D0017062

PRAISE FOR *THE BILLION-DOLLAR MOLECULE*

"A truly masterful combination of science writing, business sophistication, and simple page-turning narrative skill. This book will help any reader better understand the fundamental scientific changes shaping the world, as well as the financial climate in with which Barry Werth has carried this project off. This is a gripping and important book."

—James Fallows, author of *Looking at the Sun* and Washington Editor of *The Atlantic Monthly*

"A complex book . . . Werth knits it all together with clarity and polish."

—*Chicago Tribune*

"Entertaining . . . a lively narrative of clashing egos, grand gambles and frantic scrambling after renown riches . . . Werth bags a host of telling anecdotes in his trek behind the scenes."

—David Stipp, *The Wall Street Journal*

"Fast-paced inside account . . . enough high drama for a television miniseries . . . colorful . . . an absorbing story about a brave new world."

—*San Francisco Chronicle*

"Revealing . . . a wealth of information. . . . You'll find the book a good investment."

—Scott Lafee, *The San Diego Union Tribune*

"A fascinating and complicated story. . . a well-researched narrative . . . interesting, revealing. The story will nurture many an entreuprenerial-scientist dream."

—Sandra Panem, *Science*

"The liveliest and most revealing story written so far about the biotech industry."

—Clive Cooksen, *Financial Times*

"An authoritative book about science and the business of science. It is also a fascinating, gripping tale about real people working under extreme pressure in pursuit of what seems like an impossible goal."

—Tracy Kidder, author of *Soul of a New Machine*

"Timely, fascinating, fast moving story . . . brilliantly written."

—Ezra M. Greenspan, M. D.
Clinical Professor of Medicine (Oncology)
The Mount Sinai School of Medicine
Medical Director, Chemotherapy Foundation

SIMON & SCHUSTER PAPERBACKS
NEW YORK LONDON TORONTO SYDNEY

THE BILLION-DOLLAR MOLECULE

ONE COMPANY'S QUEST FOR THE PERFECT DRUG

BARRY WERTH

SIMON & SCHUSTER PAPERBACKS
Rockefeller Center
1230 Avenue of the Americas
New York, New York 10020

For information about special discounts for bulk purchases,
please contact Simon & Schuster Special Sales:
1-800-456-6798 or business@simonandschuster.com.

Designed by Karolina Harris
Manufactured in the United States of America

20 19 18 17 16 15 14 13 12

The Library of Congress has cataloged the hardcover edition as follows:
Werth, Barry.
The billion-dollar molecule : one company's quest for
the perfect drug / Barry Werth.
p. cm.
Includes bibliographical references and index.
1. Pharmaceutical biotechnology. 2. Vertex Pharmaceuticals
Incorporated. 3. Pharmaceutical industry—United States.
I. Title.
RS380.W47 1994
615'.19—dc20 93-32566

ISBN-13: 978-0-671-72327-9
ISBN-10: 0-671-72327-8
ISBN-13: 978-0-671-51057-2 (Pbk)
ISBN-10: 0-671-51057-6 (Pbk)

FOR KATHY

CONTENTS

PART ONE
THE STORY

CHAPTER ONE

Squatting between the twin towers of the World Trade Center in New York City, the Vista International Hotel is a minor cultural monument of the 1980s. Elsewhere it would seem a standard modern luxury hotel—contrived public spaces (schooners in the Tall Ships Bar, sail sculptures on the mezzanine), penthouse pool, chrome chandeliers—but here, dwarfed by shimmering behemoths at the foot of Manhattan, it looks like a chunk of brushed aluminum wedged between the prongs of the world's biggest tuning fork. Few of the hotel's guests seem to notice. Overnight visitors to the Vista come to New York not for the city's cultural and aesthetic smorgasbord, three miles uptown, but because, when it opened in 1982, the Vista was the first hotel in 155 years to be built near Wall Street. Less gaudy than the Taj Mahal in Atlantic City or Circus-Circus in Las Vegas, but equal to them in purpose, the Vista was constructed expressly so that guests can roll out of bed and be where the money is.

In the fall of 1989, the money on Wall Street was famously skittish, and first-time pilgrims to New York's financial markets most often came away disappointed, making the Vista a kind of Heartbreak Hotel for entrepreneurs. The stock market, despite record highs, was still shaken and defensive two years after the crash of 1987. With a recession looming, investors had withdrawn to the safety of big companies with solid earnings. Worried about liquidity, they "lightened up," especially on new companies. Such companies were too risky, it was said. They burned money. It could be years—decades—before they paid out, which by Wall Street's myopic perspective was past the vanishing point. As with many collective judgments, Wall Street's gloom was rapidly self-fulfilling. As investors retreated, stock prices sank, weakening the new companies and making their need for money all the more dire. The situation was widely considered a national tragedy by those who thought America's future competitiveness was being squandered in Wall Street's unwillingness to invest in emerging technologies. Of course, the same was said by the architects of the new companies themselves.

Yet still they came. On a warm morning in mid-October, the chief executive officers (CEOs) of more than forty new biomedical companies took their places behind several rows of long cloth-covered tables in the Vista's ballroom, unhopefully partitioned for the event. Of all the emerging fields Wall Street was cool about, biomedicine was by far the most worrisome. It spent the most money, took the longest time to pay out, and even its successes like Genentech, whose hysterical debut on Wall Street nine years earlier had driven the company's stock price from $35 to $86 in the first hour of trading, were wanting. The conference itself was an attempt to revive interest in the field. Throughout the morning, the CEOs each would have five minutes to introduce their businesses to an audience of about 150 presumed investors, although, as many of them already had discovered unhappily from perusing nametags around the coffee urn, there were few real investors present. Sellers outnumbered buyers at about the same rate, and with the same dissonant hopes, as girls do boys at an afterschool class in social dancing.

In such a market and with so little time at the podium, most of the speakers shed all pretense; it was impossible to be too bald here. Many had already filled out forms detailing their company's most intimate financial needs that owed more to newspaper personal ads than *Barron's*. "MARKET OPPORTUNITY:" flashed a California man's slide, "Thrombosis—Leading Cause of Death in the Western World."

Joshua Boger sat impassively through the morning session,

silently rehearsing his speech. As founder, president, and chief scientific officer of Vertex Pharmaceuticals Incorporated of Cambridge, Massachusetts, Boger's titular weight exceeded most of the others' on the program. But his plight was identical. Ten months earlier, he and a partner had launched Vertex with a coast-to-coast tour of venture capital companies, racking up 100,000 frequent flier miles in three months and raising just under $10 million. Vertex had no products, no revenues; it would be years, if ever, before the company would know if it even had something to sell. Yet it already was spending $75,000 a week even though its unfinished labs were crammed to the ceiling with unopened crates; even though, as the history of such companies showed, it would take up to a dozen years and more than $250 million to develop its first drug. None of this had discouraged Boger, who was thirty-eight, from coming to New York at a time when hundreds of decisions clamored for his attention back home. On the contrary, Vertex's harrowing financial need was the sole reason for his making the trip.

Six feet five inches tall, Boger (pronounced with a hard *g*) was dressed in the uniform of the day—dark pinstriped suit, jacket buttoned below the sternum—though he appeared just rumpled enough to suggest that such clothing was neither his preference nor his habit. He was long boned though not ungraceful, sitting with his legs characteristically intertwined and his torso bent forward from the waist, like a dancer's. Only his hands, which were large and torn at the cuticles, moved, thrumming a clean white pad on the table in front of him or twisting a ball-tipped marker. His face was a mask of serenity, a broad oval rising to a wide glimmering forehead and capped by thinning stick-straight brown hair that Boger parted anomalously on the right. Though he wore thick, rimless glasses and a beard that was not trim, it was not hard to imagine him as a ten-year-old—a caricature of a young scientist—grin, bones, and cowlick all askew.

Speaking just ahead of Boger was an irrepressible woman in her thirties who four years earlier had started a home delivery service for prescription drugs and whose candy-apple lipstick offset a cumulus of frosted hair. Her company, American Prescription, Inc., had grossed $185,000 in its first year, $900,000 in its second, and was now projecting sales of $70 million by 1993. "It's been an absolutely incredible history as the company continues to ramp," she gushed.

Boger, mindful of Vertex's situation, ringingly agreed. "I hate," he said, taking the podium, "to follow someone with sales."

A chemist by training, Boger had come to selling solely as a requirement for doing science. But there was no mistaking, as he began to speak, that he seemed born equally to both tasks. He had grown up in a thriving, prosperous household in Concord, North Carolina, twenty miles northeast of Charlotte, a scion of the German and Scotch-Irish pioneers who had dominated the area since before the Revolutionary War and are known to be self-reliant, industrious, unemotional, opinionated, and cold, although, as North Carolina historian William Powell points out, loyal to family and friends. His English ancestors on his mother's side, the Sneads, who descended the Piedmont by way of aristocratic Virginia, go back to the Domesday Book.

Correctly Southern, both the Bogers and Sneads have a history of carefully cultured, small-town respectability; both have long been prominent in their communities. Boger's paternal great-grandfather, a farmer in Concord, was wounded four times in the Civil War, including once at Gettysburg; his grandmother was active in the Daughters of the Confederacy until her death in 1960. His father's father, after serving as superintendent of the segregated Cabarrus County school system, took over the primitive Stonewall Jackson Manual Training and Industrial School in Concord and made it into a model reformatory, largely by preaching a stern progressivism and persuading mill owners from around the state to donate heavily for new buildings.

Boger's father, Charlie, a tank commander in World War II, was self-employed as a yarn broker. Inheriting from his own father a genuine interest in people and a practiced Rotarian appreciation for knowing what others want and how to give it to them, Charlie Boger was an exceptional salesman. With the textile mills of the Piedmont expanding headlong in the years after World War II and the explosion of new synthetic fibers and dyes, mill operators were drawn into the briskest competition they had ever known. Boger, who had a degree in chemistry, often went to sleep at night studying chemistry texts, so that when he drove to the mills in Charlotte, Kannapolis, and Winston-Salem, occasionally with one of his four sons on the front seat next to him, he could explain exactly what chemical steps would be necessary to capture the fashion industry's latest colors on any type of thread. Buyers were impressed, and Boger was popular, especially with the product men, who valued his expertise and the considerable work he saved them.

But Charlie Boger sold better than he managed. Thirty years later, Joshua would remember making customer calls with him during the summer and being shocked to find that he had signed flat-rate annual contracts. The busier his father was, Boger realized, the higher the percentage of his labor that went to others. Petulantly, Joshua grew disdainful of people who allowed themselves to be exploited this way. Though the family lived comfortably, Mary Snead Boger, Joshua's mother, never ceased to worry about money—a permanent cloud on an otherwise untroubled and equable horizon. "Charlie could sell you anything," she would recall years after her husband's death, "but he didn't know bat brains about business."

Shortly after Joshua was born, the family, riding the fullness and confidence of the postwar boom and the promising start of Charlie's new business, moved from a duplex in the center of town to a large, custom-built Georgian Colonial on a newly subdivided dead-end street on the outskirts of Concord. With two-story wooden columns grafted onto a brick facade, the house, eventually stuffed with a hefty flotilla of solid English reproduction furniture and 10,000 books, sat on a small rise, like a prow, facing a country club to which the Bogers belonged. It was nestled by a broad lawn dotted with magnolias and encircled by fields and pine woods with a stream that the boys could fish in—the complete New South idyll.

Boger grew up in the house in a whirl of boyish overachievement. He and his three brothers each excelled in school and in sports. Each was tall, thin, and high-spirited and had voluminous interests, making the household a kind of boisterous, high-expectation boarding school with Boger's mother, an outspoken and dramatic woman, as headmistress. With her liberal encouragement—later to become an award-winning theatrical director, she once marshaled other women to block the razing of Concord's antebellum courthouse by standing in front of a bulldozer—the boys relentlessly prodded themselves and each other. Together they became famous in town, the "Boger boys" being Concord's answer to the precocious "whiz kids" of the 1950s and 1960s. When Joshua was ten, he had progressed so far in a year of piano lessons that his teacher pronounced him a prodigy and contacted the local newspaper. In an article picked up by the Associated Press, Boger explained that he practiced every morning at 6:30, before school, and had begun charging forty cents a week to teach his brother Ken, who was five years older. "I keep him behind me," he said, "so I can keep up the lessons."

Diligent at everything he tried—he handed in a 400-page paper on Africa in the fourth grade—Boger gravitated early to science. By age seven, he was spending hours, and sometimes days, in a laboratory that his father helped him build above the garage. It was a low-ceilinged barnlike space with unpainted rafters and plank floors—"more space per scientist than we have at Vertex," he would joke—and it reflected the eclecticism of his world, a world that young Boger self-assuredly considered *the world* and thus strictly within his power to control. In one corner there would be a potassium permanganate crystal the size of a dinner plate suspended on a thread over a vat; elsewhere, a stack of microbiology plates, the result of a recent experiment in which he had swabbed the throats of the other kids in the neighborhood in order to compare the germ-killing potency of mouthwashes. There would be plants growing, animals in cages, rocks that Boger had chipped out and mounted on boards, chemistry experiments from a Time-Life project series that his father had ordered through the mail, a microscope with a fly impaled under its lens. Whatever order there was reflected the materials on hand and what Boger called his "dumbcat curiosity." Other than the rows of shelves groaning with the family's extra books, the space was his alone. His father might donate reagents—he once came home with twenty-five pounds of mercury, handing it over without a word—but neither of his parents ever visited.

As a young experimenter, Boger simply went his own way. Once, when he was eight, he spent a long Saturday shuttling between his lab and an open red-clay field next to the golf course. Knowing that water, Drano, and the tinfoil from a milk bottle produced hydrogen, he filled several balloons with the volatile gas. Heedless of the fate of the Hindenberg, he then ran a mouse through a maze he'd constructed, built a gondola to take the hapless animal aloft, and sent it soaring. When the mouse came down, Boger again subjected it to the maze to measure its disorientation. Being eight, he spent the next day playing baseball.

To Boger, science was the most natural way of apprehending a world that could otherwise be maddeningly obscure; he was enthralled with its precision and power. It was also fun, just as lying under twelve feet of water in the diving pool at the country club was fun. But science had its imperatives. In a school essay that he wrote at age thirteen in which he traced his academic career from kindergarten, which he recalled enjoying "except for rest time," up to the eighth grade, he concluded purposefully, "Recently my interest in chemistry has turned me towards the field of medical research. My goal in life is . . . to help rid man of the burden of disease and hunger, and to help man get along with man."

It was this trajectory, launched in puberty and accelerating more or less along the most favored path ever since, that brought Boger now to the speaker's platform at the Vista. Valedictorian in high school, he went on to Wesleyan University in Connecticut, where he trained under the legendary Max Tishler, one of the most important and prolific figures in the history of drug research, and where he again finished first in his class. One of only eight students nationally to receive a full four-year National Science Foundation fellowship for graduate school, he then went to Harvard, which then as now had the best organic chemistry department in the world, where he got his Ph.D. He ultimately went to work for Merck and Company, the world's premier drug firm. By his mid-thirties, an age when most chemists are still making compounds at the bench, he had become Merck's senior director of basic chemistry. He held seventeen patents, although none for an approved drug, and was considered among those who knew the company best as a favorite eventually to head Merck's vaunted $1 billion annual research effort—perhaps among the most powerful biomedical posts in the world. They were stunned—some furious, others relieved, even overjoyed—when Boger abruptly left the company in early 1989 to found Vertex.

Boger was never breathless as a public speaker, but he had trained himself, when talking to businesspeople, to be low-key and earnest. His years in the Northeast had purged all but the last traces of southernness from his voice, which remained sonorous and steady. Yet what set him apart now, as he described his company, was the strength of his pedigree. Everything in his past had led to his being a prince of the industry he now hoped to revolutionize, and it gave him a powerful mien. It was the aspect of the favored son, the smartest kid in the class, in the school, maybe in the school's history. Even at the Vista, Joshua Boger had valedictorian written all over him.

Vertex, Boger said, was not only about to create powerful new drugs, but also to change the way all drugs would soon be created. With only five minutes to speak, he could hardly explain the scientific rationale for such a statement. He referred fleetingly to Vertex's "unequaled scientific staff" and the "most impressive set of . . . technologies in the world." Beyond that, he quickly summarized the company's first project. It was an attempt to improve upon an experimental drug called FK-506 that suppressed the immune system. The drug had been shown to be highly toxic in some test animals, but it was still thought to have extraordinary potential in humans in facilitating organ transplants and curing autoimmune diseases.

"We will redesign the molecule," Boger concluded matter-of-factly, "and eliminate its undesired properties."

Afterward, Boger left the ballroom as soon as he could politely pull himself away. He had never expected much from this forum; Wall Street's "promiscuous imagination," as biotechnology writer Robert Teitleman called it, had long since grown impatient with stories like his, and the shortage of real investors at the Vista had proven that. But Boger knew something else. Vertex was in a historic position. The company was attempting something so bold that most people in the drug industry questioned whether it could be done at all. It would design drugs—not merely appropriate them from nature and tinker with them, as was the rule, but design them, atom by atom, as one designed a skyscraper or a computer.

As even Wall Street might have recognized, if Boger was right, the most consistently profitable legal industry in America during the past forty years (besides, perhaps, cigarettes) was on the verge of an upheaval in the way it went about discovering new products, an upheaval that would vastly increase the utility and variety of those products and the oceans of money that flowed from them. Over the next thirty years, Boger believed, drug research would become vastly more refined, more rational. Those who led the way would be heroes. Vertex, he knew, or some company like it, could well become the new Merck, which besides being a paragon of medical science had recently become, as measured by virtually every magazine executive poll, the most admired corporation in America. It was a prize of rare stature and importance.

And for that, Boger had decided, no task was too onerous, even putting himself out for as unpromising and unseemly a cavalcade as the one at the Vista.

"A meat market," he described it later. "We're talking fishnet stockings. I mean it just doesn't get any lower than this."

CHAPTER TWO

Ever since Harvard survived its first headmaster, Nathaniel Eaton, who beat students with a walnut cudgel "big enough to kill a horse," and became what historian Richard Norton Smith calls the "epicenter of American education," Cambridge, Massachusetts, has been a place where a disproportionate number of the world's smartest people come to prove how smart they are. Against this elite testing ground chafes another Cambridge, a minor, graying northern city in which generations of immigrants and African-Americans have lived crammed in underheated triple deckers and toiled in scores of shoe and candy factories, foundries, and machine shops. Until World War II, the city was roughly split: Harvard and MIT on the east and west and along much of the Charles River, working-class Cambridge in between and in the industrial flats across from Boston and Charlestown. Then history lurched, and Cambridge tilted. The universities, supported by the federal government's ambitious research programs, began pushing relentlessly outward. Manufacturing died or moved away, along with those it employed. Most critical, knowledge became a big business like any other. As Harvard's Sumner Slichter has observed, "The discovery that an enormous amount of research can be carried on for profit is surely one of the most revolutionary economic discoveries of the last century." It was during this period that the people arriving in Cambridge to prove themselves, particularly in the sciences, added to their prerogatives the takeover of the city's industrial real estate.

Along lower Sidney Street, an area of low-slung factories and warehouses abutting a necrotic rail yard, the overlay of the new Cambridge and the old is striking. On one short block, across from each other, are the Boston Pipe and Fittings Company and American Foundry, Inc.; within one hundred yards, in similarly drab two- and three-story brick buildings, such futuristically named companies as ImmunoGen, Bioprocess Technologies, and Holometrix. The barrackslike former St. Johnsbury Trucking Company depot, until the early 1980s a hub of grinding gears and hissing airbrakes, now produces X-ray telescopes in sleekly refurbished anonymity. Despite its new association with Harvard and the Massachusetts Institute of Technology (MIT), the presence of so many exceptional scientists, the restless incubation of so much profit motive, and the influx of so many Saabs and Acuras, the area remains dolefully nondescript, a temporary address for the new companies and a final one for the old.

Vertex started leasing 10,000 square feet in a former construction company warehouse at the corner of Sidney and Allston streets in April 1989, six months before Boger's outing at the Vista. In keeping with his ambitions, he began to look at once for more space. The building is brick, one story, nearly square, and the occasional target of graffiti. When it was built in the 1920s, it had mullioned shop windows and ersatz Corinthian columns. Sixty years later, the columns were stuccoed over and the windows replaced with thermopanes, giving the building an air of cheap recycling, like a motor vehicle bureau that once was an armory. In fact, the windows are as superfluous as the columns. Because anything of use to a legitimate drugmaker is of substantially higher value to an illegitimate one, the company prefers not to advertise the contents of its labs. Its blinds are all but permanently drawn.

Boger was still living in New Jersey, near Merck's giant central research campus in Rahway, when he decided to locate his then-unnamed company in Cambridge. He intended Vertex to be highly visible from the start—to the international elites of business and science, if not to pedestrians—and for that, he thought, Cambridge offered a powerful showcase.

Businesswise, it was a singularly unpromising time. During the previous decade, nearly 200 biotech businesses had sprung up, yet only one, Genentech, earned a regular profit, and even that was disappointingly small. Most of the companies had simply gone on hemorrhaging money, blindly, with no end in sight. Dozens were now failing or scrounging for buyers.

Add to that a billowing national recession and a comatose New England economy, and Boger's decision to leave Merck and set up in borrowed offices in Cambridge in the dead of a lightless New England winter seemed fateful. Boger was anything but. The previous fall, he'd been recruited by an irresistible California venture capitalist named Kevin Kinsella, an embodiment of that flamboyant breed and originator of Vertex's concept, and together—Kinsella on the West Coast, Boger on the East—they'd plunged ahead. Working from a ninety-page business plan that Boger had composed in less than four weeks, they knocked on doors relentlessly, talking with investors, scientists, vendors, developers, lawyers, contractors, regulators, and potential partners, leveraging commitments pyramidally. "Don't you think this is five years too early?" Boger was often asked, to which he answered, brimming with impatience, "Yes. But five years from now it'll be five years too late." It was a determinedly Cambridge answer, smug, marbled with arrogance and risk. But by then, they'd enlisted perhaps the one academic collaborator who could match Boger's pedigree, ambition, intellectual firepower, and cachet, Harvard wunderkind Stuart Schreiber. How, Boger and Kinsella wondered, could they lose?

This was Boger's other reason for choosing to be in Cambridge. Every young biomedical company needs in the absence of its own science the association of big-name researchers—a scientific advisory board (SAB). Most SABs are ballast for the letterhead. Boger professed to want an SAB that was more than that. Having identified as his optimal SAB five senior faculty members at Harvard and having gotten them all, most notably Schreiber, he intended to use them. Being in Cambridge meant having them within courier distance.

• • •

On the Saturday morning before Boger went to New York, Vertex's SAB and its staff scientists huddled for the first time in the company's makeshift lunchroom. Billed as an all-day strategy session, the meeting was also a critical first test of Boger's determination to use the SAB. As with the start-up of many high-minded adventures, there was the usual air of self-selection reinforced by deprivation—a mercenary, albeit ragtag, flavor. It was the first time some of them met.

In aggregate, they were the kind of people Boger felt most comfortable with—young, male, irreverent—people like himself. Of the twenty researchers on hand, just two were women; only five were over forty. Yet despite their relative inexperience, each had sacrificed something to be here, as the setting reminded them. For weeks, jackhammers had rocked the building, leaving a pall of cement dust on books, boxes, clothes. Overhead, ceiling panels had been left out by workers, and skeins of unattached pipes stood exposed. The screen on which Boger presented a fuller version of the slide show he would take to New York was gray, steel, battered, and part of a short-term leasing agreement, as was the furniture in his office, which opened onto the lunchroom and had mounds of books and catalogs splayed chaotically along every wall. Many of the scientists were accustomed to being pampered at such meetings: Merck, where some of them had previously been associated, picked up visiting researchers by limousine and toured them around by helicopter. Lunch today at Vertex would be pizza and Greek salad served on paper plates.

In fact, the close involvement of the SAB was an unpopular idea of Boger's that would require considerable selling within the company. Scientists in industry and scientists in academia tend to be brutally dismissive of each other. Academic researchers thrive on publication, attention and credit being oxygen to their careers. Yet to industrial scientists, whose own success most often depends on keeping their best work secret and who are less well known, most academics are recklessly, inexcusably self-serving—loose cannons. Boger had brought to this room some of the best industrial and academic researchers in their fields. Getting them to talk openly would be another matter.

The problem had first surfaced a week earlier, pungently, ominously, not a surprise, but sooner than most expected. Schreiber, a slender, enthusiastic thirty-three-year-old chemistry professor, had mildly proposed at a smaller meeting that everyone discuss what experiments they were planning.

Coming from Schreiber the suggestion was hardly as innocuous as it seemed. He, more than anyone else at Vertex, was Boger's equal, his other: a fast-rising star who, with the backing, position, and control he had long conceived of and only just won, was beginning to make his mark on a world stage. There were other similarities. Like Boger, Schreiber is a chemist and an avenger for the Harvard legacy, long in disfavor, of exalting chemistry above all other life sciences. He is a quick, copious thinker who can see past his own field and direct a swarming, multifrontal research effort. Schreiber worked seven days a week, had a big group of the world's most ambitious graduate students and postdoctoral fellows, published furiously, and could smell a hot idea. "Stuart is fearless," Boger once said admiringly. "He has a killer instinct for doing the right experiment."

He could also be disarming. Like Boger, Schreiber exudes an easy border-state affability: He grew up at the high end of a semirural gun-and-dirtbike culture in east-central Virginia, a few hundred miles from Concord, and partied his way through high school before discovering chemistry in college. He wears imported loose-fitting tweeds and soft loafers and commutes to Cambridge from his five-story townhouse in the Back Bay in a gunmetal gray Porsche 911 with a car phone. With a smooth and eager face, respectful manner, and large swimming eyes magnified aquatically by round wire-rimmed glasses, he looks more like a successful young art dealer than one of the two or three most promising organic chemists in the world. "Eddie Haskel," one Vertex scientist calls him.

Like Vertex, Schreiber's group at Harvard was studying drugs that suppress the immune system, a field that was rapidly heating up in large part because of Schreiber's own work. Angling sharply for what academic researchers want most in such new areas—priority, acknowledged leadership in the field—he was concerned about being slowed down by overlapping effort.

"I think it's best that we consider what we'd like to do immediately and what Vertex would like to do," Schreiber said.

There was a palsied silence, the Vertex scientists all looking tentatively at one another or at their shoes. Finally, Boger brushed aside the question by saying how many people he planned to hire and in what disciplines—a coded message that indicated the general direction of Vertex's research but no specifics. Though Schreiber was being paid $25,000 a year to attend perhaps a dozen such meetings, owned 150,000 shares of Vertex stock, and had been recruited largely for the benefit of sharing information and materials with his lab, it was clear he was not going to be fully trusted as a collaborator, not even by Boger. Sensing he would get no further, Schreiber said, "OK then, on the table, anybody who gets to an experiment first should do it." With everyone agreeing, the conversation moved uncomfortably on.

With the labs still unopened, it was too early for the threat of such competition to arise among Vertex's own scientists, but here, too, were tremors. Boger had recruited an exceptional group of researchers; of the company's ten most senior people, all but one had worked at Merck, Harvard, MIT, or Yale. Moreover, Vertex planned to integrate the most advanced disciplines of molecular biology, which deals with function, and of chemistry, which addresses structure and mechanics—whose practitioners, like behaviorists and Freudians, have little good to say about each other. Already the company had more submicroscopic disciplines—medicinal chemistry, X-ray crystallography, nuclear magnetic resonance spectroscopy, molecular modeling, computational chemistry, protein engineering, protein chemistry, enzymology—than a small university, and competition over hiring and lab space had grown fierce. As in war, victory in science is measured in bodies, territory, and materiel, and Vertex, it seemed, would be no different. Coupled with the personal ambitions of those who saw Vertex as a major drug company in the making and themselves growing in power and influence along with it, a secondary ambience of intramural squabbling had already begun to poke through the initial looseness and camaraderie.

Now, in the lunchroom, Boger moved to unite all sides. Far from being disturbed by the general testiness, he considered it affirmation that his ideas about corporate culture—a culture of enlightened self-interest—were taking root. Boger wanted people who were unbowed by competition; people who, like himself, insisted upon being best. He wanted an orgy of bristling, militantly selfish creativity of the kind he grew up with. "Arrogance doesn't disturb or impress us," he once said in another context. "We understand arrogance." As with much of what Boger said during this period, the remark seemed at least partly calculated, like a short man's swagger, to compensate for certain disadvantages: Vertex, despite its talent, would be competing against labs that were vastly richer and more experienced; outsized, even outrageous, boasts were good for morale.

And yet Boger also believed, or seemed to believe, every word he said. Devoutly irreligious in his personal life, he had a faith in himself and in science that was Himalayan, towering over most other people's. Boger's convictions were huge, and he expressed them with such confidence that it was hard not to agree with him.

Boger chose not to make the case for cooperation himself. Instead, he turned the meeting over to Rich Aldrich, Vertex's vice president for business development. Aldrich, a tall, curly-haired thirty-five-year-old with an M.B.A. from Dartmouth, was the group's sole layman. His ancestors arrived in Plymouth in 1630, ten years after the *Mayflower*, and have been ensconced in the state, in law and banking, ever since. Within those staid Yankee confines, Aldrich's decision to put his career in risky biomedical start-ups marked him as something of a family rebel. But that didn't grant him instant acceptance among the scientists in the room. On the contrary, many of them, even if they didn't know his background, viewed him as a political and cultural nemesis, a "suit." Aldrich, who was dressed today in khakis and a blue oxford shirt, enjoyed turning the disparity between business and science back on them directly. "Design any drugs lately?" he'd ask.

Despite their differences, everyone at Vertex had one thing in common. Under the terms that enable impoverished, unknown companies to recruit top scientists and expect them to work Saturdays, they'd all begun to amass large amounts of stock—from 10,000 shares for a junior scientist to, in Boger's case, 780,000 shares. These holdings were at present worthless but would likely make them all rich if and when the company went public, and extravagantly rich if Boger was right and Vertex became a major drug company. In Boger's view, this shared fortune was so obviously compelling that no one need be reminded of it; it should automatically restrain even the most rapacious ego. And indeed, as Aldrich now began to discuss the company's plans for raising the tens of millions of dollars it would need over the next couple of years, the scientists' collective attention focused as sharply as a team of accountants'.

Aldrich told them that Vertex was considering a range of options, but the most promising were its discussions with other drug companies. Vertex had approached eight other companies about the possibility of their underwriting part of its research in return for certain "downstream" development rights. In other words, he and Boger were aggressively talking with potential competitors about the company's science even before it had unpacked its first test tubes. Standard practice, the discussions nevertheless startled some of the group's academicians.

"Isn't one in danger," drolly interrupted Jeremy Knowles, a brilliant and much admired enzymologist who'd come to Harvard from Oxford and had been Boger's thesis advisor, "of giving away all we've got before we've got anything? I mean, yes, there are some splendid ideas here, and some superb people, and we will do it. But what's to stop boring old Glaxo [a British firm that had jumped from twenty-fifth to second among the world's drug companies on the strength of the world's best-selling drug, the antiulcer agent Zantac] from saying, 'Oh, oh, we see. Maybe we can do what you're telling us ourselves.' "

"But they *can't* do it, Jeremy," Boger interjected.

"But Merck can."

"No," Boger paused resolutely. "Merck can't either."

It had already become an article of faith at Vertex—as at most start-up companies—that large corporations were dinosaurs: too unadaptable and slow moving to compete at the forefront of research. But Knowles was not alone in suggesting that daring these companies to try might be an act of fatal arrogance, especially from a firm without a single hard scientific lead.

If there was a danger in Boger's intellect, Knowles knew this was it. Boger was too smart and too competitive to ignore what he could learn from others. But because he was so certain of himself, he often underestimated them, turning against them with a haughty disdain. He especially liked to tweak the mighty. This had been the case when he and Kinsella spent months searching for a name for their new company and decided on Veritas. It was one thing for a fledgling business to draw on its contact with Harvard faculty members, another to appropriate the school's 350-year-old motto. More than most schools, Harvard anguished publicly about being involved with outside businesses. Knowles began receiving rueful, cautionary phone calls from senior administrators, including the university's legal counsel. Though Boger clearly relished Harvard's angst, Knowles quickly persuaded him of the "sensitivity, the horror, the absolute unacceptability" of Veritas, and the name was changed to Vertex.

Peculiarly, Vertex was at or near the apex of the science now being discussed in the lunchroom despite having done no experiments. In February, when Boger had considered what project to undertake first, his decision to improve on the experimental drug FK-506 had seemed prescient. Now, several developments had put Vertex at the center of one of the most promising areas in drug research.

Drugs are molecules. They attach themselves along critical points in the pathway of a disease. Since not all molecules are drugs, the difficulty, from a drugmaker's standpoint, is discovering those that are. There are other challenges. A drug molecule must be sufficiently unique to patent and must be capable of getting to its relevant target, another molecule within the toweringly complex molecular universe of the body. Raquel Welch, in the 1960s movie *The Fantastic Voyage*, discovered the extreme hazards of this. She and a miniaturized team of doctors undertook a harrowing repair mission inside the human body. Solubilized in an infinitesimal submarine, they tumbled through billowing plasma, dodged the clinging death grip of chainlike antibodies, and breached greasy cell walls to fix a remote area of the brain. The journey, lasting an hour, approximates the life cycle of at least one class of drugs—those delivered by injection into the bloodstream.

Those that come in pill form run an even riskier gauntlet. It's the job of the gut to dismember chemical compounds, atom by atom, so that their constituents can be used by the body. Like machine tools in a vast automated recycling plant, enzymes in the digestive tract and liver facilitate—at speeds of up to 10 billion operations per second—the stripping and reassignment of incoming atoms. Some are saved, some reconstituted, some burned up, some discarded as slag. Because molecules are groups of atoms chained together like pop beads, a drug entering the body orally must be small, durable, and extremely resistant to being crushed or picked apart. An hour in the gut, and the nuclear sub in *The Fantastic Voyage* would look like a car left overnight on the shoulder of the Cross-Bronx Expressway.

Prior to World War II, only a handful of drugs worked, and most of those derived from some combination of luck and empiricism. Since then, however, the search for new drug molecules has narrowed to where they are most likely to be found: in soil and sludge. By far the most prolific producers of the sort of small carbon-based molecules that make the best drugs are those microorganisms that seethe invisibly underfoot. Thus the front end of any drug discovery effort at most of the world's great pharmaceutical companies consists of lab-coated scientists cooking obscure dirt samples in fermentation broths and screening them for activity. If some constituent of this foul black chemical soup is active against a disease target in the laboratory, then the search begins for the active molecule.

Besides the staggering cost and dependence on luck (you may have the right compound but the wrong target), the biggest problem with screening natural products is the molecules themselves. Though they work, in many cases astonishingly well, their activity is most often happenstance. There would seem, for instance, to be no logical reason for a molecule made by a fungus—a molecule that has evolved structurally over 4 billion years to perform some function that is eons removed evolutionarily from human cells—to reduce cholesterol, and yet most of the leading cholesterol-lowering agents, including Merck's Mevacor, a $1.6 billion seller, have been discovered this way. The best explanation is that the molecules mimic something else. And yet because they are approximations and not perfect, they may also fit with other targets or include toxic elements superfluous to their function, inciting other, unwanted activities—side effects.

It was the imperative of screening that Boger and Vertex now sought to dethrone. Quantum gains in the molecular understanding of disease and in computer technology have recently suggested another approach for finding drugs. Called *rational* or *structure based*, it presumes to design them—atom by atom—based on a precise understanding of how molecules interact. Drugs work by selectively sticking to discrete molecular *receptors*, or targets, which usually are within cells. Like pieces of a jigsaw puzzle, they interconnect—scientists use the word *bind*—based on complementary conformations, or fit. Thus the rationale for structure-based design: to optimize the shapes of drug molecules. "Connecting the dots," Aldrich liked to call it in a heroic oversimplification that made some of the scientists at the tables wince. In effect, the goal is the very opposite of screening: building the molecules one wants rather than fishing for approximations in nature.

The advantages of such drugs presumably would be enormous. Because they would be more specific, they'd be safer; there would be fewer side effects. And because they would be safer, they could be used far more widely, at higher doses; this meant untold new uses (and not incidentally market opportunities and profits). Structure-based drugs, as Boger and others have pointed out, are the industry's Holy Grail, although "like many holy objects," Boger says, "they have been more often referred to than taken seriously."

As a first project for demonstrating structure-based design at Vertex, the makeover of FK-506 suited Boger's hubris perfectly. The molecule belonged to another company, Fujisawa Pharmaceuticals Company of Japan, which had only recently begun testing it in humans; a powerful immunosuppressant, it appeared to stop transplant recipients from rejecting their organs better than any other agent. Immunosuppression occurs when some but not all of the body's defenses are disarmed. As a therapy, it is especially crucial for transplant patients, who risk having their grafts destroyed by hyperactive immune cells (the "immunological conscious," Sir Peter Medawar called it). Yet that was perhaps the least of FK-506's powers. Cyclosporine, which acts similarly to FK-506 and is the only selective immunosuppressant to be licensed, also helps dramatically against those diseases where the immune system mistakenly starts killing the body's own cells, but is much too toxic for general use. Multiple sclerosis (MS), juvenile diabetes, rheumatoid arthritis, Crohn's disease, psoriasis, lupus: perhaps dozens of autoimmune diseases could be cured with a similar, though more specific and thus safer, molecule—a molecule that Boger was convinced could now be designed at Vertex. It wasn't lost on him that the potential market for a drug of such sweeping effectiveness might be as much as $5 billion per year.

FK-506 and cyclosporine are conventional drugs: both were discovered in dirt samples (cyclosporine near the Arctic Circle in Norway, FK-506 on a mountainside in Japan) and were found serendipitously to fight disease. They're also poisonous. Many of those taking cyclosporine suffer kidney damage so severe that they require further transplants. FK-506, though it looked promising in humans, had been fatal in some dogs—a discrepancy that the Food and Drug Administration (FDA) would want resolved before approving it for general use. Thus, these two tantalizingly powerful molecules were, ultimately, limited.

Here was Vertex's presumed edge. Structure-based drug design is most often compared with making a key to fit a lock. Using a model of the cylinder, you design a device to touch only the tumbler pins. What is essential—what would give a company its advantage—is knowing the conformation of the cylinder, how it works.

And sitting now at opposite ends of Vertex's lunchroom were Schreiber and a bearded, moon-faced, mild young immunologist named Matt Harding. Schreiber had been the youngest full professor of chemistry in Yale's history at age twenty-six before he was wooed to Cambridge by former Harvard President Derek Bok; Harding, whom Boger had recently hired away from Yale Medical School, was Schreiber's main collaborator in the area of immunosuppression. The locks for drug molecules are nearly always proteins, the working molecules in cells, and between them Schreiber and Harding knew more about the protein receptors for cyclosporine and FK-506 than perhaps any other scientists. Harding had shared in discovering both targets, which because of their affinity for immunosuppressive compounds were called *im-*

munophilins, and together he and Schreiber had just produced the first available quantities of the receptor for FK-506 (FK-506 binding protein, or FKBP). To possess this receptor, a virtual carbon copy of the protein through which FK-506 presumably works within the body, was no small matter, because even a few thousandths of a gram of a reagent can give a company a significant advantage in its experiments. Along with Merck, which had discovered FKBP independently and had its own minuscule quantity, Schreiber now controlled the world supply of the protein.

Once a drug target is identified, the next task is to solve its structure—to reveal the lock's inner workings with its tumblers exposed. Steeped in fifty years of advanced science, this remains something of a black art. To calculate to within one ten-billionth of a meter the precise location of every atom within a protein (most proteins are floppy and have thousands of atoms) requires a minimum of $1 million in equipment and, generally, a highly specialized and, by tradition, temperamental researcher known as an X-ray crystallographer. Crystallographers are the most prized of all protein scientists. Association with one of the rare few who have actually solved the structure of a protein can make a company's name. Hiring one is tantamount to a franchise. And sitting in the room with Schreiber and Harding was Harvard's Don Wiley. Wiley's recent discovery of the structure of a key immune system protein was considered so critical to understanding autoimmune diseases that it had raised, for the first time, the hope of eventually curing such diseases with drugs. Wiley would not be trying to solve the structure of FKBP for Vertex himself, but that wasn't a concern. Within weeks Boger would announce that he had hired perhaps the most famous crystallographer in the pharmaceutical business, Manuel Navia. In 1988, in the record time of three months, Navia had led a team that found the structure of a major protein that causes the acquired immune deficiency syndrome (AIDS) virus to replicate, a feat of such stunning general interest and public relations value to his employer that it landed him improbably on the *Today* show. Navia, Boger liked to gloat, was coming from Merck.

Boger had decided alone what project to undertake first and had recruited the scientists accordingly. Yet the most critical development making immunophilins a ripe area for Vertex had occurred elsewhere, beyond Boger's control. That immunosuppressants like cyclosporine and FK-506 made remarkable drugs had long been known, though how they worked remained a mystery. Yet two recent papers in the scientific journal *Nature* suggested an appealing answer. They identified cyclophilin, the apparent target of cyclosporine, as an enzyme, the most complex and active of all molecules. According to the authors, cyclophilin seemed to accelerate the folding of other proteins into active shapes by catalyzing a key reaction.

It was a hugely promising observation. Proteins are nothing without folding, just chains of atoms. Yet loop them and coil them into precise, genetically ordained conformations and they snap to life. With various enzymes grabbing and splicing and assembling atomic subunits, high-speed protein folding is at its height during the manufacture of new cells, protein being half of all living matter. Thus the implication of the *Nature* papers: that cyclosporine worked like a well-thrown monkey wrench, invading a key part of the assembly and bringing the operation to a halt. No new protein folding meant no new attack cells, which meant no graft rejection or autoimmunity. Boger, like many others, was tantalized by the simplicity.

The *Nature* papers appeared in February, midway through Boger's three-month slog after seed money. Choosing a first project was undoubtedly the most critical decision he faced, since the company would have to support it on its own, without—as Vertex's competitors would all have—other programs to pick up the slack. The economics of drug discovery were dicey enough—only one in ten projects yields a drug—not to have a hedge, and Boger had already stretched the odds dangerously. Unhesitantly, he threw all of the young company's resources into immunophilins.

His reasoning was persuasive. The protein-folding hypothesis suggested a straightforward goal: to build a better monkey wrench. If structure-based design was to work, Boger believed, it would have to start with simple, well-understood biochemistry, which the immunophilins now appeared to have. Also, the theory provided what had been missing from the search for new immunosuppressive drugs—an assay, a simple laboratory experiment for testing new compounds. As Vertex began making its own molecules, it could screen them initially on their ability to bind to and block the protein-folding action of FKBP, the enzyme discovered by Harding and controlled by Schreiber. The company was unlikely again to be so strongly positioned in so wide open an area.

"FK-506 is the only exciting molecule in the pharmaceutical industry now," Boger told the scientists. "Within a year we have the opportunity to be a leader in this field."

Whether Boger was simply drumbeating or this was true, the idea that a company that still existed largely on paper might compete at the fore of what was potentially one of the most profitable and scientifically most rewarding areas in all of drug research was intoxicating for the scientists. It placed them precisely where they wanted to be.

And yet the project was far from ideal. There were the scientific questions. Could the toxicity of cyclosporine and FK-506 be separated from the drugs' activity, or were the two linked inextricably? More critically, was FKBP the correct target or was there another, yet undiscovered molecular interaction that caused the drug to work? A wrong answer to either question could kill the project—and, conceivably, the company—in a day.

There also was the competition. FK-506 was a new enough molecule that no company hoping to beat Vertex had an insurmountable lead. But many companies, most notably Merck, had already recognized the opportunities that Boger had seen and were moving rapidly ahead. Indeed, Boger had launched Merck's program, creating possible legal problems now. Though the project appeared well suited to a small, highly focused effort like Vertex's, as Knowles pointed out, it could be foolish for Vertex to challenge Merck and the other big firms.

Boger raised these issues only to brush them aside. More than anything, he was a rationalist. He believed ineradicably that no decision was ever wrong. A decision based on incomplete information might be, in his word, "suboptimal," but it couldn't be incorrect, not logically. The key was to have the best information possible. Given the data now available—about the protein-folding activity of cyclophilin, about the similarities between cyclosporine and FK-506, about the state of the art of drug design and the people he had assembled to do it, about who else was in the field—Boger was convinced that Vertex had as good a chance as anyone, including Merck, of designing the next great immunosuppressive drug, one that would not only capture the $800-million-a-year transplant market, which cyclosporine now dominated, but open up the autoimmune market as well. Now that he had assembled everything he needed in order to, as he would put it at the Vista, "redesign [FK-506] and eliminate its undesired properties," he had no choice but to believe that he would do it. As the afternoon went on, those in the lunchroom came to believe more and more, as Boger had planned for them to, that they might do it as well.

One of the day's final speakers was a thirty-one-year-old Australian protein chemist named John Thomson. Chiefly responsible for supplying Vertex with FKBP for its other experiments and attempts to solve its structure, Thomson was, in effect, the company's lead-off batter, a pivotal, high-visibility position he wouldn't have had any other way.

Thomson's work was as unglamorous as it was vital. One of the few researchers of his generation to prefer extracting protein from animal tissue to the more modern recombinant methods, he revels in an earthy image. As a graduate student and postdoctoral at MIT, he spent several years isolating a protein from the lenses of fetal eyeballs. Now, in the lunchroom, he presented a price list from a research supply house for human organs—"Igor," Boger code-named it.

As FKBP was concentrated primarily in the adult spleen, Thomson explained, he would be buying those first. The base charge for an organ from a brain-dead cadaver was $360. Those discarded during transplantation of other organs were discounted at $200, although, as Thomson pointed out, there might be other charges: $25 for sterile or undiseased specimens, $25 for snap freezing.

"If we want same-day delivery," Thomson said in his thick Melbourne accent, "that'll be about $85."

Surprised by the efficiencies of the marketplace, many of the researchers groaned. But Boger picked up the theme. If Vertex was to solve the structure of FKBP—indeed, if it was going to design a drug—then nothing was scientifically more important now than developing an abundant supply of the protein. Even having Schreiber would be of little help if a collaboration with Harvard for his recombinant FKBP couldn't be worked out soon. Thomson, Boger explained, would first attempt to isolate protein from the thymus glands of unborn calves as a model system for the scarcer and more expensive human spleen. Boger didn't like to set priorities in stone, but this one he did.

"If we get to a situation where we have to bring a cattle truck up to the back door and start unloading calves," Boger said, "we'll do that."

The daylong session lasted until five, and though it was a luminous fall Saturday at the height of the New England leaf season, no one rushed out. The fifteen staff scientists, nine of them Ph.D.'s, had been sitting around telling themselves for months that they would be doing the best science of their lives here. Now, hearing each other and the SAB and, most important, seeing Boger's expansiveness played out on a larger scale in front of senior people who might, but didn't, dent his enthusiasm, they felt their own optimism vindicated. Boger's promise that Vertex could do what no one else could suddenly looked more real to them. And so they left, as Boger had intended, with a renewed degree of confidence, although nothing compared with the feeling of rare preparedness, of triumph before the fact, that now seemed to consume Boger and still transported him several days later at the cattle call in New York.

CHAPTER THREE

Not perhaps since cortisone, forty years earlier, had a molecule arrived promising so much. Chemists and transplanters, pathologists and cloners; specialists of the liver, kidney, skin, joints, eyes, bowels, pancreas, nervous system, and immune system; drug companies, insurance companies, ethicists; xenografters (people experimenting with replacing human organs with animal organs), oncologists, microbiologists, yeast specialists; people suffering from dozens of chronic, incurable diseases or at the precipice of death; and basic biologists, at the farthest remove from suffering, exploring the molecular essence of life—all were drawn to FK-506. They all wanted it—to examine, to take apart and reassemble, to experiment with, to treat with or be treated with—even though it remained largely untested, even though the first scientific paper documenting its effects on humans had yet to be published. In the world of modern medical research, where the rigors and jealousies of specialization act like water-tight bulkheads, compartmentalizing knowledge and separating those who pursue it from one another and from those they might benefit, and where the opportunities for hype are galactic, interest in the drug was broad, consuming, immeasurable, and in the early fall of 1989 focused irreducibly on one man, Dr. Thomas Earl Starzl.

Best known as one of the pioneers of transplant surgery, Starzl ran the world's largest, busiest, most messianic and—at about $100 million per year in billings—most successful transplant center, the only medical center offering FK-506 to patients, at the University of Pittsburgh. Mercilessly driven, Starzl had directed the rescue and development of the drug after it was initially deemed too toxic for humans, and he hadn't stopped there. Like a fight manager or impresario, he had also groomed the drug, choosing how it would be tested, with which patients, and under what circumstances. He had controlled what the world knew about it and when. What the spectrum of researchers and patients now clamoring for FK-506 all wanted was in fact something that only Starzl and the scores of surgeons and researchers around him had witnessed up close, a siren's song. "A miraculous drug," Starzl repeatedly called it. "A wonder drug. One of those drugs that comes along once in a lifetime."

In September, Boger received an invitation to a meeting in Barcelona where the results of the first human clinical trial of FK-506—Starzl's results—were to be presented. The session had been tacked on to a regular meeting of the European Society for Transplantation in late October, apparently in some haste, since the invitation was not formally printed, but came by fax.

It was hardly a good time for Boger to be away. Work on the labs was proceeding glacially. He still had several key positions to fill. Though much of what he was doing could be done by others, he insisted on making even the most minute decisions himself. "This is the easiest time to set things up right," he said. "A year from now things will be twice as hard to do or impossible to fix." And so he did everything. He designed the company's computer network, selected the fonts for new slides, interviewed every candidate, reviewed every purchase. Working most nights until ten, he then stayed up past midnight reading scientific journals, like his father. He worked every Saturday in a den off the kitchen at home. When he left the house now to drive to Logan Airport, as he often did, his two-year-old son refused to kiss him good-bye.

Because small biomedical companies are years away from having any income, time equals money for them in a perfect sense. A company's lifeline is computed by its "burn rate." Six months after Vertex started writing checks, it was burning $15,000 a day. In a year the figure would double. Boger never forgot what Vertex's burn rate was or what it demanded of him. On the day he and his family moved from New Jersey to Concord, an immaculate suburb far enough from Cambridge to be in another area code, he tripped down a flight of stairs; for the next two weeks he dragged himself around on crutches. Amy, his wife, seldom saw him anymore. They had met when she was at Radcliffe and he was tutoring in exchange for room and board at Harvard. A pediatrician, she was now taking time off to care for their two sons and was pregnant again. Resigned and supportive herself, she was having a harder time with the boys. On a night soon after the SAB meeting, Boger was in Vertex's lunchroom at about 8 o'clock when the phone rang. It was Zachary, his five-year-old. "OK, I'm leaving," he said. "I'm coming home right now." Hanging up the phone, Boger shook his head admiringly: "He knows the area code." A half hour later Boger raced out of the building, listing from a foot-high stack of journals under his arm. He flew to Barcelona the next day.

Since the goal of most researchers is publication, not lecturing, most scientific conferences are torpid rehashes of old work punctuated by tantalizing previews of forthcoming articles. Those with nothing new to say speak too much; those with real news, too little. But Starzl had published sparingly on FK-506. And so as the cavernous auditorium at the University of Barcelona began filling shortly after noon with some 500 researchers from around the world, there was a rare sense of anticipation. Starzl himself, a handsome, graying figure at sixty-three, six feet tall and thin on the verge of being gaunt, remained in the background, chewing a nicotine substitute incessantly though gritted teeth. However, there was no mistaking his role. Of the thirty-one papers to be presented, twenty-six were from his group. Only one of those had already been in print, a summary of the first sixty cases using FK-506 that had appeared in the previous week's issue of *The Lancet*, a British medical journal.

Starzl's findings were breathtaking, defying belief. FK-506 had first been given to liver transplant patients who either were rejecting their new organs on cyclosporine or couldn't tolerate the drug's harmful side effects—so-called rescues. Not only did most of them improve dramatically with FK-506, but many of their rejection episodes simply stopped: The patients didn't have to be retransplanted. There was some evidence (Starzl called it "minor league") of nephrotoxicity—kidney poisoning—but none of the hirsutism, massively swollen gums, or tremors that occasionally made transplant recipients in their teens so distraught that they quit taking cyclosporine despite being told that doing so could kill them. Overwhelmingly, patients on FK-506 felt better, recovered sooner, left the hospital quicker, and needed fewer other drugs. Their hospital bills were cut almost in half—from $244,863 with cyclosporine to $134,169 with FK-506.

One after another, the members of the Pittsburgh group built a powerful clinical case for FK-506. Yet to Boger the most tantalizing talk was that of a young Japanese surgeon, Dr. Nukio Murase, who never saw human patients. In halting English, she reported on recent animal experiments in which the entire lower viscera of rats—liver, kidneys, stomach, duodenum, pancreas, large and small bowel, everything but the spleen—were successfully transplanted using FK-506. Such experiments had been tried previously with cyclosporine, but none of the rats had survived more than thirteen days. However, Murase's animals had lived as long as seventy-two days (some would eventually live more than seven months), with no evidence of rejection even after she had discontinued the drug. Incredibly, the surviving rats had all put on weight.

Boger marveled at the implications. If FK-506 was potent enough to keep the immune system from rejecting such a forbidding mass of tissue, it could probably be given in small enough doses to cut down substantially on side effects; therefore it would win not only the transplant market, but perhaps the autoimmunities market as well. And yet it was also inconceivable that Starzl had ordered the rat experiments simply to confirm the drug's potency. If Starzl's people were doing multivisceral grafts in animals, it could only be as a prelude to attempting the same operation in humans.

The session, which began at 1 P.M. and was supposed to end at 7, went on with only a single fifteen-minute break until 10:15. No one left. It was a landmark meeting, one of the very few most of them would ever attend. And yet for Starzl it was more than that. It was the apotheosis of a career that spanned practically all of modern transplantation and clinical immunology—a career of spectacular highs and profound lows played out against some of the most dramatic events in experimental medicine of the past forty years. Starzl's heroic rescue of FK-506 had brought him to a central place in the world of scientific medicine, and it had enlarged him. And yet now, in Barcelona, he was also forced to concede, however unintentionally, the great paradox of his triumph. If FK-506 was all that Starzl said it was, it would be criminal not to give the drug to anyone who might benefit from it. Yet no drug—especially one as powerful as FK-506—is ever approved without careful comparison with those agents already available. It was lost on few people in the room, least of all those like Boger who were from the drug industry, that FK-506 might be *too* good a molecule in the hands of someone as daring, as unyielding, as messianic, as Tom Starzl.

"Once we started switching patients over to FK-506, we couldn't get people to take anything else," Starzl told the audience. "We were faced with a practical and ethical dilemma in continuing to work in a controlled manner with this drug. By summer we were experiencing a patient revolt as word spread in the hospital on the success of FK-506."

Starzl has always been a figure of superhuman perseverence, determined to choose the most difficult problems and attack them with the most murderous acts of will. He was born and raised in LeMars, Iowa, a heavily Catholic county seat in the hog and corn country near the South Dakota border, where his mother was a nurse and his father owned a newspaper. Rome Starzl, a steel-eyed second-generation German-American, inherited the paper from his own father, who was tried and acquitted for sedition during World War I for editorializing against the inhumane treatment of soldiers en route to France. The stain of the episode—Rome Starzl, then attending officer's training school in Texas, had actually been the piece's author—never left the family, and it embodied for the young Starzl the suffocating narrowness of small-town life.

Starzl's recollections of his father, like Boger's, are flavored with disappointment—a man, despite hard work and a good mind, roaming through life frustrated and less than successful on his own terms. Rome Starzl ran the family paper out of obligation. His real love was science. He was an inventor whose innovations were ingenious but failed to catch on commercially and, during the late 1920s and early 1930s, a science fiction writer of certain but limited success. His first published story, "Out of the Subuniverse," remarkably foreshadowed "The Fantastic Voyage." It was about people who shrunk themselves to explore a microscopic cosmos—a genre that the elder Starzl would continue to pioneer until he was forced, midway through the Depression, to abandon fiction for the safer middle distance of running the *Globe Post*, and which his son would ultimately consider a metaphor for the life he chose and that he, young Tom, dreaded above all else.

"My father stayed in that tiny universe within a universe but never was reconciled to its limitations," Tom Starzl would write in his memoirs. "When my time came, I wanted to escape. The fear of failing and being forced to return defeated for a lifetime of regret made trivial all other fears, even death. Like a grim watchdog, this feeling stayed until the long course was run."

World War II catapulted Starzl out of LeMars. He joined the navy, graduating from Westminster College in Fulton, Missouri, where he'd been assigned for officer training and where, as a Latin scholar with ideas of becoming a priest, he was a gofer for Winston Churchill during his famous Iron Curtain speech in 1948. From Westminster he went to Northwestern University Medical School in Chicago. Starzl was indefatigable. In five years at Northwestern, he finished an M.D./Ph.D. in neurophysiology under the brilliant and imperious brain surgeon Dr. Loyal Davis, Nancy Reagan's stepfather, while working almost every night at an all-night surgical clinic in one of Chicago's worst slums. He went on to do an internship at Johns Hopkins University. Though Hopkins was widely regarded as having the best surgical program in the country, even Starzl found the school "ruthless." Interns were on duty twenty-four hours a day, every day of the year except for one week off. The system was "pyramidal," with students being culled after each rotation so that only one in nine made it through the entire program. Four years after he began in a class of eighteen, only Starzl and another student remained. He was thirty years old.

The fury with which Starzl left LeMars didn't abate at Hopkins; it intensified. He ate and slept haphazardly, smoked three packs of cigarettes a day, and pushed himself beyond his physical and emotional limits. Despite his endless hours in the hospital, he had no money, as Hopkins interns were not then paid. By his own description, he was turbulent and confused. In 1955, he left Baltimore with his wife and infant son for Miami, where he worked in one of the busiest and most notorious hospitals in the world, Jackson Memorial. "Nowhere have I seen such a parade of sorrow," he would later recall, ". . . drowned children, raped and murdered women, blond suntanned muscle builders with neat bullet holes in their heads."

For two years Starzl operated slavishly, performing some 2000 operations—three a day. When he wasn't operating, he worked in a primitive animal laboratory he'd set up in an empty garage across from the emergency room, experimenting on dogs that he'd gotten from the pound. In part because of the types of injuries he was seeing in Miami—gunshot wounds to the gut, massive internal bleeding due to cirrhosis—Starzl began to concentrate on the liver. Typically, his frustrations were huge. He was dissatisfied with the limits of abdominal surgery, which then was mired in academic discussions over the best techniques for arterial repair, and with physiology, which offered few new solutions for saving lives. Even more, he was dissatisfied with himself for impoverishing his family and not yet, at thirty-two, having chosen a life's work. Pent up, he developed an ulcer. "I felt," he would write, "like a missile looking for a trajectory."

The next fall Starzl returned to Northwestern, having decided to stay in experimental medicine. Cancer and open-heart surgery were then the promising fields, and Starzl planned to perfect the techniques of the heart and lungs. Yet that, too, was not enough. To Starzl, thoracic surgery looked much the way LeMars had looked to his father: safe, conventional, and ultimately stultifying. "The allure of cardiac surgery had faded for me," he wrote. "Cancer research was a possibility, but the optimistic literature of that period suggested that a cancer cure was close at hand. I thought I was too late."

In contrast, what most appealed to Tom Starzl was the struggling field of transplantation, still dawning and considered hopeless by most experts. "The literature on transplantation of the kidney and other organs was uncompromisingly pessimistic, and therefore, paradoxically attractive," wrote Starzl, who, it will be recalled, feared failure more than death. "This looked like the vacuum I was seeking." As if to ensure that nothing in the path he chose for himself would be even remotely easy, Starzl decided to concentrate again on the liver, the body's largest and most complicated glandular organ. It was 1958. The only other team seriously in the field was at Harvard, where four years earlier the first successful kidney transplant had been achieved between identical twins. Starzl could hardly have asked for a more challenging competitor. The Harvard group was directed by Dr. Francis Moore, chairman of surgery at Peter Bent Brigham Hospital, the chief clinical laboratory for the medical school. Moore, at age forty-five, was already a titan of academic medicine; the Brigham, one of the two or three best research institutions in the world. As in Miami, Starzl, yet to receive his first academic appointment, found a place to operate near Northeastern and began carving the livers out of dogs.

It was a sacrilege rooted in the ancient past. The earliest descriptions of animal hybrids were of the monstrous fire-breathing chimeras of Greek mythology—a lion's head, a goat's body, a serpent's tail. In the shifting Middle Ages, the miracle of a leg graft by two saints, Cosmas and Damian, was a favorite subject of Renaissance painters. Though reports were sporadic, isolated attempts at transplanting whole organs began in Europe in the late nineteenth century and continued until the early 1920s. Technically primitive surgeons grafted the kidneys of sheep, pigs, goats, and lower primates into humans with abysmal results. None of the organs functioned for more than a few hours, and the patients all died within days. Nature, it appeared, abhorred the fusion of animal parts as much as the Greeks did.

Though none of the organ recipients lived long enough to reject their grafts, the search for the biological barrier to transplantation focused on the immune system. Ever since the 1870s, when Louis Pasteur first showed that invading germs provoked specific defensive chemical responses in the body, scientific immunology had been on the rise. In the 1890s, Pasteur's work was advanced spectacularly by another chemist-turned-biologist, Paul Ehrlich. Studying how dyes bind to wool, Ehrlich ushered biology from the level to which Pasteur had brought it—the cell—to its ultimate arena: molecules. He showed that immunities were triggered by certain molecules on the surface of cells "recognizing" others. Unlike Pasteur's work, this couldn't be seen under a microscope. But Ehrlich's theory that molecules bind according to specific affinities and that their interactions make up all life instantly became the touchstone for all subsequent biomedical research.

For transplanters the question was, What molecules caused the body to abhor foreign tissue and could they be disarmed? For fifty years, the problem addled immunologists. Most surgeons had long since given up. But World War II, with its "improved methods of inflicting wounds and burns," as historian Arthur Silverstein points out, revived interest in at least one type of transplantation: skin grafting. Returning to the problem, researchers soon discovered the long-sought immunological barrier to transplantation. Molecules on the surface of a class of immune cells called T cells distinguished between substances that were native to the body and those that weren't—between self and nonself—and initiated the production of new cells to track and kill the latter. Intriguingly, these molecules appeared to be shared genetically by some, but not all, family members.

The observations helped rationalize the first successful kidney transplant between identical twins at the Brigham in 1954—a surgical procedure that had first been devised by surgeons in Paris with organs from guillotine victims but had always ended in rejection. But the kidney was an unusual organ: There were two of them. And few people had identical twins with matching tissue types. Almost all other transplants would require taking organs from dead donors almost certainly unrelated to the recipients. Unless a way could be devised to lower the immunological threshold by suppressing the immune system, the prospects for transplanting hearts, lungs, and livers were unremittingly bleak. "On the whole," wrote one of the fathers of modern immunology in 1961, three years after Starzl began experimenting with liver transplants in dogs, "the present outlook is highly unfavorable to success."

This was Starzl's "vacuum," his "trajectory." The physical removal and resection of the human liver, an organ about the size and shape of a boxing glove wedged inconveniently against the diaphragm, was daunting enough. It had half the body's blood pushing through it at any given time; all its major vascular connections and ducts, tying it to the body's largest vein as well as other organs, were buried out of sight; and it started to die almost instantly upon removal. The logistics of getting a liver out of someone who had just died and into someone else who would die without it were nightmarish, prohibitive. And yet the far larger problem, as molecular scientists had now shown, lay ahead with keeping the recipient's immune system from destroying the graft.

As a surgeon, Starzl knew nothing about controlling the immune system, but then, hardly anyone did. Indeed, few therapies have begun more blindly than immunosuppression in the years after World War II. With the goal of simply knocking out T cells, the first transplant patients received full-body irradiation. It was like fixing a watch with a hammer; the procedure, similar to being exposed to a nuclear blast, destroyed their immunities entirely. Patients were like the "bubble boy," who lived in a Houston hospital for twelve years before dying of massive infection within weeks of being released. Azathioprine, a powerful cell-killing anticancer drug, was also used, but it proved far too toxic over the long term. (Because the threat of rejection remains constant, transplant patients must take immunosuppressants as long as they live, lowering their tolerance for side effects, like kidney poisoning, that are cumulative.) One approach, developed later by Starzl, involved installing a shunt at the back of the neck to drain the immune system of billions of white blood cells, then pumping the patient full of antibiotics and antifungal drugs: exchanging artificial immunities for natural ones. That, too, had to be abandoned.

Starzl attempted the world's first human liver transplant on March 1, 1963, on a three-year-old boy named Bennie Solis. By then he had performed more than 200 transplants on dogs in Chicago and in Denver, where he had moved to continue his research at the University of Colorado. On the theory of lowering doses to increase tolerance, he had come to favor a "cocktail" approach to immunosuppression—radiation, azathioprine, and cortisone—and had planned such a therapy for Bennie. It was a moot issue. The boy bled to death on the operating table.

Two months later, Starzl grafted a new liver into a forty-seven-year-old janitor who was dying of liver disease. The man lived twenty-two days—longer than two of his next three patients. Though the surgery had been a success and though the man had not rejected his graft, Starzl was scorned and rebuked. An editorial in *The Annals of Internal Medicine* condemned his work as "cannibalization." Another journal accused him of "grave robbing."

Starzl returned undeterred to the laboratory. He consumed in one year 10 percent of all the research dogs in the country. During the next two decades he would perfect many of the surgical techniques that would make the mechanics of organ transplantation more routine. He refined a bypass system that allowed blood to be diverted to the lower half of the body during surgery: His patients no longer bled to death. He developed preservative solutions that extended the time the liver could survive outside the body from four to ten hours, making it possible to ship organs by air between cities. But the defining challenge, as ever, was in immunology. The inadequacy of the available drugs resulted in a therapeutic knife edge: "Use too much and the patient doesn't survive," said a surgeon at the time. "Don't use enough and the transplant doesn't survive." With the failure of Starzl's shunt therapy in the late 1970s, the field appeared at a dead end. Heart transplants, which had captured the world's imagination a decade earlier, all but stopped. By 1980, the year Starzl moved to Pittsburgh, survival rates in transplant patients were plummeting, and even Starzl had to concede that without a more specific drug the procedure was likely to die out from its own cruel ineffectiveness.

The Hardanger Vidda, a vast, forbidding highland plateau in Southern Norway, is nearly the size of Connecticut, yet so unrelievedly barren that the only buildings are climbers' huts and the summer shacks of herders. There is no permanent population. Though a portion of it has been declared a national park, the Vidda (waste) is considered by most Norwegians appallingly inhospitable, a primeval terrain of lichens, mosses, and treeless grasslands dotted with glacial outfalls—enormous boulders and plunging, frigid lakes favored only by Nordic trout fishermen. In 1943, after Norwegian saboteurs destroyed a secret German heavy water plant in the nearby town of Rjukan, the brigade's leader fled to an isolated hut in the Vidda, where he was tracked for two years before being killed by Nazis. Twenty-five years later, in the summer of 1978, a vacationing microbiologist working for the Swiss pharmaceutical company Sandoz toured the Vidda. During his stay, he routinely scooped up a spoonful of its alkaline, calcium-rich soil and placed it in a sealed petri dish to bring back for the company's natural products screen.

There are in any fingernail of dirt between 50 million and 100 million living organisms, representing 3000 to 4000 species and living in a constant state of chemical war. To ensure their own survival, these microbial colonies develop molecules that are lethal to one another. Thus it was with the sample from the Vidda. Screeners at Sandoz discovered it contained a new molecule, which they named cyclosporine, that was fatal to a broad range of fungi. As Sandoz was screening for antifungal drugs, the compound appeared promising. But it turned out to be useless against those parasites that attack humans. For two years the drug was shelved until it was routed to an immunologist named Jean Borel, who discovered that it was also a potent immunosuppressant. Borel's story—because immunosuppression was then considered a small, unimportant market, Sandoz repeatedly tried to kill the program, forcing Borel ultimately to test the drug on himself—quickly became famous within the drug industry, although interpretations of it vary. Screeners believe it exalts screening. Antiscreeners, like Boger, believe it shows the hair-thin luck on which screening ultimately rests, and the lunkheadedness of most big drug companies. Borel himself is more sanguine. "I'm afraid the definition of a scientist," he has said, "is a man who can take frustration without end."

Cyclosporine more than resuscitated the field of organ grafting. After a decade of failure, suddenly there now was a drug that not only disarmed T cells, but didn't fatally undermine the rest of the immune system. How it worked, what the molecule bound to—those were secondary questions to be answered later in biology labs. Now, in the late 1970s, the overriding question for transplanters was toxicity. Could the drug be tolerated? The initial trials on humans revealed a terrifying medley of complications: diabetes, gout, neurotoxicity, tumors, mood swings. The worst of these from a clinical standpoint was kidney poisoning. Up to 80 percent of those taking cyclosporine eventually developed nephrotoxicity so severe that in many cases they required additional transplants.

The first human trials of cyclosporine were conducted by Sir Roy Calne of Cambridge University and were dismaying enough to dash the hopes of most transplanters. Starzl, however, had always believed that toxicity could be controlled by reducing dosage. He got the drug and immediately began administering it in a cocktail with steroids. The result was adequate immunosuppression at a therapeutic price—a wider range of diminished side effects—that most doctors and patients found tolerable. Survival rates of transplant patients suddenly soared. Transplantation units proliferated. News stories about people snatched from death with other people's organs became nightly staples. "We've gone," Starzl announced, "from the unattainable to the routine."

Starzl had not developed cyclosporine; by the rights and rules of medicine, the credit belonged to Borel and Calne. But cyclosporine made Tom Starzl the most famous and influential transplanter in the world. Transplantation, which had been macabre and dismaying, now gleamed with optimism, and no one was more emblematic of its fearless new image. Starzl's all-consuming determination, his stamina, his daring, his obsession—as well as his long-standing friendship with the Reagans, who helped qualify liver transplantation for reimbursement—advanced the field dramatically throughout the 1980s.

There seemed to be no limit to what Starzl could—or would—now do. In 1984, during a grueling sixteen-hour operation, he replaced both the heart and liver of a six-year-old girl, Stormie Jones, who within two weeks was skipping around the hospital. More than once he performed back-to-back liver transplants, working up to seventy-two hours at a stretch without sleep. When he did sleep, it was in a flannel-lined sleeping bag in the aisle of a chartered jet on the way to procuring organs or in blood-spattered scrubs on the floor of his office, underneath a round wooden table piled high with unfinished paperwork. Calls from all over the world, as many as a half-dozen a day, now flooded into Starzl's office, which he began to call the "court of last resort." With cyclosporine, Starzl and his team could now extend indefinitely the lives of people who otherwise would surely die—if only organs could be found for them, if only they could get to Pittsburgh. Shortages of organs and critical care beds, not antirejection drugs, now imposed the severest barriers on grafting, and it was against those that Starzl most frequently railed. As for cyclosporine, only the drug's nagging toxicity—and the fact that Starzl himself hadn't developed it—sustained his dissatisfaction.

In August 1986, Starzl flew to Helsinki for a meeting of the International Transplantation Society. Cyclosporine dominated the proceedings, which included a promotional side trip to the Hardanger Vidda sponsored by Sandoz. Starzl, as ever, was looking forward, not back, and was more interested in the conference. Like many others, he'd heard that a Japanese surgeon named Takio Ochiai had data on a new immunosuppressant that was one hundred times more potent than cyclosporine. Rumors of new drugs were not unusual at such meetings, and Ochiai, a middle-level professor at Chiba University, was assigned a small room to speak in. Quickly, it filled to overflowing. Borel was there. So was Calne. People attracted by the crowd pressed three- and four-deep at the doorway, straining to hear.

Ochiai's data were tantalizing. Experimenting mainly with beagles imported from the United States, he showed that the new compound, FK-506, worked in much the same way as cyclosporine by slowing the proliferation of T cells. Like cyclosporine, the molecule had shown up in a soil screen. It had been discovered in a dirt sample taken from the lower slope of Mount Tsukuba, an hour by train from Tokyo and just a few kilometers from the central screening facility of its discoverer, Fujisawa, the third-largest drug company in Japan. (FK-506 was an abbreviation of the molecule's identification number, FK-506009.) Like cyclosporine, FK-506 also appeared to be toxic. In dogs given less than immunosuppressive doses of the drug, Ochiai said that fully all had developed vasculitis, a weakening of the blood vessels of the heart.

After Ochiai's talk, Borel stood up and pronounced both the presentation and the drug important new developments. Calne added that he too had the drug—although he hadn't begun testing with it—and also considered it promising. Once again, early reports of toxic side effects were damning, but anything that appeared better than cyclosporine was considered hopeful by transplanters, who found Sandoz's product, despite its effectiveness, unpredictable and hard to control.

Starzl, never tentative, leapt. Two weeks after returning from Helsinki, he flew to Japan to ask Fujisawa for exclusive rights to test FK-506. Starzl had never before directed the sort of basic research required for bringing a drug to clinic. Nor was his characteristic hubris suited to the impartial requirements of the job. Even if Fujisawa was looking for such a partner, the situation was complicated by a cross-licensing agreement the company had with another pharmaceutical firm, Fisons. It was through Fisons that Calne had obtained his own FK-506. Fortunately for Starzl, when Calne began testing the drug, he also found it caused vasculitis in dogs. More, Calne believed it had been fatal in some baboons. Starzl spent two weeks in Japan while Fujisawa deliberated, eventually returning home with less than a gram of FK-506—enough to begin cell assays and some experiments in rats—and a tentative commitment from Fujisawa not to give other transplanters the drug.

By the end of the year, Calne concluded that the drug was too toxic for humans and left the field. Starzl at last had what he had long desired—complete control over the testing and development of a promising new molecule. That it appeared to be poison barely intruded on Starzl's sense of inevitability. As with himself, he was determined to make the molecule succeed.

"We were like a human machine to bring the drug through," Dr. Mike Nalesnick, a pathologist pressed into service by Starzl to measure drug doses, would recall. "You wanted to make sure you didn't blow it. It was your ass out there."

Starzl had been influential at the University of Pittsburgh Medical Center before; he was its star, its first international figure since Jonas Salk, and Pitt was rising in prominence because of him. Now his demands and the reaction to them among those who were less favored multiplied furiously. Insisting on independence, he refused all financial support from Fujisawa—a decision that eventually would cost the university up to $8 million a year. Even before the Helsinki meeting, Starzl and a small group had began meeting

Monday nights to discuss new immunosuppressive therapies. Now the group burgeoned to nearly one hundred—surgeons, oncologists, organ specialists, animal toxicologists, pharmacologists, technicians. Where Starzl couldn't find an expert, he willed one, as he did with Nalesnick, by importing someone from another field or, as with an Italian pancreatic specialist, Dr. Camillo Ricordi, simply ordering the defenseless man to leave Milan for Pittsburgh. He commandeered labs, operating rooms, scarce intensive care beds. Characteristically, he stopped at nothing.

During the next twenty-nine months, Starzl's team performed hundreds of studies on mice, rats, pigs, baboons, and dogs, eventually showing that Calne's dog studies were at least ambiguous and should not stop the drug's progress. ("Drugs kill dogs," says Boger about the animal's well-known proclivity for showing side effects unobserved in other species, "and dogs kill drugs.") Working closely with the FDA, which sanctions all such clinical trials in the United States, Starzl's team produced compelling enough evidence to warrant testing the drug in people. Calne and others still claimed FK-506 was too toxic for humans. Criticism of Starzl's ad hoc research methods and army of medical conscripts was considerable. But by now, Starzl was absolutely convinced that the FK-506 was the best immunosuppressive therapy ever developed.

"In every animal model, in every organ, FK-506 won," he said. "It wasn't bold to give the drug to people. It was the most responsible thing we could have done."

As a first candidate for a powerful medication of unknown risks and benefits, Robin Ford was a natural choice. Twenty-eight years old, she was dying. Her third liver graft in three years was failing, she had lost a kidney, and her remaining one was so damaged by cyclosporine that it, too, was shutting down. On February 28, 1989—less than a month after Boger chose immunophilins as Vertex's inaugural project—doctors prepared Ford to receive cyclosporine and FK-506 together in a last-ditch attempt to save her life.

Starzl aside, concern over the new drug remained such that Ford was required to sign a release detailing those side effects most prevalent in animals—vomiting, weight loss, elevation of blood sugar—that read: "Other side effects, including death, are possible in humans, but cannot be anticipated." As in the first stage of all clinical testing of new drugs, the initial goal was to establish that FK-506 was not prohibitively toxic. Officially, at least, Ford was not expected to improve on the drug, only to show whether it could be endured. A full complement of emergency equipment, including a resuscitator, was wheeled into her room.

Ford spent much of the first day under critical observation. Starzl flew to Paris to attend an emergency meeting to discuss the feasibility of trans-Atlantic organ sharing. Those left to supervise Ford's treatment—three surgeons, John Fung, Ashok Jain, and Saturo Todo, and a pharmacologist, Raman Venkataramanan—were "very anxious," Jain recalls. "We didn't know the optimum dose. We didn't know how to treat her."

On the third day of taking both drugs, Ford began vomiting. She complained of a severe headache and nausea. "I said, 'Oh God, maybe the drug really is no good,' " Jain recalls. Distraught, he and others began to argue in favor of stopping the FK-506, though Ford herself wanted to stay on it. According to Starzl, his staff came within "a hair" of pulling her off FK-506—a decision that not only would have caused her to reject her new liver but almost certainly would have killed the drug's progress outright—before he, returning from Paris, examined her. Starzl believed Ford was reacting to the two powerful immunosuppressants in her system. Since she couldn't tolerate cyclosporine, he reasoned, she had nothing to lose by coming off the drug. Consulting with the FDA, Starzl decided to take Ford off cyclosporine to see if she could survive on FK-506 alone.

Forty-eight hours later, Ford's nausea dissipated. She was able to hold food. Within two weeks, biopsies showed, her immune system stopped rejecting her new liver. More startling, her liver function began to improve: From this one case, FK-506 seemed actually to revive dying tissue. "It was like a miracle, like a dream," Jain says. Because severe kidney damage, unlike serious liver damage, is irreversible, Ford eventually required another kidney transplant, but with the FK-506, she tolerated that, too. After about a month she was home, returning eventually to work and a normal life.

• • •

As Boger would guess in Barcelona, Starzl's attachment to FK-506 went deeper than wanting to advance the cause of liver transplantation—a cause he'd more or less helped bring to its climactic stage with cyclosporine. Starzl was more relentless than that, his goals hungrier, more personal. The testing of FK-506 in humans was necessarily secretive—no drug developer wants to prejudice the FDA, or inflame the public prematurely—but Tom Starzl was not a reclusive man.

As at many other times in his life, Starzl in the first half of 1989 was embroiled publicly on several fronts at once. Six weeks before giving FK-506 to Robin Ford—Boger had sensed correctly here, too—Starzl was forced to halt a new series of multiorgan transplants at Pittsburgh. Two three-year-old girls who had each received five new organs—stomach, liver, pancreas, and large and small bowel—had died the previous year. Though Starzl blamed cyclosporine, the deaths invited yet another round of public questioning of his methods. "I believe," Harvard's Francis Moore, Starzl's primary competitor in the early days of liver transplantation, wrote in a leading medical journal, "that this procedure shouldn't be performed again until it has been shown . . . that there is a palpable likelihood of success."

Meanwhile, Starzl had begun to talk openly at the medical center and, in a few cases, publicly about FK-506. By the time the next rescue patient was given the drug—a thirty-eight-year-old New Orleans contractor named Lester Wilson, who'd previously received five new livers—Starzl was so convinced of its superiority that he no longer considered it ethical to prescribe cyclosporine. The position inflamed the University of Pittsburgh Medical Center's Institutional Review Board (IRB), which was responsible for protecting patients—and the hospitals themselves—from overzealous experimenters and overstated claims. "Starzl says, 'Trust me, I'm the expert,' " complained Dr. Richard Cohen, head of the thirty-member IRB. "We've had to protect him from himself."

That anyone should think that Starzl would need such protection only drove him to more militant extremes. He responded by attacking in the darkest, most unyielding terms those who insisted that he compare the two drugs blindly in clinical trials, even though, as he knew, the FDA would eventually insist on such comparison testing. "In this cruel world there have been too many examples of people who've conducted unconscionable experiments on human subjects because they said their superiors ordered them to," Starzl said in a speech at the medical school. "The worst of them ended up in the dock at Nuremburg. I don't want to be put in that position."

Starzl received a temporary reprieve from his travails in April, when the U.S. Justice Department, after years of investigation, decided not to prosecute him for violating the National Organ Transplant Act—an act he helped foster. A Pulitzer Prize-winning series in the *Pittsburgh Press* had shown that in the mid-1980s Starzl's unit transplanted a disproportionate number of foreign nationals, who paid higher rates and were not often as ill or hadn't waited as long as other patients. Normally skillful with the media, Starzl had dug himself deeper by telling a reporter that the organs reserved for foreigners were "crumbs" and "bottom of the barrel." Eventually published, the remarks had been impossible to wash away, Pittsburgh and the world being, in that respect, no larger or more forgiving than LeMars.

Assailed in the press, entangled with the medical center review board, barred by the lack of a more potent immunosuppressant from developing new transplant operations, stung by peer criticism, Starzl was explosive. A notorious taskmaster, he began to drive his people even harder in anticipation of the meeting in Barcelona, where the world would see their work and where he expected they would be vindicated.

In September, Starzl flew to Minneapolis to give a talk on "cluster" transplants. These involved removing the liver, stomach, spleen, duodenum, pancreas, and large and small bowels and replacing them just with a liver infused with pancreatic islet cells. Another of Starzl's experiments, the procedure made it possible for transplant recipients who'd had their pancreases removed to live without depending on daily insulin shots. Animal studies with FK-506 had shown surprising success with the procedure, which Starzl now suggested might be used to cure people whose abdomens were ravaged by metastatic cancer. The talk was vintage Starzl—low-key and dignified and startlingly heretical. Transplanting for Starzl had always been a straightforward therapeutic trade-off: remove and replace the sick parts of an otherwise healthy person to save the whole. The more potent the immunosuppression, the more parts that could be removed. After the conference, Starzl was approached by Lawrence Altman, a reporter from the *New York Times*, who said he'd like to visit Pittsburgh and write about the cluster procedure.

For credibility's sake as well as for the planting of a firm flag scientifically, Starzl did not want his clinical work in FK-506 announced publicly before it appeared in *The Lancet*. But from the moment Altman arrived in Pittsburgh, it was clear something more was happening than animal studies. Whether or not it was the "patient revolt" Starzl would describe in Barcelona, Altman, a physician, couldn't help but inquire. He wasn't alone. A reporter from the *Pittsburgh Post-Gazette*, Henry Pierce, also had begun hearing about Starzl's new miracle drug and was preparing to write about it.

Starzl tried to delay the newspapers from publishing until the conference in Barcelona, but without a guaranteed exclusive, neither paper was willing to wait. Then, in mid-October, Starzl heard from Pierce that the *Post-Gazette*'s editors feared being scooped and were planning to print his story. He immediately called Altman, thereby crafting the release date—October 18, 1989—himself. It was the day of the San Franciso earthquake, but the stories, ready in time for the early editions, both made the top of the front page. In the case of the *Times*, the placement guaranteed wide distribution by the news services, the television networks, and Cable News Network (CNN). By the end of the day, calls began flooding into Pittsburgh from around the world—journalists, doctors, patients, drug industry analysts—all wanting breathlessly to know more about Tom Starzl's miraculous new drug.

At Vertex, several hearts sank at the news. Scientists don't rely on the *Times*, or any newspaper, for scientific information. But to see the *Times* give the story rare above-the-fold coverage—headlined "Great Success with Drug in Transplants of Organs," the piece ran sixty column inches, more than the paper devotes to the election of most European heads of state—was unnerving, especially as it touted the drug's safety. "FK-506 has shown little evidence of toxicity in humans so far," Altman wrote. It wasn't long before chemist

David Armistead, speaking by informal proxy for several others, wondered aloud at a rump meeting of the Vertex scientific staff: "What if this is the wonder drug of the century? Why should we go after it?"

Boger, who is bemused in general by the symbiosis of the medical community and the press and in particular by "breakthroughs" that are announced to the public before other scientists have reviewed them, was not in serious danger of being second-guessed because of FK-506's sudden publicity. The drug still had a long way to go, and everyone agreed that Vertex's rationale for chasing it remained sound. But some members of the board of directors already had called anxiously that morning, and how Boger responded was the first real test of his leadership. It was not too late to launch another program. With any drug project, it was always better to cut one's losses early. Characteristically, Boger was staunch and unequivocal.

"This is good news for us," he told the staff. "It raises the stakes. It shows that cyclosporine is a beatable compound, and it heats up the area. This whole way of going after autoimmune diseases looks more reasonable because of [FK-506's] reduced toxicity." Indeed, his first comment after reading the *Times* piece had been: "I'd be happy if by the time we have a compound to go into clinical trials, FK is doing $3 billion in sales. I want Fujisawa to succeed."

As with so much else about Boger, the iconoclasm at first seemed brash. But as the others began to hear him analyze the situation, they quickly came to see that he was right. FK-506 was probably going to be a blockbuster drug: It would steal the transplant market from cyclosporine and would probably be safe enough to open up the autoimmune market. But, the *Times* article notwithstanding, it was unlikely to be safe enough to capture that market as well, so the greatest prize would be left for the next generation of compounds. Vertex couldn't have asked for a better scenario.

Of course, every major pharmaceutical company in the world would soon come to the same conclusions and rush ahead with programs in immunosuppression, clogging the field. But that burst of attention was likely to be less than it seemed. Most of them had already spent the better part of a decade screening for new molecules to replace cyclosporine, with only a single contender emerging—FK-506. Hundreds of chemists at Sandoz and elsewhere had tried for years to redesign cyclosporine to make it safer without a single clinical candidate to show for their efforts. Without their own leads, many of the companies would now likely end up looking to underwrite a small specialized research partner, which played directly into Vertex's strategy. Either way, it appeared, Vertex won.

There were only two problems. The first was Merck, which already had a sizable program in the area. Merck had set the industry standard with its billion-dollar hypertension drug, Vasotec, for capturing a growing market with a second-generation compound. Confronted in the mid-1980s with Squibb's development of Capoten, a drug based on a molecule found initially in pit viper venom, Merck had put scores of chemists on the task of improving it, then followed up with a withering sales campaign so effective that it ended up beating Squibb in the market even though Capoten was launched first and was much the same drug. Huge resources could be brought to bear if Merck suddenly got more serious.

Which led to the second concern: Vertex's unfinished labs. As Merck's most spectacular defection in years, Boger knew as well as anyone the limits of its science. As he told Knowles at the SAB meeting, he thought Vertex could defeat Merck going head-to-head. But that presumed an equal start. Merck, which had the greatest research capacity of any drug company in the world, had been working on improving FK-506 for more than a year under a program set up largely by Boger himself. Vertex, meanwhile, had fifteen scientists sitting on their hands. On the day the *Times* piece came out, they began the morning bristling with competitiveness, but with no place to put their fire, they were soon back at their rented steel desks, rechecking their equipment orders and wondering when they would start. For all but Boger, it was another slow day.

CHAPTER FOUR

"This is supposed to be a delicacy."

Matt Harding pinched a thymus filet, his hands encased like a surgeon's in blood-smeared latex gloves. "Sweetbread. I've never had it, but I think I've seen too much of it to give it consideration." He cut the filet into thin strips with a pair of surgical scissors.

Further down the lab bench John Thomson snipped aortas, looking like clam necks, from the pale meat. He plunked the slices of thymus into a double-walled steel cylinder roiling with liquid nitrogen. The pieces bobbed up dappled and lustrous.

"Looks like crabmeat, doesn't it," he said. "We'll track down some lemon for you."

Not a cattle truck but a thirtyish man in running shoes now came to Vertex each Thursday, fast-walking through the lunchroom toward the biochemistry lab with two sallow-looking bags of meat, an entrepreneur from a suburban abattoir doing a steady side business in research supply. Two months after the Barcelona conference and the opening of the labs, Thomson, Harding, and a lanky, rail-thin assistant named Matt Fitzgibbon spent the rest of the day and part of Friday cutting meat, the first stage of a protein "prep." The setting, the job, the ethos evoked a chop-house kitchen.

"If you dipped your 'and in it," Thomson said, nodding to the bubbling liquid nitrogen, gas spewing down the side of the cylinder, "it would turn into glass. You could smash it to dust." He shook his hand once toward the hard epoxy countertop, then jerked it back. "The 'frozen frog' effect," he said ghoulishly, recalling an undergraduate prank.

It was Thomson, the Australian, who would attempt to isolate from the stringy thymus meat the scarce, gossamer protein molecules of FKBP, the putative target for the drugs Vertex planned to design. But Harding, as codiscoverer of the protein and of cyclophilin, pitched in with the cutting, a familiar ritual to both that in Harding's case had led to a highly promising career in immunology, and in Thomson's, to a prized mastery of a dying art coupled with a mordant resentment of how other scientists viewed him.

On the day Harding arrived at Yale nearly a decade earlier as a postdoctoral in pharmacology, the first thing he noticed was other postdocs cutting thymus. Starzl's recent rescue of cyclosporine had revived interest in the molecule, and Yale was among those places looking for its protein target within the body—the next step in understanding how the drug worked. The effort had been partly successful: Students in the lab of a pharmacologist named Robert Handshumacher had been able to find several possible candidates. By linking cyclosporine to a chemical tether, they'd used the drug as a magnet, fishing proteins one by one out of a crude thymus "soup." The problem now was to find which receptor was biologically relevant and determine its chemical makeup.

It was daunting work. Proteins, as any cook knows, are promiscuous and unstable. To be active, they must be looped precisely, yet they're held together with the molecular equivalent of spit—the combined power of their weakest members, hydrogen atoms. Heat up an egg white, and the hydrogen bonds in ovalbumin, its most abundant protein, fly apart like the rivets in a submarine that has gone too deep; the individual proteins, having lost their internal bonds, flop open and glom together in a gel. Beat the same unheated egg white with a whisk, and the surfaces of the protein molecules explode into a frothy meringue. Extracting protein without destroying it requires the precision and delicacy of an obstetrician combined with a steely equanimity and bottomless capacity for loss.

Harding, a marathoner at the time, developed a steady routine.

He came to the lab early, ran six to eight miles out to the New Haven waterfront and back in the late afternoon, then worked until well after dinner. Quietly anxious, agreeable to a fault, he rarely took a day off in six months. Handshumacher's people had brought the protein to within a couple of steps of purification but hadn't been able to isolate it from a group of contaminants. Methodically, Harding experimented with different solvents and conditions until he was able to produce trace amounts of the pure receptor.

Doing more experiments, he and his labmates were able to calculate its molecular weight and, ultimately, characterize its sequence of amino acids, the variable links of any protein chain. Proteins are identified chiefly by their size and amino acid sequences and by other characteristics such as their behavior in water. Curiously, the protein they had discovered, cyclophilin, so named because of its affinity for cyclosporine, was highly soluble: it existed in that part of the cell between the membranelike sack and the nucleus. This would be consequential later on, as it meant that the protein was not one of those well-understood surface molecules that immunologists considered central to organ rejection, but something new, some part of the cell's inner workings.

Harding's paper, his first as a postdoc, appeared in *Science*, which ranks with *Nature* as one of the two most widely read journals in the world for early reports of biological breakthroughs. Though its significance was debated—"We had a structure without a function," he explains; "a lot of people said, 'So what?' "—the discovery launched him professionally, leading to a faculty appointment at the medical school and the opportunity, harder and harder to come by for a young scientist, to do independent research in a competitive field. Yet to lure the kind of money that would enable him to support a serious lab effort and to thrive academically, Harding needed a success independent of Handshumacher. He needed his own molecule.

Stuart Schreiber, though more secure than Harding professionally, also was interested in new molecules for similar reasons. For fifty years the great achievements of organic chemistry were in synthesizing biologically active molecules, and Schreiber, still in his twenties and already a full professor at Yale, was conceded to be one of the coming masters of the art. Like Boger, who had just had his own first great success at Merck, Schreiber was drawn to cyclosporine not for what it did, but for how it did it—its structure. Yet he also was blocked. Major work on the synthesis of cyclosporine had already been done in the lab of his department chairman, Sam Danishevsky. Danishevsky was Schreiber's chief sponsor. Much as he was ambitious, Schreiber was careful not to be thought uncollegial.

A leader in studying cyclosporine, and now cyclophilin, Yale pressed ahead. All universities covet such leading positions for the money and prestige they invite, and Yale, long Harvard's inferior in the sciences, relished especially any project on which it had an edge. It began collaborating with Merck, taking money in exchange for reagents and data and, not incidentally, bringing into contact Boger and several of those, including Harding, whom he would soon hire. The university also sought large-scale federal funding of the type that grants an institution instant benediction as a center of a particular field. As junior members of the team, Harding and Schreiber often found themselves sitting together in the back at group meetings, talking about science and falling into friendship.

"In October 1988 there was an NIH [National Institutes of Health] site visit," Harding recalls. "I already had my faculty position—actually I was half an assistant professor, and half a postdoc, which served everybody's best interests but mine—and the Warty paper had come out." (A Starzl conscript named Vijay Warty was the first to speculate publicly that FK-506, like cyclosporine, worked through a discrete binding protein in the body.)

"Stuart had become interested in FK, which was natural, and we were at this meeting where Bob [Handshumacher] was talking about the derivative of cyclosporine we had used to discover cyclophilin. I don't know if Stuart was aware of how easily you could pull protein out of a crude mixture, but when he saw this little reagent we had, he started drawing, and his eyes lit up. He kept saying, 'Fujisawa. Fujisawa. Fujisawa has this compound.' Then he said, 'We're making this, and I'm pretty sure we could put something on to make a reagent you could couple to this column.'

"He knew I needed to do something new, and he said, 'If we could make something, would you be interested?'

"I said, 'Yeah. In one experiment we could find Fujiphilin.' "

Schreiber by then had been "called" to Harvard. Harding's first experiments, meanwhile, were a "dismal failure," which, he says, "was good. I always like experiments that don't work the first time, because if they do, they never work again." In February 1989, the *Nature* papers about the role of cyclophilin in protein folding and the burst of attention that followed suddenly lifted Harding's standing as codiscoverer—his molecule now had the cachet of an important biological function. Then, in March, Harding ran a thymus extract through a one-inch column and in two days came up with what he was looking for—a bright band on a photographic gel that indicated a protein with a strong molecular affinity to FK-506. He sent the gel to Yale's protein chemistry lab, which conducted a quick search of the data bases. Harding's protein was brand new.

For someone in Harding's position, the discovery was a bonanza, provided he and Schreiber published their results first. All that was likely to follow—important publications, funding, patent royalties, more lab space and students, international recognition, wealth, travel, star power, and, perhaps, after all that, tenure—hinged utterly on not being beaten to press. Yet almost as soon as he was done with his experiments, Harding began hearing rumors that Merck had also discovered a high-affinity binding protein for FK-506. In science, Harding knew, winner takes all; there are no silver medals. The only saving grace is that, because of the stakes, there are an inordinate number of ties.

"The race was on," Harding recalls. "I was trying to get some of the material, purify it, and prove that it was really a specific binding protein. And of course the obvious thing to ask was whether [like cyclophilin] it catalyzed protein folding. If you didn't ask that question, you didn't deserve to be a scientist."

With Schreiber calling every day from Cambridge, Harding raced to finish his experiments, then drove to Harvard with a draft of the paper. In Rahway, meanwhile, Boger's former collaborator on FK-506, a Canadian-born M.D./Ph.D. named Nolan Sigal, was similarly pushing hard to finish up. Sigal's group submitted its paper to *Nature* on June 16, three weeks ahead of Yale and Harvard. Though the two articles were finally published together that October, as Boger was leaving for Barcelona, the Merck paper appeared first in the magazine, formally signaling its priority. The distinction was noted carefully only among scientists. To the world at large, copublication signified a tie, and Harding, listed as first author, would thereafter be known as a discoverer of FKBP. Because universities customarily own all discoveries in their labs, Yale and Harvard filed jointly for the patent.

Harding was elated. He was on his way. But if he was happy in his career, he was miserable in his job. He was working at the medical school under surgeons who depended on him for test results involving their transplant patients. To them, as for Starzl, a junior faculty member in immunology was prized for his technical ability first and his ideas second. Far from aggressive, Harding found himself running assays for doctors rather than doing his own experiments. Frustrated, and encouraged by Schreiber about the possibility of further collaborations, he wrote Boger in July asking for an interview at Vertex.

A small tableau graces the bookshelf above John Thomson's desk at the end of his lab bench: a jar of Kraft Vegamite, the bitterly salty soy paste Australians favor over peanut butter, and three bottles from a chemical supply company labeled, in plain black type, Caffeine, Nicotine, and Ethyl Alcohol. The bottles, though meant to be ironic, represent Thomson's muses. He gulps sludgelike coffee that no one else at Vertex will touch, smokes unfiltered Camels, and drinks hard. In the lab, in winter, Thomson wears Ray-Bans, battered running shoes, tight blue jeans, and even tighter T-shirts; a favorite one bears a picture of Calvin, the boy in the comic strip *Calvin and Hobbes*, and says: "I hate everybody. As far as I'm concerned, everyone on the planet can just drop dead." The introduction to his doctoral thesis contains the prologue to Goethe's *Faust*, in which Mephistopheles mocks the existence of a good soul.

Like Faust, who sought redemption in applying science for a larger good, Thomson is a creature of the laboratory. It is his sanctuary, his crucible, his cave. Yet his attachment to it separates him from his peers. With science drawing more and more from the norms of business, individual success now most often correlates with how much time a researcher spends on the phone, on airplanes, and in meetings, not in the lab. As Harding had discovered at Yale, doing experiments one becomes the "hands" of others whose work is moving ahead of one's own.

Boger, who had hired almost exclusively people who were on the verge of moving up sharply in their careers, was determined to reverse this trend. To get full value from his scientists and to get them talking to one another and help block the formation of hierarchies, which he considered a dull waste of talent antithetical to good science, he insisted that even his most senior researchers make a commitment to working at the bench. Boger personally designed the labs without private offices so that even those making $80,000 a year had to lug their briefcases from one communal desk to another to write letters or make phone calls. Boger's "social experiment," as the scientists called it, was another unpopular idea, like making use of the SAB, that most of them tolerated begrudgingly. They assumed it would fail and hoped that it would be sooner rather than later—except, notably, Thomson. "I'm proud to be called a blue-collar scientist," he said.

Fortunately for Thomson, Harding and Merck appeared to have discovered the same FKBP, which saved him from the abyss of a thoroughly blind search. They agreed on a number of key points—how tightly it bound to FK-506, its chemical sequence, that it accelerated protein folding—which Vertex could translate into analytical tools. Eventually, Thomson would use those tools to determine if what he had was what he was after. But he didn't have them now, and Harding and Merck had also had their differences. Harding's protein had a molecular weight of 14,000—it weighed 14,000 times as much as a hydrogen atom—while Merck's weighed 11,000. In the first stage of Thomson's search, FKBP's known characteristics were the only signposts. That some were confusing and others unavailable hardly cheered him.

Thomson plunged ahead with what he knew: the approximate size of the protein. Like Harding, he had chosen the thymus because it was the organ in which the preferred receptor of a drug whose chief effect was immunosuppression was likely to be most abundant. Shrunken and vestigial in adults, the thymus is draped illustriously over the heart in infancy and is one of the seats of the immune system, churning out those cells that orchestrate the body's defenses, the T cells (T being for thymus). Of the tens of thousands of distinct proteins in each T cell, only a handful were likely to be in the 11,000 to 14,000 molecular weight range. By creating a clarified protein soup and running it through a series of submicroscopic filters, Thomson initially hoped to capture FKBP solely on the basis of its size. In other words, he would pan for it.

In a dozen years of wet-lab experiments going back to the University of Melbourne, Thomson had come to know animal tissue, which is coarse and matted, and protein, which is lacey and ephemeral, and could separate the two without harming the protein as well as perhaps anyone in the world. It was grueling, unglamorous work. *Tierchemie*, runs an old German expression, *ist nur Schmierchemie* (Animal chemistry is just the chemistry of slimes and messes). Observing once the aftermath of a small prep in Schreiber's lab at Harvard, Thomson had taken note of the flecks of wet tissue covering the walls.

Now, at Vertex, he decanted the snap-frozen pieces of thymus into an industrial-strength blender, which pulverized them without spraying. Then, adding water, he reconstituted the mixture, producing a salmon-colored bisque. He poured the bisque into plastic jars, carrying them into another room where he placed them in a high-powered centrifuge, which spun out much of the fat and some of the larger, heavier proteins. As with every step, Thomson had to weigh the necessity of treating certain constituents harshly enough to remove them from the broth while gently leaving the FKBP intact and unperturbed. To this end he used as few solvents as possible and kept the mixture, which could change even at room temperature, refrigerated for all but short stretches. This required him to work long hours in a forty-degree Fahrenheit cold room, which he did in his T-shirt, slipping out only for an occasional swig of coffee or a cigarette.

Once Thomson started something, he didn't rest. When the first stage of the prep gave way to the more difficult step of clarifying the extract, he immediately set in with a new and more demanding series of experiments. Proteins tend to be colorless. If Thomson was going to obtain a mixture from which he could single out individual proteins, he had to eliminate everything else, especially the fats and waxes that would gum up his microfilters and gave the bisque its pinkish cast. Generally, these can be pulled out with organic solvents like ether and chloroform, but so can protein. Over a period of days, Thomson, working mostly in the cold room, washed and rewashed the extract until it was almost clear.

To manage the widening effort and also because it suited him, Thomson began staying at Vertex two and three days at a stretch, then three and four, then four and five. As the work expanded, so did his compulsiveness. During the day he raced between his bench and the cold room, running several preps simultaneously, he and Fitzgibbon, his assistant, becoming a two-man factory. At night he kept up his vigil while washing glassware, a purifying ritual that often lasted several hours and was still going on when others began arriving in the morning for work. His hands became chafed and swollen from being in detergent and in solvents. His feet swelled from his being on them for twenty-four hours or more between those occasions when he would collapse in a settee in the lunchroom and, with his Ray-Bans still on, sleep for a few minutes sitting up. His face assumed a fluorescent-induced pallor.

A complex mixture of choice and necessity propelled Thomson ahead. More than anyone else at Vertex including, possibly, Boger, he was a true believer in structure-based design—he imagined Vertex in utopian terms. He passionately admired Boger and the all-for-one, lab-centered, egalitarian social ethic he espoused within the company, and he believed Vertex would be limited in its success only by a shortage of pure, high-quality protein: protein that he would now provide. At the same time, Boger and Aldrich had begun negotiating a possible collaboration with Glaxo, Inc., the giant British pharmaceutical company, which planned to send a delegation in mid-February. Boger had called this a "camel's nose" visit, with Glaxo sniffing Vertex's goods, and was pressing hard to have "as much on the tent wall as we can." Thomson had vowed publicly to have pure FKBP to show Glaxo. He had five weeks. "I'd stuck my neck out," he says.

As with almost all of Thomson's actions during this time, there was a Faustian motivation to his resolve. Unlike Harding, he had not come to Vertex borne on a succession of triumphs. The previous two years in America had been a downward spiral, punctuated by impulsive acts he would come to regret. Though he had done superb work resulting in several important publications, his postdoc at MIT ended badly with a falling out with his lab chief. Thomson, who was married and had two young children, began seeing another woman. His marriage broke apart, with his family returning to Australia. He then was hired by Boger at Merck, but Merck's lawyers were unable to arrange a work visa for him, and he ended up taking an academic post in Wyoming for a year. His girlfriend dropped him soon after.

Thomson came to Vertex "in a very angry sort of a situation" and "determined to make a new start." But after a smooth beginning, his troubles resumed. He met another woman, but she left him for Jeff Saunders, a Vertex chemist who had become Thomson's closest friend, in a reversal that many others at Vertex apparently knew about before Thomson did. Blaming himself for the breakup of his marriage and his stalled career and now feeling mocked and betrayed, Thomson began drinking heavily. On a night in early November, he got drunk, climbed on his motorcycle, and roared off from an east Cambridge bar on his way back to Vertex. He went most of a block on one wheel before skidding on a puddle and careening out of control.

Thomson totaled the bike and tore up his hands, but his rage was undiminished. Three weeks later he went out drinking again with another chemist, John Duffy. This time Thomson drove his car. "I got shitfaced," he recalls, "real rip-roaring drunk. I don't remember leaving. I don't remember the accident at all. I just smashed into another car head on. I went through the windscreen, smashed my face up, and totaled the car. I was looking through my eyelid."

Thomson was taken to Massachusetts General Hospital, then to the neighboring Massachusetts Eye and Ear Infirmary, where doctors sewed up his eye and removed dozens of glass slivers from his face. "I remember waking up," he says, "and saying, 'I need to call someone at work and tell them I'm not coming in.' " Two days later he was back at the bench, resuming his pursuit of FKBP and seeking redemption, as Faust had, in the one thing he hadn't destroyed—his work. His right eye was swollen and bandaged behind his Ray-Bans, and his vision was blurred. For months afterward, throughout his ever-longer sieges in the lab, he continued to pick pieces of glass from the stigmata on his face. He avoided all but perfunctory contact with most others in the lab, keeping to himself.

"I was real screwed up at the time," he says. "Plus I didn't have a vehicle. I couldn't afford one and I didn't want to drive: I was afraid I'd hurt someone.

"I just wanted to make things work and I knew what needed to be done. I didn't have wheels. I didn't have a lot of friends I could trust. And I had a lot of work to do. So I stayed here and I did it."

To Boger, the impending Glaxo visit meant two things, money and benediction, each more vital than the other. Vertex needed a corporate partner to carry the immunophilins project forward and to help salve its burn rate. It also had to show the drug industry and investors that someone (it didn't especially matter who: Nissin, the largest Japanese noodle manufacture, was another early prospect) took the company seriously enough to put up a stake in it. That Vertex's suitor was Glaxo, the most dramatically successful company in Europe throughout the previous decade and the strongest challenger to Merck's hegemony in drug research, sweetened Boger's calculations dramatically.

The drug industry, powered by the huge profits of the 1980s, had lately embraced the concept of "strategic alliances." Fervently in vogue, they were thought to solve a common problem: in a fractionated marketplace spanning countless diseases, an explosion in knowledge, and thousands of laboratories, no big company was big enough and no small one clever enough to go it alone. Kissingeresque marriages of convenience were a necessity. For Boger to be entertaining overtures from Merck's chief rival, a company with more than $2 billion pouring in annually from the world's best-selling medicine, automatically enhanced his position. Glaxo had never done a large research deal before—a "virgin," one former manager called it—but Boger was unworried. He was sure it would come around.

But Glaxo also was uneasy. Boger's ambiguous relationship with Merck troubled some company officials, who fretted a Merck "backlash." Had Boger's separation been clean? they asked: as one of them put it, "Could Merck ever come back at Josh?" It was a reasonable concern. Boger had gone to great pains to "lobotomize" himself against what he knew about Merck's research, refusing to discuss anything that wasn't published. Yet by choosing as his inaugural project a target molecule he had previously pursued at Rahway, he invited inevitable questions. Merck's senior management was known to be furious about the decision. ("Does Vertex have a single project that wasn't first formulated here at Merck?" Ed Scolnick, Merck's head of research, would fume more than three years later; he refused to elaborate.) And so Boger remained, at least for now, a defector, highly prized yet suspected on all sides. As much as Glaxo and other big drug companies might like to know what Merck was doing in immunophilins, they preferred not to know it from an impoverished prospective partner whose bills they were being asked to pay and whose fortunes they were considering hitching to their own.

In fact, Boger knew little. He knew Merck's approach in such situations, which was to take a molecule like FK-506, synthesize it, and have its chemists begin making molecules that were similar, testing them for activity. But that required a method of testing, something no one in the world had prior to the discovery of FKBP several months after Boger had already resigned. His own ideas for improving the molecule were considered radical and went untested while he was still at Rahway. He was confident they would not have been pursued since.

Boger and Aldrich visited Glaxo during a week-long swing through Europe in December, and Boger had come away impressed enough to consider the company a possible competitor—another reason for seeking a collaboration. Yet he knew Glaxo was probably uninterested in Vertex's theories about drug design, being even more firmly committed than Merck to discovering its drug leads through screening. What seemed to interest Glaxo most about Vertex, besides Boger himself, was what Vertex might know about the biology of FK-506 and, in particular, Vertex's relationship with Schreiber—a relationship that, unknown to Glaxo or anyone else outside the company, was now starting to unravel.

After the first SAB meeting in October, Aldrich had confronted Boger about Schreiber's ambiguous relationship with the company. Prior to the meeting, Schreiber had insisted he had no interest in making drug molecules at Harvard. But in exploring how FK-506 blocked FKBP, he and his group had now begun making compounds that mimicked parts of the drug. Aldrich, as head of business development, failed to see the distinction between Schreiber's molecules, which Schreiber claimed were only for experiments, and intentionally designed drugs. To him, Schreiber represented only severe, multiple contaminations. If one of Schreiber's graduate students made a molecule similar to one discussed at Vertex, the company could be dragged into a ruinous patent fight with Harvard. If a Vertex chemist made a molecule similar to an idea Schreiber claimed he had while driving to work and mentioned accidentally at an SAB meeting, Vertex could be forced into paying the university steep royalties. Either way, Aldrich told Boger, Harvard now had a lien on the company.

The combination of Schreiber's personality and ambition posed a more immediate threat. He brimmed over. Like Boger, who was more circumspect but just as loquacious when it served his needs, Schreiber loved to talk about himself and what he was thinking, and he now had the world's ear. "I'm not concerned that Stuart will find a compound that will compete with ours," Boger said. "But I'm tremendously concerned he may tell everyone in the world what we're doing. Two of the companies we saw in Europe said they had telephone calls from Stuart within the past couple of weeks. His phone bills must be enormous."

Boger feared that Schreiber was "persistently naive" about the need for secrecy; Aldrich thought him calculating and opportunistic. A "good Stu/evil Stu" polemic ranged sporadically between them for months while they tried to negotiate an agreement with Harvard that would give Vertex exclusive rights to any discoveries relating to immunophilins in Schreiber's lab. Schreiber, for his part, favored such an arrangement, since it also included a collaboration he coveted. The big prize with any new protein was solving its three-dimensional structure. But Schreiber had no X-ray crystallography lab of his own and no ready access to one. For the chance to work with Manuel Navia and Vertex's biophysics group, Schreiber was happy to grant Vertex sole rights to whatever he produced.

The obstacle was Harvard. Ethically, the commercialization of academic science had given the university, as the standard bearer of American education, an immense institutional heartburn. No entanglement was harder to swallow than the specific type of alliance that Vertex and Schreiber were now proposing, an exclusive license between a professor and a company in which he or she owned stock. Such arrangements slashed at every presumption of liberal education—free and impartial inquiry, pursuit of learning for its own sake, the sanctity of knowledge—and replaced them with a tangled system wherein a faculty member stood to profit directly from the work of students. Harvard, though it had negotiated two or three such deals over the last decade, recoiled ritualistically each time one came up and was balking now at Vertex's proposal.

In December, Boger, Aldrich, and Boger's brother Ken, Vertex's lawyer, met at Harvard with officials of the Harvard patent office to discuss a compromise. In preparation for the meeting, Schreiber had sent them the terms of a $10 million research agreement he and two professors at the medical school had recently signed with Hoffman-LaRoche, the large Swiss-owned drug company. In his cover letter, Schreiber had said he saw no conflict between his work with Roche and his commitment to Vertex. Now, however, Joyce Brinton, Harvard's head of licensing, was reviewing a proposal that Schreiber had written earlier, in which he listed, among those research interests that Roche might want to support, his group's work in chemistry and immunology. In other words, FK-506.

Boger was stunned. Schreiber, who had a major equity stake in Vertex and was privy to its science, had licensed his work in immunophilins to a direct competitor; Vertex, Boger thought, might as well have invited Roche to its lab meetings. What's more, Schreiber hadn't told him. Schreiber considered the overlap the result of "miscommunication." Moving quickly to correct it, he agreed to restrict the Roche agreement to work in another area, AIDS, in which he had suddenly become interested. Still, as word spread among Vertex's scientists, their anger and confusion soared. Pressure mounted on Boger to act.

Boger was caught in a contradiction of his own making. He had wanted Schreiber involved in the company, a decision that had already paid off with the recruitment of Harding and others and was now of defining interest to Glaxo. Schreiber's marquee value, which remained Vertex's alone despite the Roche deal, was indispensable for the sort of joint venture Vertex now needed to survive.

Yet Vertex had nothing if not its ideas for designing a novel immunosuppressant, and Schreiber, who knew most of them, was acting like a free agent. Demonizing Schreiber had its advantages. Having a famous academic collaborator who appeared to care only about himself reinforced Vertex's culture of industrial chauvinism, and it helped to close ranks behind Boger, who, despite his success in recruiting, still had to sell himself internally by proving he could play in the blood sport of big-time science. But Boger was walking a fine line: making good enemies was one thing when they loomed at a distance. It was another when they were in your huddle.

No other recent situation so rankled Boger, who, ever since, as a child, he had watched his father squander his business dealings, had developed a conservative's heated loathing for the expropriation of intellectual property. Boger didn't think Schreiber would contribute significantly to Vertex's science, yet Schreiber was moving so fast it was impossible to rein him in. That Harvard, which had done nothing toward creating the company, was also now in a position to share in Vertex's future appalled Boger as much as if it were some new, especially heinous form of street crime.

Resisting Aldrich and the scientists, Boger determined that Schreiber would stay but in name only. From now on, he would be brought out for such goodwill displays as the Glaxo visit but otherwise was to be treated like any other competitor. Boger ordered all lab books updated and signed daily to document the conception of ideas—a defensive measure anticipating future lawsuits. As for direct contact with Schreiber, Boger counseled his usual rule for negotiating with rivals: "Tell them only what they need to hear so that they'll tell you what you need to know worse."

"I think," Boger told the scientists, "everybody should be very friendly to Stuart and listen to what he has to say."

Boger didn't inform Schreiber of the change. He worried that Schreiber might bolt, negotiating with Glaxo on his own. He had other reasons for keeping Schreiber in the dark. Now that Schreiber and Vertex had both begun making compounds, the stakes for each of them had risen exponentially. Despite his denials, few things troubled Boger more than the possibility of Schreiber's group developing a promising drug lead before Vertex—a prospect Schreiber believed he already might have achieved.

In December, Schreiber's group had synthesized a new molecule that he called FK-506 binding domain, or 506BD, which was central to Boger's concerns. FK-506 is what chemists call a *macrocycle*—as its name implies, a large ring. It has 126 atoms and a pronounced, hornlike gaff. Drawn conventionally (Schreiber, practicing a different aesthetic, draws it upside down), it looks like this:

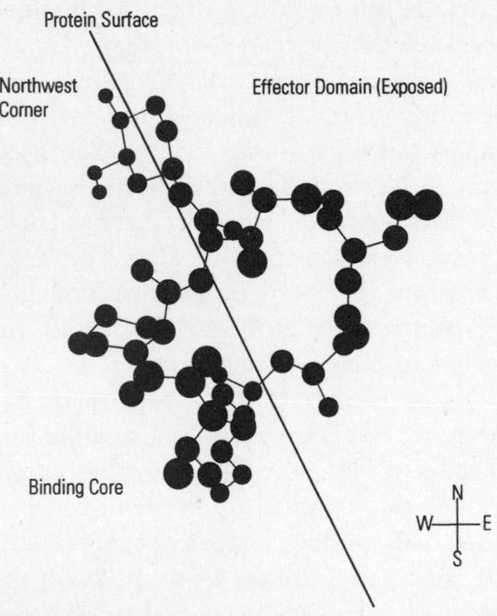

As shown in the figure at the left, FK-506 has two regions. At left is the binding core, which connects the molecule to its target. The right side, the effector domain, remains exposed. It extends from the protein surface (the northwest to southeast axis in the drawing) to interact with other molecules. The balls in the picture depict atoms.

As a chemist, Schreiber cared little that FK-506 was a powerful drug. What intrigued him, indeed what drove him to a kind of solipsistic rapture, was something else: its extraordinary resemblance to another microbial compound, rapamycin, which not only was immunosuppressive but also inhibited cells from reproducing.

Isolated in 1975 from a soil sample from Easter Island, rapamycin was considered a possible anticancer agent, but again Schreiber was uninterested. What consumed him was the almost mystical fortuity: two microbial compounds, sharply similar structures, yet apparently different modes of action. As with any molecule, the big prize academically was figuring out how it worked—that and, of course, being first. Here, quite possibly, was a three for one opportunity. Schreiber couldn't help but be enthralled.

Thus 506BD. It contained only those structural pieces common to FK-506 and rapamycin, and Schreiber and his colleagues theorized that 506BD was the part of each drug that bound to its target. However, when they tested it, they discovered that although it blocked protein folding as well as FK-506, it was "dead"—biologically inactive. In other words, something else appeared necessary to make FK-506 a drug.

Schreiber was ecstatic. He began to theorize that what accounted for FK-506's activity was not its ability to block the protein-folding action of FKBP (as Vertex and everyone else still assumed), but some other function. This he attributed to an *effector region*—that part of the molecule not associated with binding. Using 506BD as a platform, he and his students began building outward to replicate the missing piece.

Boger maintained that he, not Schreiber, had first proposed making 506BD at a meeting at Vertex in September, although getting credit for the discovery didn't concern him nearly so much as what would happen now that Schreiber had begun announcing 506BD to the world. Though Boger discounted the molecule as the basis for drug design and dismissed Schreiber's conclusions, he believed other drug companies would be drawn to the theory. And that, Boger feared, could be shattering for Vertex.

Boger had little but contempt for what he called the "monkeys with typewriters" approach to medicinal chemistry practiced by virtually all major drug companies. Legions of chemists systematically changing every atom and subgroup on a molecule in order to improve it was not Boger's idea of intelligent science. Like screening, the approach was largely statistical—how many chemists making how many derivatives for how long—and Boger didn't like gambling when the odds were against him.

But conventional medicinal chemistry could also, Boger knew, be extremely successful, and it favored those companies with the most and best chemists. Merck had routinely been able to bludgeon competitors simply by throwing more chemists at a project.

This was what troubled Boger now. Without its own legions—Vertex had only seven chemists making compounds; Merck, which guarded the information as if it were a state secret, more than

400—and without any structural data, Vertex was left with its own risky strategy for drug discovery, a strategy that Boger had devised. The approach was to explore not miniscule variations in the structure of a molecule, but to carve it up wholesale in an effort to discover those minimal portions that accounted for its activity. Such "small-molecule" derivatives—attempts to reduce a drug to its minimal size and weight while still retaining its effectiveness—were controversial among drug developers, but Boger believed in them categorically and counted on others not to do so. Small molecules were Boger's religion, the intellectual basis for his faith in structure-based design, the critical fault line in his break with Merck and the industry. Indeed, if Vertex had a "secret formula," it was the half-dozen small-molecular blueprints that he and the chemists—some in discussions with Schreiber—had designed.

Boger believed that 506BD, although dead, would broadcast Vertex's strategy to the world. It would point toward those other small molecules that Vertex had made and that, although probably novel, would not be hard to imitate and expand on—something Boger believed no company would try without prompting from someone as effusive, as influential within the field, as Schreiber. Boger couldn't believe Schreiber didn't see that. And yet Vertex couldn't afford to lose him. Other drug companies would be pounding down his door. And Schreiber, in his ebullient way, would want to accommodate them.

On the second Friday in January 1990, Schreiber visited Vertex for the first time in weeks. A meeting with the full SAB was scheduled for the next day, and Schreiber had come with one of his postdocs to report on some new work—"trying to put some cookies back in the jar," suggested Harding, who had begun to be torn by shifting loyalties and was growing increasingly uncomfortable because of it.

Schreiber seemed no less ill at ease, for other reasons. Although he had helped recruit Boger to lead the company, he was increasingly unimpressed by Vertex's scientific production, especially Boger's decision to try to make drugs that were inhibitors of FKBP. "A nonstarter," he called the effort. (What else Vertex might have made compounds to do, given that the true mechanism was unknown, remains unclear.) He was convinced that Vertex's scientists were concentrating on the wrong parts of the molecule, and he saw no gain in being associated with their problems. It wasn't that he distrusted them, as a growing number of them distrusted him. He disrespected them—in Schreiber's view, a far harsher judgment.

What Schreiber and Boger thought, they colored the world with, and their mutual displeasure now set them irremediably apart. Black was the other's white. Not only, would Schreiber say, had he not disclosed the company's secrets as Boger believed and feared, he would have been embarrassed to do so. "Other people's studies were irrelevant to me," he says. "It's absolutely absurd that I would talk about Vertex because they never had any work to talk about." (In fact, Schreiber by now had been taken unawares out of Vertex's communications loop and didn't know what the company was doing. His disdain reflected accurately what Boger and the scientists were telling him.)

Meeting now with Harding and David Livingston, Vertex's chief enzymologist, in the cramped conference room, Schreiber seemed tentative, awkward. He repeatedly reached for something in his open briefcase, then pulled his hand away as if reminding himself that he couldn't touch it. The temptation grew, with Schreiber muttering "dammit" at one point before bringing back his hand sharply. After forty-five minutes, Schreiber finally pulled out two documents—a preprint of a recent article he had written on cyclophilin and one by a competitor addressing many of the same issues. After a minute, he returned his own paper to its file, apparently feeling free to share his competitor's work, which was in press, but not his own.

After Schreiber left, the tension of the past weeks suddenly erupted in a round of gripe sessions that stalked Boger wherever he went. The Roche deal, 506BD, Harvard's hard line, the next day's SAB meeting, the decision to keep Schreiber on—Boger was called sharply to account for each.

"Digging up your neighbor's yard because you want a bigger garden is one thing," groused Harding, standing in an angry knot with three or four others, "but Stuart's rototilling the whole neighborhood."

"The problem is he tells people whatever he's thinking," said chemist Jeff Saunders.

"The problem is he's making compounds," Boger said. "We can control Stuart's life when he's here, but he has another life we don't control." Boger paused. "The problem . . . the problem is that they're exactly the same."

"We should follow Stuart to every meeting he goes to," Saunders said.

Hal Meyers, a chemist and one of Schreiber's first doctoral students at Yale, nodded approvingly. "It might work. Stuart doesn't document anything."

"Good," Boger snorted. "We document, he doesn't. At least we don't want to change Stuart's method of operation in that regard."

"Can't he be informed that there's fame and there's fortune?" Saunders asked.

There was a collective sneer. "He thinks he can have both," Dave Armistead, another chemist, said.

"When we can walk into a major drug company and do a half-hour seminar on our own findings," Boger said, "Stuart's name won't mean anything to us anymore. But until then, his capital is too valuable to squander.

"Stuart's value and his liability to us are on approaching curves. Unfortunately, his liability factor is not as great as his value right now. But those curves aren't flat. They're going to cross."

For months Boger's stock answer to questions about competition from Schreiber was that Vertex would soon surpass him. Schreiber would cease to be relevant, Boger said, the day Vertex was first to solve the structure of FKBP—when Vertex had vital data that no one else, including Schreiber, had. Boger had no doubt this was true, but now, in the gloaming of a late Friday afternoon in January, after weeks of mounting frustration, he turned impatiently on his heel and headed back to his office.

"The world is competitive enough," he said, "without competing with yourself."

Two hours later, during the customary Friday afternoon beer hour, John Thomson, stumbling and glazed, approached Boger in the lunchroom.

"We've got the pure protein," he said.

CHAPTER FIVE

It was *not* the pure protein. Thomson had isolated a substance that had the same molecular weight and abundance as Harding's FKBP. But as he analyzed it, he discovered a second constituent—assumably another protein—almost identical in size to the first. Thomson was distraught. He believed he was jeopardizing the company and that others blamed him for it. Typically, Boger consoled him that any information, however discouraging, was "money in the bank," but Thomson refused to be cut any slack. He would now go back and find out what was in his mixture—another dark, slippery descent, as Vertex still had no specific assays to guide him.

Science, as Nobel laureate David Baltimore has observed, proceeds "not from truth to truth, but from suggestion to suggestion," and Thomson was buoyed instinctively by the suggestion that his two molecules might be linked by some other feature besides size. The chemical pathway that included FKBP remained unknown, and it was still uncertain what the protein did besides help other proteins invert themselves. If Thomson's second constituent bound chemically to FKBP and if it was new, he conceivably had a molecule as important, as rewarding, as FKBP. This and a fear of failure to equal Starzl's kept Thomson from relaxing his pace.

Thomson now moved into the lab altogether. He abandoned the waist-high stack of take-out pizza boxes in his apartment and a refrigerator empty but for the world's supply of fetal eye lens protein and some beer and began those experiments that would delicately tear his two molecules from their embrace. What he needed was something that bound to FKBP but not the other component—a molecular butcher's hook—and that Boger himself now provided. Like fly-fishermen and safecrackers, molecular researchers like to improvise their own tools. Boger, in the earliest days of the program, had designed several antibodies to FKBP, compounds with known affinities to parts of the molecule. Now, working night after night, Thomson used them to try to detect what was in his mix.

"When I had a little bit of the second isolate pure enough to get sequence information and a physical photograph of it of any kind, I'd know what I had," he recalled.

Thomson stayed in the lab for five days, "on the hop," as he put it, moving all the time. His feet swelled, and he was limping. Frustrated and exhausted but determined to make good on his promise, he became his own demon so that no one, not even he, could imagine doing any more. All of the scientists had known people in graduate school who never left the lab, and at one time or another, most of them had slogged through intense periods at the bench. But none of them had ever seen anything to match Thomson's stamina or resolve. Boger, who saw the value of his scientists in competitive terms and was still crafting the Vertex mystique, said, "Not even Merck has someone like John. If they did, I wouldn't have had to hire him when I was there." Added Aldrich, "One of our chief goals for the next year is to keep John alive."

Thomson eventually isolated the contaminant. He'd been right: There was another protein in his mix. But it wasn't new, it had no obvious connection to FKBP biologically, and there was much more of it than FKBP. Indeed, Thomson's red herring appeared to be one of the most universal—and universally understood—of all proteins. From what Thomson could tell, he had discovered ubiquitin, which, as its name implies, is found literally in every living cell, from amoebas to carrots to people. Just as proteins are constantly being produced, they're also being destroyed, and ubiquitin leads the kill. It's everywhere. And now it was overrunning the FKBP in Thomson's sample tubes.

Thomson was undismayed by his failure to produce an important new find: Dry holes are the rule in science, and he had trained himself, like most researchers, to expect nothing even as his hopes soared. But he shuddered at what the extraction suggested about FKBP. He now feared that Harding's protein was much scarcer than he or the Merck group had originally reported—so scarce, in fact, that if its relative abundance had been known at the time, Boger might have been forced to cancel the project outright. Suddenly, Thomson's goal became much more daunting. Even if he was able to isolate FKBP, there was so little of it, even in thymus, that he might not be able to produce enough of the substance to crystallize, which was nearly essential for determining its structure. At the very least, larger preps and much higher yields were now critical.

He was back at square one. The Glaxo visit was less than a month away and, though no one would say so, the lack of protein meant that the other scientists, who were just as eager to prove themselves as Thomson and were falling behind their competitors elsewhere, were forced to sit idly by. It also meant that Schreiber, who was thought to have nearly a half-gram of recombinant FKBP, became a double threat. Without its own protein supply, Vertex needed more than ever to retain Schreiber and cultivate a deal with Harvard. Yet Schreiber was becoming increasingly impatient. Few goals were as critical to him as finding a crystallographer who, using his protein, would be first to solve the structure of FKBP. Stock or no stock, Schreiber was hearing the same rumors as Thomson that Merck had now produced sufficient protein and was trying to crystallize it. He could hardly be expected to forfeit his priority for the company even if he had been so inclined.

Thomson went home and slept for twenty-four hours, showered, shaved, and changed his clothes before returning to the lab to start again. He was determined this time not only to get pure protein, but to devise an extraction process that yielded one hundred times more FKBP than either Harding or Merck had gotten. The decision meant not only doubling his standing order for thymus from five kilograms per week to ten, but experimenting with new methods at every stage of the process, going all the way back to tissue cutting. Every feature had to be optimized. "Basically I threw both papers in the trash at that point," Thomson says. The search for new protocols would no doubt slow him down sharply, but Thomson considered it essential for producing the quantities of protein that Vertex would ultimately need. Despite the rising anxiety of some of the other scientists, Boger supported Thomson implicitly, placing himself as a barrier to criticism and ensuring that Thomson would not have to deal with any haranguing but his own.

By working slavishly, Thomson dispelled most criticism without Boger's qualification, but he now was so unforgiving of himself that others began to worry. On January 22, 1990, for instance, less than a week after sorting out the apparent contamination with ubiquitin, Thomson entered the lab and didn't leave—except to go home twice to shower and change—for eight days. At the end of that time, he had extracted a batch of what appeared to be pure FKBP, but it was dead—inactive. Thomson concluded that what killed it might have been his use of chloroform to segregate out the fats early in the process or perhaps the snap freezing, which he was now ready to abandon completely. "Worst comes to worst," he said, "tissue day around here may be a bit more nasty." His matter-of-factness did not strike anyone who knew him as a good sign. They laughed nervously when Thomson told them that as soon as he got the protein, he planned to reward himself by buying another motorcycle.

The Glaxo visit was now set for February 20, and though Vertex had other intriguing developments to show, the possession of a method for isolating protein would guarantee that Glaxo would have to take Vertex seriously not only as an ally but as a threat—a coveted position in any negotiation. Thomson and Fitzgibbon now went flat out, racing to beat not only that deadline, but an earlier one. Boger had stuffed Vertex's labs with so much high-end equipment that the company had recently outstripped its power supply. On February 16, the labs would be shut down and the building sealed for most of the day while a utility crew installed a 50,000-watt transformer. Al Vaz, Vertex's lab manager, had already told Thomson he would be taping shut the cold rooms, and Thomson anticipated having to stop work. With Glaxo coming four days later it would be hard enough for him to gear up again, much less salvage any experiments that might be lost when the power went off. Whatever he would do, he would do by the 16th. He had two weeks.

As Boger could have expected, Harvard grew more rectitudinous by the day. Though the licensing office favored a deal, Joyce Brinton now told Aldrich that Schreiber's relationship with Vertex was raising red flags in other quarters. She advised him to talk with Schreiber. It was an open secret that Harvard offered tenure in chemistry only to those it believed would win a Nobel Prize. This bestowed on Schreiber a status quite incommensurate to his age or scant seniority within the university, which put a high premium on such prizes and had more of them in biomedicine than any other institution in the world. Perhaps, Brinton suggested, Schreiber could use his influence to move the deal along.

Brinton advised that Schreiber meet with economics professor Jerry Green, chairman of the Committee on Professional Conduct, explaining what Aldrich called the "party line": that collaborating with Vertex, now an established leader in immunophilin research, was vital to his work and that of his students. Not to do so, he was to imply, would deprive him, his students, and ultimately Harvard of an important opportunity to compete at a vital forefront of research. Brinton also suggested that Schreiber visit President Derek Bok. Bok had lured Schreiber to Cambridge in a tough recruitment battle with Yale and was required, because of Schreiber's equity stake and overlapping scientific goals, to rule personally on whether his relationship with Vertex met the university's strict conflict-of-interest guidelines—guidelines that Bok, a former dean of the law school, had promulgated amid one of the most fraught and anxious battles of the university's history. Aldrich needed no reminder that Bok had been deeply offended by Boger's decision to name his company Veritas and had pressed for it to be changed.

"It all comes back," Boger said, hearing from Aldrich of Bok's personal interest in the case.

"The circle is closed," Aldrich said.

Boger had hoped to present Glaxo on February 20 with a signed deal with Harvard that secured Schreiber's services once and for all. Now, without it, he needed more than ever to present an appearance of harmony. Schreiber still didn't know of the depth of feeling against him at Vertex, and Boger, who still believed Schreiber might be naive, was determined to exploit whatever loyalty remained between them.

"Stuart will do whatever we ask him to do," Boger told Aldrich in his office. "We just have to give him the right script."

Aldrich was skeptical. "The question is, Do you want to use the carrot or the stick?" As with all matters with Schreiber, Aldrich favored intolerance; he wanted Schreiber reprimanded and reminded that he was at risk of losing all his company stock.

"I want to use the carrot for now," Boger replied.

"I guess it depends on where we use it," Aldrich said.

Aldrich's distrust of Schreiber, based on his experience at Vertex, had become immense and personal. But his cynicism about Harvard was more roundly historical. As for many scions of old Boston families, the university was an essential part of Aldrich's heritage, nearly a bloodline. His father, mother, and both of his father's siblings had gone there, and his father and uncle had graduated from Harvard Law School. They had attended in the years around World War II, when Harvard was in the final throes of transforming itself from a Brahmin redoubt to an academic superpower and its highmindedness was unencumbered by profit motive or self-doubt. It was a time—extending ultimately into the early 1970s, when the younger Aldrich was in high school—marked by steeply ascending federal support for the sciences and an attendant virtuosity about the goals and consequences of research. For years Harvard's patent policy was simply, "No patents primarily concerned with therapeutics or public health may be taken out . . . except for dedication to the public," and there was little reason to suggest that it be otherwise.

Partly because of himself ("I never found school enough of an incentive," he says) and partly because of Harvard (underachieving young Brahmins were no longer so easily absorbed as the university rose in stature), Aldrich ended up at Boston College. From there he went to work for the headiest of the big business consulting firms of the 1970s, Boston Consulting Group—a Harvard Business School spawn that introduced such buzzwords as *learning curve* and *cash cow* and led the way toward recasting American business as a shovefest over market share. Young and brash, Aldrich was a self-described "punk in pinstripes," researching ways for companies to squeeze out new profits and typifying an old joke about consultants—that they know a hundred ways to make love but don't know any women. After getting an M.B.A. at Dartmouth, where he supported himself by marketing beer caddies at football games, Aldrich took a job developing business deals for one of the two earliest leaders in the infant biotechnology industry, Biogen.

Biogen brought Aldrich face-to-face with the new, changing, commercially minded Harvard—a Harvard that Bok was straining to keep from going astray or, perhaps worse yet, becoming uncompetitive. Biogen was founded by Walter Gilbert, perhaps the quintessential Cambridge figure of the postwar era. A brilliant biochemist and Nobel laureate, the brindle-haired Gilbert left Harvard to run Biogen, raised $125 million, came to earn $285,000 a year and own 580,000 shares of stock, then almost lost it all before surrendering the company to experienced managers and returning to the university. He later would attempt to raise capital to compete privately with the government's effort to map the human genome, a plan that, if successful, would have given Gilbert and his cohorts the patent to the human blueprint. In Gilbert, Aldrich saw the prototype of the new, big-name academic scientist: brilliant, yet "mercilessly egotistical" (to use Jeremy Knowles's phrase) and driven by a profit motive to equal anything on Wall Street. Aldrich greatly admired their ability to create value—Gilbert's, especially—but found them as a rule "cutthroat pirates" when it came to appropriating their share. "I have no qualms about throwing any of these guys overboard," he said. As for Harvard, he resented that the moral certitude of his father's era—when the school knew what it was and what it stood for—had given way to ambivalence and rank greed. A political and fiscal conservative, Aldrich, in his mid-thirties, was in this way like a man much older.

Still, using Schreiber to motivate Harvard held a certain grisly fascination for Aldrich. He felt they deserved each other. But he was not comforted by the prospect of having to depend on

Schreiber for so important a mission, and he tried to calculate Harvard's incentive for making a deal with Vertex regardless of Schreiber's entreaties.

"If Harvard decides this is a positive net to Harvard," he told Boger, "they'll have a hard time not making this deal."

"Yeah," Boger said, "but Joyce Brinton knows that come May, she may be able to walk to the table insisting on a 5 percent royalty on anything we do and say, 'Take it or leave it.' We're stuck. We can't wait.

"I've got Stuart all ready to go. I'm going to tell him that whether we can get all or part of this resolved by the 20th will determine if the company expands or not."

"Expands," Aldrich suggested, sharpening the thought, "with or without Stuart involved."

The morning of February 16 broke over Boston with deep-winter, arctic clarity, freezing the undulate white fumes from the smokestacks in midcurl and impaling them against an ice blue sky. A bone-slicing wind slashed in from the harbor, bowing commuters. The unplowed whalebacks of ice on Allston Street were as hard and gray as plank steel.

Such frigid weather played havoc with Vertex's chemistry lab, at the center of the building, and with the rest of the facility. Chemists do most of their work inside glass-enclosed safety hoods that rapidly remove gases from the environment. The hoods also remove the environment. Several high-powered roof units continuously heat the air in the lab, then suck it out. A new atmosphere comes and goes every three minutes, which in the early predebugging period of the company regularly produced wild thirty-degree temperature swings on cold days. A young crystallographer who had begun working almost as much as Thomson, Mason Yamashita, took to napping on the floor next to the X-ray generator because, he had found, it was the warmest place in the building. The glass front door, several corridors away, slammed violently from the backdraft whenever someone opened it. For all but a few employees, Vertex's extreme technical demands now accounted for a welcome Thursday morning in bed while the electric company came to triple the company's power supply.

Alone in the labs, John Thomson drained a cup of coffee and rushed back to the protein sequencing room, an alcove adjacent to his cold room, which contained a small hood and six feet of benchtop. A single computerized instrument covered the bench. Approximately every eighteen minutes the machine spat out a three-letter code . . . Gly . . . Val . . . Gln . . . Val . . . Glu . . . Thr. Each code was an abbreviation for one of the twenty amino acids from which all proteins are assembled—Gly for glycine, Val for valine, Glu for glutamic acid—and which are themselves bundles of atoms: a carbon atom in the center of a small, tightly configured constellation of oxygen, nitrogen, carbon, and hydrogen atoms. Like fingerprints, no two protein sequences are the same. Thomson was looking for the sequence that Harding and the Merck group had both identified as belonging to FKBP.

Thomson had been in the lab for six days. He was beyond exhaustion or relief, feeling simultaneously awake and asleep, dead and alive, crippled and immortal. He stared at the machine, unshaven, his hands and feet raw and red and pumping nervously. The codes came; they were indifferent to him. Yet with each syllable he knew that what the computer was analyzing could only be pure, isolated, biologically active FKBP. It was a trace amount, nowhere near enough to supply crystallography. But it was there, as he had pledged, four days before the meeting with Glaxo.

Thomson stood at the machine, recording the sequence, and was still standing numbly when the power went down and there was nothing left for him to do. With no one to share his triumph, he celebrated by returning to the lunchroom and, in the sullen half-light and unfamiliar silence, sitting down to close his eyes.

Benno Schmidt's twenty-third-floor office in Rockefeller Center looks directly down upon the twin gothic spires of St. Patrick's Cathedral, rising somberly across Fifth Avenue. "God's view," Boger calls it, less because Schmidt is Vertex's board chairman and one of its largest stockholders than because, as a founding partner in J. H. Whitney and Company, the world's prototype venture capital firm, and ex-chairman of the federal government's War on Cancer, the eighty-one-year-old Schmidt is the senior gatekeeper of biomedical innovation in the United States. Ensconced in the office since he became Whitney's managing partner in 1959, Schmidt, a lawyer, arguably has influenced the course of medical research during the past twenty years more than any other nonscientist. He looks the way God would look if he were an unreconstructed Texas Republican who came to New York in the muscular years after World War II and, working with one of the great private fortunes of the time, became hugely rich and powerful by merging old-boy eloquence and an instinct for making deals with charm, charisma, a flair for agenda building and an exceptionally well-placed network of friends and associates. Along the walls, amid faded floral drapes and surrounding a titanic desk, are pictures of Schmidt—his looming frame, robust jowls, sleeked salt-and-pepper hair, and extravagant white eyebrows—with every U.S. president since Richard Nixon.

Schmidt was not born to such company. He grew up in Abilene, in the western Texas hill country. His mother, widowed when he was twelve, worked as a secretary at the county welfare association. Attending public schools, he graduated at the top of his class at the University of Texas Law School, then went on to teach at Harvard. He enlisted in the army two days after Pearl Harbor, eventually rising to the rank of colonel and joining the State Department after the war. In 1946, he was approached by John Hay Whitney, heir to what once had been the largest individual estate ever assessed in the United States, to start a company to invest in new technologies. Schmidt didn't know Whitney, but it was hard not to know of him. "Jock" Whitney was one of those men of vast personal resources and style who defined the nation's attitude in the decade surrounding the war, a man for whom, a friend once observed, money had three purposes: "to be invested wisely, to do good, and to live well off." Whitney, dashing and competitive, was famous for all three: he bankrolled the filming of *Gone with the Wind* and was ambassador to Britain; he spent $40 million trying to rescue the *New York Herald Tribune* and gave millions more to victims of discrimination; he played high-goal polo, bred racehorses, and amassed one of the world's finest private art collections, including a number of Matisses that traveled with him as he moved among his eight residences, among them a town house on East 63rd Street, a 500-acre estate on Long Island, and a 19,000-acre refuge in Georgia. As America was preparing to exploit its success in war and men like Whitney resumed their places of prominence, Whitney invited Schmidt to sit beside him.

Schmidt quickly adopted Whitney's penchant for doing good by doing well. Coining the phrase *venture capital*, he and his partners began underwriting small companies that Wall Street and other lenders considered too risky. They were gamblers, speculating on new talent and new ideas and reflecting the rising expectations and romance of the time. "We don't live so much on our batting average as our slugging average," Schmidt once said. "We live off our extra-base hits." Within a year, the company had invested in two small firms—Minute Maid, which invented frozen orange juice, and Spencer Chemical Company, a midwestern fertilizer concern—that paid off so spectacularly that Schmidt had his fortune. Like his benefactor Whitney, he began a life of conspicuous public service: chairman of the Bedford-Stuyvesant Development and Services Corporation, the Fund for the City of New York, the Welfare Island Planning and Development Committee, and, most presciently, president of the Memorial Sloan-Kettering Cancer Center, one of the world's premier cancer research facilities.

Laurence Rockefeller, who worked with Schmidt on the Sloan-Kettering board and was an old friend of Whitney's, recommended that Schmidt be appointed in 1971 to a commission studying the federal government's role in fighting cancer. Not since World War II had there been so much pressure to put research in the service of a distinct medical goal, and Schmidt, known for getting things done, was put in charge. He remained as chairman of the President's War on Cancer through the administrations of Nixon, Ford, and Carter, and though no cure for cancer was found, the surge in directed research and outpouring of federal money gave rise to a new, entrepreneurial era in biomedicine—an era in which Schmidt, as a businessman and investor with incomparable connections among research directors and in Washington, was consummately positioned to become a major broker. Coming full circle, J. H.

Whitney and Company, under Schmidt's management, began investing heavily in those new companies spawned to capitalize on gains in basic research that grew out of the War on Cancer. "Making plays," he called it.

One of Schmidt's most successful plays was in a company called Genetics Institute (GI). Founded by two Harvard molecular biologists, GI was launched in 1980 at Harvard, and for a time, before Wally Gilbert and others registered such an outcry that the university was forced to withdraw, Harvard intended to buy most of it. Harvard's retreat, coming just as the initial fever over biotechnology was spiking, worked in Schmidt's favor. He and William Paley, chairman of CBS and Jock Whitney's brother-in-law, presented the scientists with an intoxicating alternative: take their labs private. Why work for Harvard when they could work for themselves? GI went on to become one of the glamour companies of the new biomedical age, with Schmidt, who became chairman, compiling for himself and Whitney a combined stake of more than 250,000 shares.

GI had already paid off handsomely for Schmidt, but the largest prize, in the winter of 1990, remained to be taken. It was a molecule called EPO, for erythropoietin, a protein that stimulated red blood cell production, which GI and Amgen, a California company, were competing for the right to market first in the United States. Here were the all-or-nothing stakes of the new biomedical order writ large. Developed initially to treat kidney dialysis patients, who become severly anemic and can't, because of their condition, be transfused, EPO was long recognized as a likely blockbuster, perhaps the first billion-dollar drug of the biotech age. It was one of perhaps two or three molecules guaranteed to catapult a new company from making bonfires of its cash to Fortune 500 profitability—while granting its early backers, like Schmidt, a more than hundredfold increase in their initial investments—and GI and Amgen both possessed the key for making it.

That was the problem. Patent laws grant drugmakers exclusive markets for their new products for seventeen years, and Amgen and GI sued to block each other's patent claims. The lawsuits, carried on simultaneously on both coasts, dragged on for more than four years, even after Amgen won FDA approval to begin selling the drug, and still had not been resolved in November when a federal court in Boston ruled that both companies' patents and infringement claims were valid. With each side now spending millions of dollars annually on legal fees and issuing press releases nearly every time they filed a brief, the ruling was appealed by both sides.

Even with the uncertainty about the U.S. market, the world's largest for pharmaceuticals, the news for Schmidt was hardly all bad. Like nearly all small, research-based companies, GI couldn't afford to develop EPO itself and in 1985 had licensed it to a partner, Chugai Pharmaceutical Company of Japan, one of a rising class of Japanese drug companies with global ambitions. Chugai had bought the rights to market GI's EPO both in Asia and as a joint venture in the United States, and though its U.S. prospects were now stalled, in January 1990 the company had received approval to begin selling EPO in Japan, which consumes more legal drugs per capita than any country in the world. The Japanese market for EPO, in which Chugai had no competition, was estimated at $200–400 million. GI would get a 5 percent royalty—not a spectacular annuity, but tens of millions of dollars nonetheless. Meanwhile, Chugai, feeling flush, had begun looking to make another play in the United States and was seeking Schmidt's counsel.

Schmidt had not meddled before in Vertex's business development; he was busy and he wanted to let Boger run his own show. But in early February he called and invited Boger to meet his "old friends" from Chugai sometime in New York, and Boger, not wanting to kill a day freezing idly in Cambridge, suggested that he and Aldrich take the shuttle down early on the morning of the 16th. That Glaxo would be coming the following Monday favored Boger's calculations neatly. Introductory business meetings with the Japanese were largely ceremonial, and deals with them most often moved glacially and took years to complete. And yet even if nothing happened, Boger would be able to state to Glaxo with unimpeachable accuracy that Vertex was now considering other options, implying that Glaxo needed to move fast or lose an opportunity. Boger had an affirmative distaste for negotiations where the power balance heavily favored the other side, and he didn't want to be held up while Glaxo shopped around. At 6 A.M., as Thomson slumped over the protein sequencer, Boger, Aldrich, and Manuel Navia climbed aboard the Trump Shuttle for LaGuardia.

Schmidt set the tone of the meeting, sweeping around the room, pumping hands like a host at a barbecue, his blue eyes boring into each charmed gaze from under the great cirrus of his eyebrows. The head of the Chugai group was a surprisingly young man in a conventional dark suit and aviator glasses, Osamu Nagayama, the company's deputy president and son-in-law of its aging chairman. Most Japanese businesspeople who are in a position to spend their company's money came of age during World War II and affect an unapproachable resolve, particularly with Americans. But Nagayama, who was forty-three, spoke English well enough to enjoy its jokes and seemed to relish Schmidt's informality. Boger, Aldrich, and Navia, mindful of the emphasis that the Japanese place on gestures of respect and on social hierarchy, dutifully presented their English-Japanese business cards and addressed Nagayama and the others by their honorifics—Nagayamasan, Ohtasan. Schmidt called Nagayama "Sam" and said that he'd had some business cards printed up in 1946 but hadn't seen them in some time.

After the introductions, they moved into the partners' dining room, Schmidt's sanctum. It was formidably paneled in butternut stripped from the Fifth Avenue mansion of Jock Whitney's parents and decorated with some of Whitney's early American holdovers and a Miró that Boger thought resembled a molecule. Boger ran through his standard slide show, which remained long on generalities and short on details. Even had he known about Thomson's breakthrough that morning, he would hardly have disclosed it, proprietary information being best saved until there is the prospect of real money on the table. He scarcely mentioned Schreiber. Then Nagayama gave a brief summary of Chugai's strategic goals. With the development of EPO now complete, he said, the company was entering the second phase of a fifteen-year growth plan in which it intended to triple its sales to $3 billion, do 30 percent of its business overseas, and become one of the top thirty drug companies in the world. The company, he added, had another, longer-range plan that ran through the year 2100.

To Aldrich, who had negotiated with Japanese companies ever since his days at Biogen, it was an unusually promising exchange—not for what was said but for who said it. Most negotiations with the Japanese begin with low-level licensing people who have no authority to move things along within the ritualized consensus-building process by which virtually all Japanese firms are run. But Nagayama was one step away from the board. He had the confidence of chairman Kimio Uyeno, son of the founder, who had parlayed the best-selling hangover remedy in Japan into a nascent global empire. Aldrich figured that by this one meeting alone they had gained a year, perhaps eighteen months, toward the prospect of Chugai's underwriting Vertex's immunophilins program.

Even more than Boger, Aldrich hated sharing Vertex's "value" with those he thought contributed nothing to it. That was why he favored research partnerships over other types of financing. He knew that if he and Boger didn't find a development partner soon, they would end up going back to venture capitalists (VCs) like Schmidt for more money, which, although potentially a boon for Schmidt and the other board members, was the most expensive route for the company. Aldrich's frequent use of the generic "bloodsucking VCs" reflected this point. Because of the high risk they take, venture capitalists expected the highest return for their money of any investors. More, as board members, they priced stock that they then sold to themselves. Aldrich wanted nothing more than to free Vertex from its indentured condition through a deal of his and Boger's own making, and he was delighted to think that the ground had suddenly been laid for one. That Chugai seemed not to care about Schreiber was a stroke so fortunate he hadn't even dared consider it up to that moment.

Schmidt, eyeing the exchange avuncularly, also thought he smelled a deal. He had two tests for doing business: Is this a field you want to be in and are these people you want to be in it with? Viewing the situation from both sides, he thought the auguries promising. Casually, he suggested a roughly even split. If Vertex produced a new immunosuppressant, the two companies would market it as a joint venture in North America and Europe, with Chugai getting the Far East and paying Vertex a sizable royalty there. From a pharmaceutical marketing standpoint, the rest of the world is a write-off; Schmidt didn't mention it.

"I outlined my ideas of a deal and asked Sam if that sounded to him like the makings of a deal, and he said it did," he later recalled. "So I wrote down a price on a little piece of paper and I handed it to him."

Schmidt's figure—for a project that was scarcely any closer to producing a drug than it was three months earlier when the company had no labs, a project whose chief selling point, Schreiber, was now persona non grata and perhaps even a double agent and competitor—was $40 million. Nagayama folded the scrap of paper and, committing nothing but a reply, returned with it to the company's headquarters in Tokyo.

CHAPTER SIX

In the weeks before the Glaxo visit, when not only Thomson but all of Vertex's scientists were working feverishly, Boger often sat at his desk cranelike and unmoving for long periods, his eyes fixed on a computer screen, a look about him of preternatural composure. Occasionally he unfurled several long fingers to type something or reached without looking into a foot-high mound of plastic slide envelopes stacked precariously at his left. Extracting a slide, he examined it against the light, then put it in a small box or returned it to its slot, his visage unchanged. He stopped eating during these times, causing Roger Tung, one of the chemists, to suggest that he was an autotroph, an organism of such exquisite metabolism that it supplies its own food and can live on air. Compared with the researchers' own anxious exertions and the towering unknowns that swirled before them, Boger's calm and assiduous attention to his slides seemed to many of them—dangerously, a few argued—aloof.

Boger thought to convince them otherwise was "an unpersuadable issue," and so he didn't try. He was in full "sell mode"—he was thinking about money—and there was no complication in his mind about what Vertex had to do, what he had to do, or what the scientists had to do. He let them think about him whatever was necessary to make a better drug than FK-506.

Boger was now acting on first principles: Long before Vertex could design and sell its first drug, it had to design and sell itself. Small companies that are years away from making a profit are called *story stocks* on Wall Street because their value is based not on products or sales, but information—information about, and generally supplied by, the company. Usually the term connotes a flagrant volatility: Company A's value soars upon winning FDA approval to test a drug in humans, then plummets as Company B files a patent interference. But there is also an element of illusion and seduction, of hype. Among investors, stories generate heat and, if one is fortunate, lust. Far from incidental, the possibility for such arousal is oxygen to emerging drug companies, because it enables them to raise hundreds of millions of dollars in the face of what appear to be wanton, savings and loan–sized losses. Thus Boger's attention to his slides. They comprised a montage of the story he was scripting for Vertex, a narrative that would have to stand up, with or without actual progress in the labs, indefinitely and against other stories that were equally compelling.

Boger knew that stories have to be accessible and that what investors want most from them is affirmation, so he molded Vertex's slide show not as a disquisition on science or business strategy, but as a quest. The grail—the object of the quest—was structure-based design and its transcendent prize of safer, smarter, more profitable drugs. The impetus, as always in such stories, was a combination of righteousness and greed; Vertex had a better way to discover drugs than screening and biotechnology (both of which, Boger would say, were terminally limited) and was intent on capturing the spoils of its victory whole. The rationale for the quest was the company's unique melding of disciplines and technologies, which he represented as a kind of circular flying wedge, and its scientists, who, he noted, all came from the world's most powerful research institutions. Harvard, naturally, was a key supporting element, as was Merck, and on the financial side, Benno Schmidt. FK-506 and immunosuppression were the story's set pieces, meant to illustrate its correctness.

That was the text. There also were subtexts that Boger didn't mention, the most intriguing being about himself. Boger never referred in his slides to his relationship with Merck, but he was seldom introduced anywhere without it being mentioned. To listeners with a knowledge of the drug industry, his defection was the most tantalizing part of Vertex's story, introducing, as it did, a whiff of patricidal intent, of vengeance. Here was Boger, a scion of America's Most Admired Corporation, the most productive drug company in history, Wall Street's gold standard, rejecting all that it had to offer because he thought he could do better. It didn't take a rereading of Genesis: Boger's saga of defiant departure was as old as Adam.

Because of his central place in the story, Boger believed he also had to be the one to tell it; as Aldrich put it, "People want to kick the tires. They want to see Josh." The decision complicated Boger's position within the company. As Vertex's chief scientist and CEO, he was responsible equally for its research and its business. But he had not stopped traveling in a year, and the rigors of selling had drawn him increasingly away from the labs. His grasp of the intricacies of each lab's work and of their interrelationship remained unequaled within the company; chemist Jeff Saunders, far from dismayed by his absences, was astonished by his "omniscience." But some of the scientists began to doubt that he could competently manage both roles for long. A few missed his confident presence and were floundering.

Boger saw no evidence that he or the company was seriously faltering, and so he dismissed their concerns. He believed he could run both the science and the business as long as he preferred, provided that everyone else did what he expected. Typically, as whenever others raised worries about events that hadn't occurred yet, he decisively deferred action. If it got to a point where he couldn't do everything, he told them, he would do something about it then. Now, his job was to sell, which in Boger's view meant a calculated attempt to master and improve the contradictory art of selling.

"Anytime you talk to people—especially a group of people—you're by definition projecting an illusion," he said some months later. "You're projecting the illusion that you're talking to each one of them individually. So you've got to picture in your mind what the conversation would be if it was just with one person. You've got to make each individual feel like they're witnessing a play in a drawing room and that they're sitting in a comfortable chair listening to lines that are intended just for them."

Boger used a variety of stage tricks to elicit such feigned intimacy—lowering his voice to make a point; speaking to the back row, which gave those in forward rows the impression that he was speaking to them. His main device, though, was a strenuously rehearsed spontaneity. As he spoke, Boger listened intently to his own words, then quickly anticipated the questions of those in the audience and tried to answer them matter-of-factly in the next sentence or two. Thinking that this was what anyone who wanted something from another person did unconsciously, as a matter of course, he was surprised that others didn't try to systematize it. And yet he did it so well that it allowed him to give the same talk over and over, never once the same, without the slightest trace of boredom.

"You can't fake excitement and you can't fake sincerity," he said. "It can't be done . . . and to do it properly you have to practice."

Practicing now, Boger believed he knew exactly what it would take to sell Vertex's story to Glaxo. He also knew that Vertex didn't have it; no company did. In the process of cross-examining himself, he heard questions that the big drugmaker would surely want fully resolved and for which there were no firm answers. What role did cyclophilin and FKBP have in suppressing the immune system? How did they function? Why should Glaxo put up a dime to design molecules for a target that, though it bound to FK-506 chemically, had no demonstrable link to the drug's activity?

Boger treated the question of FKBP's biological relevance with a studied pragmatism; it mattered, he would argue, but not nearly as much as generally believed. Because FKBP bound so tightly to FK-506, a drug of rare and proven efficacy, it either had to be the right target or was close enough in molecular structure to serve as an interim template for drug design. It was a chemist's view, to be sure, and Vertex's biologists worried especially that Boger gave the question short shrift. They suspected him of whistling past the graveyard. Boger, however, saw no immediate downside and so persisted.

Appealing to Glaxo on strategic, rather than scientific, grounds was for Boger a no-brainer. Glaxo, thoroughly and cantankerously British, had developed the world's biggest-selling drug, the anti-ulcer agent Zantac, through a traditional combination of screening and medicinal chemistry; it had a well-earned and lordly disdain for unproven methodologies like structure-based design. Yet in Boger's reckoning, the company couldn't fail to be deeply interested in Vertex. Best-selling drugs seldom retain their position for more than a few years either because their patents expire or new blockbusters come along and push them aside. The treatment of autoimmune diseases with immunosuppressants, meanwhile, represented a virgin market of perhaps $4 billion per year. Implicit in Vertex's story was that FKBP, discovered by Harding and controlled by Schreiber, was the best target for immunosuppression yet found and that Vertex knew more about it than anyone else, including Merck. Glaxo, needing to stuff its pipeline in its product war with Merck, could not but be extremely attentive.

Of course, there were at the time other stories about other small companies also working in immunosuppression that Boger, though he disdained most of them, was forced to take into account. For instance, a California company named Cytel had recently signed a $30 million deal with Sandoz to develop drugs for rheumatoid arthritis, juvenile diabetes, and multiple sclerosis—the same autoimmune diseases targeted by Vertex and practically every other new company working in the area. Cytel planned to design molecules that blocked receptors on the surface of T cells that falsely recognized the body's own tissue as alien. Seattle-based Icos, the most talked-about biotech story of the year, had raised $33 million while still on paper by saying that it would block "cell-adhesion molecules," which act like miniature Velcro tabs. The company intended to gum up the molecules so that they couldn't snare errant white blood cells and divert them to areas where they result in inflammation.

Boger was unimpressed. Perhaps these molecules were legitimate targets, but whether anyone could block them with drugs remained unproven. On the other hand, FKBP bound to a molecule of clear therapeutic value. To Boger's mind, the distinction was immense. Cyclosporine and FK-506 had already vouchsafed Vertex's approach. Cell-adhesion blockers? Boger thought perhaps in a decade the world would know better whether such molecules had even a chance of working.

And yet Boger couldn't deny that companies like Cytel and Icos made great stories. Preparing his slide show, he factored them in now. He would not resort to their extreme speculation, but if Glaxo wanted biology, biology he would give them. If it wanted proof that FKBP was the optimal target for immunosuppression, he would make a powerful case. If the company wasn't, as Chugai was, interested in structure-based drug design, rejecting Vertex's very reason for being, he would still find a way to represent a deal between the two companies as being in the best interest of both. Stories, Boger knew, could be edited to suit a variety of audiences. It was the key to successful selling.

Unlike the meeting with Chugai, which was informally arranged and anointed by Schmidt, Glaxo's visit to Vertex's labs was steeped in protocol and strained by nuance. Lab coats were cleaned and pressed and, for the first time in many cases, worn. Every penny of the $4 million Boger had spent on equipment was ceremoniously engaged, every machine and computer "pinned" for maximum display value. Shepherding Glaxo's six-man delegation was its North American licensing chief, Rick Hammill, who had done the advance work for the visit and stood to gain most if it presaged a deal. But it was a staid Briton, Leslie Hudson, that Boger knew Vertex must impress. Hudson was Glaxo's chief of immunology, which by definition meant he would lose by a deal with Vertex. Not only would it imply a failure within Hudson's department, but he'd have to pay for it out of his own budget. Boger detected no light behind Hudson's visage, nor did he expect to.

As Jeremy Knowles had cautioned at the October SAB meeting, impressing a potential scientific collaborator without revealing so much that you become dispensable is often impossible. Thus Vertex began at a disadvantage. Hudson, speaking first, predictably said little beyond the obvious: that Glaxo was launching a screening program in immunosuppression and was looking for targets to screen against. But Vertex couldn't afford such generality. Boger had warned the scientists that they had to risk telling Glaxo all they knew about FKBP in order to create a strong negotiating position. Now, as they began detailing the work of the past several months, a few of them felt, despite the decorum, that they were about to be reamed.

David Livingston would recall that ten minutes into his slide show, two or three members of the Glaxo team were scribbling so furiously in their notebooks that the table shook. Others noted the same breathless interest in the comments of Debra Peattie. Peattie, a tall, thin biologist whom Boger had hired away from Harvard Medical School, suggested a way to improve the use of FKBP to screen compounds, based on an unpublished observation. Sensing Glaxo's keenness and Aldrich's horror—"Rich was turning white," she says—Peattie quickly returned to the lab to record the idea in her lab book, having her assistant, Judy Lippke, cosign the entry.

Hudson was politely enthusiastic about Thomson's sample of pure bovine FKBP and about Manual Navia's disclosure that he had coaxed from it what appeared to be the first known crystals of the protein. Privately Navia called the slender, needlelike crystals Mickey Mouse, but the ability to grow them just three months after the opening of the Vertex labs supported his and Boger's boast that Vertex would generate and use structural information faster and better than anyone else. Reluctant to take credit for so preliminary an accomplishment, Navia acknowleged his debt to Thomson for providing him protein of exceptional purity, a blandishment that Thomson underscored by working again through the night.

Several times Hudson importuned for more data, but otherwise he seemed impatient: It was Schreiber he'd flown across the ocean to see and hear. Boger hadn't misled Glaxo about the delays with Harvard, nor had he told Schreiber of his internal exile. He wasn't dissembling. But his plan was to exclude Schreiber from any discussion of chemistry or biophysics, then cede him the major part of the meeting to talk about biology.

Schreiber performed flawlessly. He made a compelling case for his theory that FK-506 consisted of two distinct regions—binding and affector—and that the necessity of both for immunosuppression suggested that the molecule worked by imbedding itself in a host, then reaching out and hooking onto something else, most likely another protein. Though Schreiber didn't claim to know what this partner was and although his hypothesis implied that it, not FKBP, was the drug's real target, the pluses in his scenario strongly outfavored the minuses in supporting Vertex's view of FKBP's critical importance. Boger was pleased. Vertex's bases were well covered. Even if there was such a molecular partner to FKBP, the clear implication was that Vertex and Schreiber's group at Harvard were already collaborating to find it.

Of distinctly more interest to Vertex than Glaxo was Schreiber's announcement that his lab had now produced 400 milligrams of pure, recombinent FKBP. Boger, sitting across from him, noted to himself the amount. The only conceivable reason for directing a highly promising graduate student or postdoc into the career abyss of grinding out that much protein was to try to solve its structure. Schreiber would not want to sit on such a treasure for long if a collaboration with Navia couldn't be arranged soon. Boger also noted Schreiber's pleasure at speaking with Glaxo. "If he gets stroked any more he'll be too shiny to look at," he said privately after the meeting.

Hudson listened noncommitally. Boger was uncharacteristically quiet, preferring to let the story unfold by itself. Neither was undiplomatic, and yet as the morning went on both were clearly becoming more and more irritated. Hudson was not hearing the proof he wanted that FKBP was the target Glaxo should be pursuing, and Boger was beginning to sense that Glaxo's disdain for rational drug design was more than academic. It was about money. "I could see they were trying to figure out how they were going to fold everything else but biology into overhead," he says. "We were going to be expensive for them, and I didn't need to have my research controlled by someone who didn't believe in my basic philosophy. I've been there before."

Finally, breaking the standstill, Hudson asked Boger for his own view on how immunosuppression worked. The room drew silent, a scuffling of chairs quenching in its own echo. Tartly put—"belligerent," Livingston recalled it—the question seemed to have only two answers, equally unacceptable. Either Boger had to concede he didn't know, deflating Vertex's plumped-up claims for FKBP, or else he had an explanation but insufficient data to support it. Either way, he seemed to be left little room to avoid being embarrassed, much less to salvage a deal.

Boger leaned forward. He told Hudson insouciantly that it was too early to have such a view nor was it important to have one at the present time. He would have one, he said, when the information warranted.

Hudson stared. He said nothing. It wasn't until after, as the two groups of scientists stood around talking, that he remarked offhandedly that he agreed.

Whether Boger had anticipated the question and rehearsed his answer or was simply reasserting his view that decisions are only incorrect when they're premature or neglect the most advanced data, Glaxo's tone suddenly now resumed its earlier friendliness. Hudson, meeting privately with Boger and Aldrich in Aldrich's office, suggested that Glaxo might be able to fold some of Vertex's structural research into a deal after all. In the next days, a flurry of follow-up calls from Rick Hammill confirmed Glaxo's interest. In the postmortems at Vertex, a consensus now emerged that the company, with barely three months of experiments behind it, had held up under direct fire. After the months of misgivings about Schreiber and Harvard, the aura of triumph, of invincibility, of big things about to happen, returned to the lunchroom. Aldrich, generally pessimistic in any negotiation, now placed the odds of doing a deal with Glaxo at 50–50. Once again, Boger seemed charmed.

Only Livingston, formerly a senior scientist with a once-promising biotech company that had run out of money and was sold amid layoffs and other misfortunes, debunked the good feeling. "I think all this hype about this being Vertex's deal to lose is a crock," he snapped. "Glaxo did exactly what they should have done. They took back everything that we gave them, and they're keeping us waiting as they evaluate it and see where it fits into their plans."

Events hurtled disproportionately, quickly now. Two days after the visit from Glaxo, on the afternoon before Boger and Aldrich were to leave on a long-scheduled eleven-day door-knocking mission to Japan, Chugai's director of research and several other top company officials arrived at Vertex as a follow-up to the initial contact in New York. To Aldrich, it was a sign of Chugai's exceptional urgency that the company would not be put off until after his and

Boger's Japan trip—a perception that grew decisively when its haggard delegation, arriving from Europe, took a cab directly from the airport to Vertex and remained cloistered in scientific discussions for six hours before finally going to their hotel. At 7 P.M., with the normal time frame for negotiations telescoping wildly, Aldrich went to the University Club for a swim, then returned to Vertex until midnight to write a formal business proposal to Chugai. A week before, the two companies had barely known of each other.

He and Boger boarded a 7 A.M. flight to San Francisco, where they would make connections for Tokyo. Aldrich called these trips "death marches," though it was during them that his status rose. At Vertex he was often excluded by the scientists, but here, in the air next to Boger, he was the company's vanguard, its elite. Aldrich called Vertex from the cabin, only to discover that Glaxo, apparently having second thoughts, was also now pressing insistently for another meeting. The company had called that morning, inviting Boger to fly to London at once for discussions with its senior officers. Caucusing in front of the first-class lavatory, Aldrich and Boger reviewed their options. "It wasn't going to be a straightforward exercise," Aldrich recalled.

Glaxo and Chugai didn't know about each other, nor could they know except in the most general terms. On the other hand, the attractiveness of using two ardent suitors as leverage against each other was an obvious no-brainer. Boger and Aldrich decided to stall Glaxo, at the risk of losing its interest, at least until they returned to Cambridge.

It was gratifying, this sudden surge of attention, and as Boger would say, wholly anticipated. And yet it also pointed up sharply the strengths and weaknesses of his position. He had assembled all the elements of a luminous story—so luminous that it had enticed two companies that saw within it prospects that couldn't be more different and represented for Boger, and for Vertex, starkly divergent futures. Glaxo was the more formidable development partner: richer, more aggressive, more capable of bringing a major new drug to market. But Glaxo, like Merck, saw structure-based design as a sideline to the main business of screening for billion-dollar drugs. It would command and take a larger share of whatever Vertex made and possibly treat the company as a biological service in return, perhaps even discarding it after Vertex came up with the information it required. Glaxo would overshadow Vertex in any collaboration. It might end up using Vertex's molecular design work in its annual reports but not in its drugs.

Alternatively, Chugai saw structure-based design as key to the future of drug discovery and wanted to buy into Vertex while it was still affordable. It would want not only drugs but technology, not only data, but proof of concept. As a classically Japanese company, it intended to swallow Vertex's example whole. Thus it would want Boger to do everything he set out to do, including ascend to the leading place in science he seemed destined for. It was the best way to ensure Chugai's own success.

Boger's ego wasn't immune to those who would aggrandize him, but he realized, in considering which company's money he'd rather take, that there was another reason he might need Glaxo's strength: as a counterweight to Merck. It had been slightly more than a year since he had left Rahway, but Boger was not easily let go. Merck, both as a competitor and as the parent of his ambitions, still dominated his thinking. It defined him in more ways than he liked. If he was to succeed, it was going have to be in spite of and measured against Merck's own singular success. He sometimes jokingly called the company "Mother Merck," aware of the implications. But in the manner that he viewed it and the broad streak of history that now funneled narrowly into his thrust for independence and keen ambivalence toward the company, the firm was much more like a father to him. In that context, Glaxo looked like an optimal partner, even if its surrogacy threatened to undercut Boger's control over his science as much as Merck had, even if, as he understood, it dismissed everything he stood for.

Manuel Navia was also a renegade son of Merck and, having solved three important protein structures by age forty-three, unique for Vertex: He was already famous when he arrived. His hiring was a coup for Boger, important enough for Aldrich to write a press release that was printed in the *New York Times*. And yet his presence as an unofficial first-among-equals rankled those who believed Boger's social experiment barred such favoritism. Navia, an outgoing, fastidious, first-generation Cuban-American and the only Vertex scientist—including Boger—to wear a tie to work each day, overcame the discrepancy initially by downplaying it. He made comic self-flagellating gestures with his tie, muttering "Ka-chunk . . . ka-chunk" as he alternately flailed his shoulders, suggesting a hairshirt as prickly as everyone else's. But everyone else didn't fly to New York to meet with Chugai, or have his experiments performed on the space shuttle, or serve as company spokesman while Boger was in Japan. Clearly, Navia, formidable and articulate, was to have a larger role to play, though typically Boger had been careful not to specify what it might be.

Navia was distinct from Vertex's scientists in other ways. He seldom drank, after-hours drinking being something of a company team sport. With his early-1960s haircut and attention to grooming, he looked like a divinity student at his kid brother's frat party, gripped by no vice save perhaps ambition. He was also worldly and likable and had a keen sense of the absurd, as when describing his stint in the army: He had spent a year during the Vietnam War firing laser beams at pigs, helping the service defend itself from wrongful injury suits by tank drivers who'd made the mistake of staring into their nightscopes.

An only child, Navia arrived with his parents in New York City in 1953 after an overnight flight from Havana and immediately became his parents' interpreter and the sole, dutiful object of their rising expectations. In the 1960s, he attended Xavier High School, an all-boy Jesuit military school in Manhattan, and then New York University, because they were close to home. He later received his Ph.D. at the University of Chicago and won a fellowship at NIH, where he solved the first structure of a human antibody. As courteous and well spoken in public as he is sardonic informally, Navia professes to be motivated only by a desire to design drugs. Despite his fame, he claims not to care about scientific competition and its reward system, insisting, "I'm not an academic." This is partly true, but no research career as visible and respected as Navia's occurs haphazardly. Navia knew an opportunity when he saw one and attacked it with a ferocious temperament. Surprised by a small dog that began barking at him once from behind a chain-link fence on Allston Street, he exploded at the hapless animal. It was not the only time that others at Vertex would see him snap.

With Boger and Aldrich in Japan and without enough protein from either Thomson or Schreiber to begin major work on FKBP, Navia used the time after the Glaxo visit to plan out experiments and catch up on some scientific literature. Along with Boger and Livingston, he tended to read more outside of his field than others at Vertex and had begun thinking broadly about other projects besides immunophilins. One article in particular quickly absorbed his attention. It was in *Nature* and was about a possible new drug for AIDS. A well-known Belgian chemist named Paul Janssen had synthesized a new group of inhibitors that in the test tube were more active against the AIDS human immunodeficiency virus (HIV) than any other known group of compounds, including azidothymidine (AZT), at the time the only drug approved for treating the disease.

Janssen's molecules were benzodiazepines, derived from the same basic chemical structure as the best-selling tranquilizers Valium and Librium. He speculated that they worked by blocking reverse transcriptase, an enzyme that enables HIV to hijack healthy human cells and turn them into factories for making new virus. The ability to convert the body's sentinels into assassins is what makes HIV such an insidious killer, but reverse transcriptase is highly mutable and thus a poor target for drugs. Because of the extreme toxicity of AZT, Janssen's more specific molecules conceivably represented a major breakthrough.

Years earlier, Navia had studied benzodiazepines closely and had committed their wedgelike triple-ring structure to memory. At Merck, he had solved the structure of HIV protease, another enzyme crucial to the proliferation of new virus. Now, reading Janssen's paper, he was thunderstruck. "The two compartments of my brain immediately folded together," he recalled, snapping his hands shut like an alligator's jaw. "I took the structure of HIV protease, and I built a model of one of Janssen's compounds, minimized it, and matched them up. And there it was. The overlays were just incredible."

Believing that Janssen's compounds inhibited not reverse transcriptase but HIV protease thrust Navia into a searing personal and professional dilemma, one that had vivid implications not only for him and Vertex but, Navia thought, the world. Navia saw AIDS as a disease of "planetary significance." He thought that unless the virus was checked it could eventually kill off the human race. These were not hysterical claims, and yet they were so remote from the terms in which AIDS was generally discussed within the drug industry as to be considered irrelevant and bizarre. There, universally, diseases are not treated first and foremost as diseases, but as markets. And the industry consensus in the winter of 1990 was that on the bottom line AIDS was going to be a loser for all but a few companies. The coldness of this calculation was reinforced by Boger, who swore practically as a matter of principle that Vertex would never work in AIDS. He was not unsympathetic, but given the odds—the large number of other companies already in the field, Vertex's embryonic lab effort and onerous finances, what was known scientifically about the virus—Vertex appeared to have little to offer and everything to lose by taking on HIV.

Navia had agreed with Boger, but now he was transformed. He couldn't prove his hypothesis—not without supplies of protease and Janssen's compounds and not without crystallizing them together to show that they indeed bound as he envisioned. But in a series of experiments using high-powered computers and three-dimensional imaging software, he was able to create visual models that gave his "hallucination" striking credibility and form.

HIV, like all viruses, consists solely of a snippet of genetic information inside a protein casing—so low on the order of life that it can't reproduce outside a host and so small, as writer Fred Hapgood has observed, that compared to a human cell it's like a basketball against the World Trade Center. HIV protease is one of a well-known class of enzymes found up and down the evolutionary scale that cleaves protein into smaller subunits. In HIV, it works during the assembly stage. Slowly, indolently, the enzyme draws fragments of protein into it maw, where two scissor-like groups of atoms cut them to the specific length and chemical composition needed for the construction of new virus particles. Theoretically, a well-placed inhibitor would fill the tunnel-like cavity where the pincers are located and occupy them, like chucks driven into lobster claws. Without raw material, replication would slow. No new virus would be produced, and the spread of infection would be checked. No such inhibitor had yet been tested in humans, but many researchers now considered HIV protease the most promising drug target in AIDS.

Well aware of Boger's admonitions, yet stirred by the prospect of an important breakthrough, Navia now plunged into designing HIV protease inhibitors based on Janssen's compounds. For several days he sat in a darkened room next to the X-ray lab at a high-speed graphics workstation that had previously been used mainly by some of the chemists to simulate aerial dogfights, maneuvering benzodiazepine derivatives into the active site of HIV protease. One after another, he positioned and repositioned them inside the enzyme. And as he imagined, and with an excitement that quickly became uncontainable, he found he could fill the contours of the site, occupy the pincers, and satisfy a striking number of the known chemical requirements for protease inhibition. Boger believed that Vertex's primitive facility for designing drugs was insufficient to fighting AIDS. But Navia's models were so convincing that he quickly came to believe in them. This computer modeling—speculative to be sure—was the germ of structure-based drug design. Bursting, Navia felt he had to act quickly toward the next step: seeing the molecules made and tested.

"I didn't think we were ever going to work on HIV so it was a moral issue," he said. "My thought was, we've got to do one of two things; either we get somebody else to work on it or we have to publish it—put it out and see who picks it up."

Concerned that no one would pursue his observations if he simply published them, Navia decided he would rather offer them to another drug company as a freebie—a simple matter, it would seem, in an area where the stakes were human extinction. And yet scientific companies are filled with their own researchers brimming with their own ideas. They're organized in such a way that at every level, from technician to senior vice president, they must compete, often brutally, for the resources to see their most promising observations acted upon. No self-respecting drug company welcomes over-the-transom proposals, especially from someone who can't get his own company to commit the time, money, effort, and faith to pursue them. Navia needed a company that had both the understanding and ability to make use of his concept and regarded him highly enough to take his suggestions seriously. Only one came to mind.

"The dilemma was," he would later say, "could I trust Merck to run with the ball? I still had a certain credibility there, and I felt as if at Merck I at least knew where all the handles were. I knew where the buttons were. I thought I could engineer it in such a way that the thing would get a hearing."

During the first week in March, with Boger still in Japan, Navia wrote a four-page letter to Merck's director of research, Edward Scolnick, detailing his work on HIV. As the company's number two scientist, Scolnick was an obvious choice: He had the authority to act unilaterally, yet if he didn't, Navia still had room to appeal to CEO Roy Vagelos. Both men had personally beseeched Navia to stay at Merck after he'd decided to leave.

The choice of Scolnick had an unintended irony. Navia had been hired at Merck in 1980 to help the company design drugs, but his assistance often was not welcome by those who were skeptical of his efforts. His departure was precipitated by one such incident. Navia had assumed that his structure of HIV protease, which had been greeted in the world at large with front-page coverage in the *Wall Street Journal* and by senior company officials with lavish praise—would become a vital part of Merck's drug discovery. But when he asked for routine information about the company's compounds so that he could do what he was now doing at Vertex—trying to predetermine their biological activity—the project's head chemist refused. Incensed, Scolnick overruled the decision, but Navia, appalled that it had been made in the first place, withdrew from the project in angry protest. Now, as a competitor, he was asking Scolnick to have Merck do what it had failed to do when he worked there.

With his letter to Scolnick in hand, Navia had only to apprise Boger before sending it off. He faxed a copy to Chugai's headquarters in Japan, expecting quick approval. Boger received the fax just as he and Aldrich were about to meet with the company's executive committee, which had taken the unusual step of canceling its own meeting to receive them. He and Aldrich talked most of the night about the matter, then again the next day while touring the Tokyo Stock Exchange. That night, Boger ordered Navia in a confidential fax not to send Scolnick the letter.

If there was something to Navia's hypothesis—and Boger believed there could be—he wasn't about to give it away. Contradicting his own sworn judgment and with barely enough science and scientists to warrant what he was spending on immunophilins, Boger authorized Navia to begin a second "protoproject": an exploratory program in AIDS.

Scolnick never learned of the letter or its contents. But two days later, on March 5, he unintentionally answered Navia's question during a speech at Harvard Medical School. Asked by Schreiber, who was now concerned about such matters, whether having the crystal structure of HIV protease had helped Merck design new inhibitors, Scolnick answered: "It gives you some ideas you might not otherwise have." Pressed, he added, "It's given us some help, not dramatic help." But by then Vertex had set out to prove him wrong.

All Boger's ghosts came ablaze at the decision. Ever since high school he had been on a seesaw course between doing better than anyone else and waging guerrilla attacks against those whose authority he believed kept him from doing still better. In tenth grade, he harshly criticized his chemistry teacher for making mistakes, then taught the class himself from October to June. At Wesleyan University, in the early 1970s, he founded CRAW (Committee for Reform of Academics at Wesleyan), which fought to keep the school from compromising its mission of "learning for learning's sake." Largely successful, its call for what Boger called "savage rage" was tolerated, in part, because of the intellectual rigor and studious presentation of his arguments and because he was the top student in his class. The situation repeated itself at Merck, where his relentless attacks on screening and on the company's bureaucracy were balanced by brilliant feats of science and a scrupulous ability to get things done. "You almost can't do better than Josh did at Merck," says former vice president of basic research Ralph

Hirschmann, who first hired Boger out of graduate school. If Boger was insurgent, he was always cautious to lead his rebellion from the front row after doing all his homework and scoring perfectly on all his exams.

But this was another order of challenge. Even more than the decision to leave Merck and go it alone, the attempt to design drugs against AIDS based on the same information that Scolnick said Merck had found only marginally useful set Boger irrevocably on a more oppositional path. He was rejecting the judgments of the world's best drug company and of his own mentors. As with immunophilins, he was trying to reach beyond screening and medicinal chemistry to a more rational way of finding drugs, yet he was doing it in an area he swore he would never enter, against one of the most complex and insidious diseases in history and against every leading drug company in the world. He was outside the classroom now, outside the school, setting up alternative classes on the playground.

He was in repudiation mode, as he might say. "Joshua's found a new religion," observed Hirschmann. "I just hope he doesn't forget that the old religion, in its way, was incredibly successful."

CHAPTER SEVEN

Like any good student's, Boger's reformation encapsulated sweeping changes in a struggle that was mainly personal. He was less intent on perfecting than replacing the old order.

His ambitions, as always, befit his precocity. Medically, the twentieth century may be remembered as a time of two great pandemics separated by more than six decades and bracketing the most productive period in scientific history. The first contagion, influenza, obscured by the horrors of World War I, swept westward with calamitous speed in the fall of 1918: Covering the globe in just two months, it killed 22 million people, almost twice as many as the war itself. Though the general nature of infectious disease had been understood for more than forty years, science was scarcely more able to identify the infectious agent or stop it with drugs than in the 1340s, when the Black Death claimed one third of Europe and even the most learned thinkers ascribed the affliction to a bizarre conjunction of planets and treated it with powdered stag's horn and potable gold. The century's second great contagion appeared slumberingly in 1980 and, by contrast, has spread much slower since, granting a vastly more knowledgeable and priority-driven research establishment a rare chance to catch up. Within four years, when nationally still fewer than 3300 were dead and less than 4500 new cases were reported, the cause of the contagion was known. By 1988, an array of promising molecular targets had been identified. The second contagion—AIDS—epitomized the challenges to the new religion of molecular pharmacology that Boger, by the mid-1980s, had embraced as one of its leading apostles. But it was the 1918 flu epidemic, one of the three deadliest contagions in history, from which the old religion—the strikingly successful combination of soil screening and medicinal chemistry—first arose.

The pandemic reached America last, coming ashore in Boston in the first days of September, three months before the end of the war. Four days later the first cases were reported at an overcrowded 45,000-man military cantonment at nearby Fort Devens, where within three weeks the slate-colored bodies of up to ninety men a day were being "stacked like cordwood" for burial. The flu itself was not the killer. With no antibiotics or other treatments available, the victims' lungs became saturated with pneumonia-causing bacteria. Many died, drowning in their own fluids, less than forty-eight hours after their first cough. In an age when American prosperity and technology had seemed capable of conquering any problem, the fact that organized medicine offered only palliatives against the plague struck many as its most apalling aspect. "Science," the *New York Times* editorialized, "has failed to guard us."

In Boston, eleven-year-old Max Tishler raced among disease-infested brick tenements, delivering aspirin to the families of the dead and dying. The fifth and next-to-last child of poor Jewish immigrants, Tishler had gotten a job washing bottles and filling them with powders for a druggist. Despite tentative advances in other areas, the pharmacopeia for acute infectious disease still consisted mostly of metals and extracts; aspirin, with its wondrous pain-killing and fever-reducing properties, was one of only a few drugs owing their discovery to science. It was derived chemically from coal tar, a noxious sludge that one hundred years earlier became the first major toxic by-product of the industrial age. Aspirin didn't cure people, but it comforted them, and only in hindsight does young Tishler's diligence seem futile.

Ministering to the sick gave Tishler "a feeling that I might want to do something in the line of disease," but his circumstances were bleak, prohibitive. His Romanian-born father left when he was four years old and would stay away more than thirty years. His mother and siblings all worked, and within the family only he and a younger sister would finish high school. Tishler was undaunted. Lean and small boned, with a helmet of wiry rust-colored hair, sprightly ears, and a piercing grin, he had a bristling intelligence and fiery resolve. He worked constantly—hawking newspapers at trolley stops before school, baby-sitting, answering phones—all in addition to holding down jobs at various drugstores. After graduating from Boston English High School, he won a scholarship to Tufts, where he majored in chemistry and graduated magna cum laude. He obtained his pharmacist's license the same year.

Eschewing medical school, Tishler opted for a career in chemistry. A professor told him that "Jews have a hard time getting placed and you won't get anywhere," but Tishler, irascible and determined, disregarded the advice. He enrolled at Harvard in the fall of 1929, the near aftermath of the most blatantly antisemitic period in its history.

As a graduate student, Tishler was drawn to organic synthesis; he wanted to make biologically active molecules, although what use such molecules might have as drugs remained much in doubt. Like many aspiring synthetic chemists, Tishler had been seduced exultantly by Paul Ehrlich's prophecy of a "magic bullet," a molecule that attacked the cause of disease but not its host. To his professors, however, what a molecule did was not as important as how it was made, its architecture. Tirelessly, Tishler immersed himself in the developing new reactions by which compounds could be broken down and put back together. He was an intrepid bench chemist. Working once in a tiny lab with wet hands, he dropped a bottle of highly flammable benzene, which burst into flames: With the billowing smoke blocking his exit, he climbed out onto a third-story ledge and stayed there until several students rescued him. "What I think bothered me about the whole thing was the fact that I caused a fire," he would recall, "and we used up all the carbon dioxide extinguishers."

Tishler excelled at Harvard and was hired as an instructor, then as now a slap on the back as Harvard gave tenure only to those who had proven themselves elsewhere. By 1936, married now and with his wife, Betty, expecting at the height of the Depression, Tishler began looking for a more permanent job. Hoping for an academic post, he applied widely but received no offers. Meanwhile, one of his sisters had recently died of tuberculosis, and Tishler, who had continued to work as a druggist through graduate school, grew even more deterimined to work in medicine. Slowly, he began considering what until just a few years earlier would have been unthinkable for a promising chemist, much less a young Harvard professor. He began applying to drug companies.

Ehrlich aside, the U.S. drug industry in the mid-1930s was still a tainted business, more likely to produce such parodies as William Radam's Microbe Killer or Wendell's Ambition Pills than a genuine therapeutic. Science's failure eighteen years earlier during the flu epidemic had not been erased by its few successes since nor by the attempts of several companies to establish their own research labs. Companies were small, good at selling old products but with little idea of how to find new ones. As in *Arrowsmith*, Sinclair Lewis's 1925 novel in which a scientist who goes to work for a drug company is given up as "gone wrong" and "dead," an academic scientist who went into industry risked losing not only his credibility but his friends.

Tishler was not averse to working in business, but he also had no choice. Barred from a significant future at Harvard, unable to find another academic post, he plunged into applying to the major drug houses. After months of trying and apparently being rejected by at least one company—DuPont—because he was Jewish, he finally received an offer. It was from a small company in New Jersey that had a reputation for doing high-quality, interesting science but had yet to develop its first drug.

The company was Merck.

Since he'd taken over his family's fine chemical company in 1925 at age thirty-two, George Wilhelm Merck had devoted himself equally to three goals: expanding the business, becoming a scientific patron in the noble style of his forebears, and assimilating himself and his company at ever higher levels of American life. His firm, Merck and Company, ensconced on 150 acres adjacent to the Pennsylvania Railroad's main line in semirural Rahway, New Jersey, was not yet a drug company, but then hardly any American pharmaceutical manufacturer was. There were drugs and companies that made them, but the idea of discovering new medicines through research—the sine qua non of the modern drug house—was still new. It was also notoriously unsuccessful and, in the baldness of its profit motive, aggressively opposed by both doctors and patients. Merck, six foot five inches and 250 pounds and brimming with postwar optimism, was undaunted. The key to finding powerful new drugs, he knew, was having the most advanced labs, which, with the extraordinary promise of allowing its scientists to publish their work, Merck and Company began developing in the early 1930s.

Tishler, happy to be employed, subsumed himself in his first project: vitamin B_2. Without its own drug leads, Merck had decided to manufacture vitamins, an uncommercialized field where the limitations were the very opposite of those during the flu pandemic. Here the cures were well known; the failure was that no one had found incentive to make them. Scientists had known for nearly twenty years that those deprived of even one- to two-thousandths of a gram daily of B_2 suffered an array of ailments: Their mouths cracked at the corners, their tongues swelled, their eyes hurt, their skin grew inflamed. Among starch-eating sharecroppers in the South, who didn't get enough B_2 or a related vitamin, niacin, the suffering was extreme: skin lesions, wrenched bowels, depression, apathy, and the so-called three D's (dermatitis, diarrhea, and dementia) of pellagra. Two years earlier, German and Swiss chemists had patented a method for making B_2, but, seeing no market, had refused to license them in the United States.

Tishler was a dynamo. He quickly developed a novel synthesis free of all European patents, then, as the company began building a $5 million manufacturing plant, commandeered its scaleup. Chain-smoking, avoiding sleep, he arrived at his lab before dawn, left at night after everyone else, then returned after hours to prod and cajole the plant's engineers into reproducing his fine-tuned organic reactions on an industrial scale. A molecule, he knew, was worthless, a laboratory curiosity, if it couldn't be made cheaply enough to sell at a profit. On the job less than a year, thirty-one years old, he now showed that Merck could produce a complex organic molecule to compete with the lordly Germans.

In 1935, while Tishler was still at Harvard, a German dye chemist named Gerhard Domagk stunned the world by announcing that he had fed a white powder synthesized from coal tar to infected mice and had cured their infections in every case. The drug, called sulfanilamide, was the first "magic bullet" to be synthesized since Ehrlich's discovery, thirty years earlier, of a cure for syphilis and the first specifically to target bacteria. Like most coal tar derivatives, it was originally made as a dye: Binding tightly to the protein cells in wool, it had been expected to stand up well to washing.

To a world whose notions of bacterial infection were forged in 1918, the sulfa drugs were nothing short of miraculous. People were now suddenly being cured of agonizing, often fatal diseases like spinal meningitis and childbirth fever that previously had defied treatment. Gonorrhea, more prevalent than syphilis, could be reversed in a matter of days with a shot or two. Twenty years after the most devastating outbreak of pneumonia in history, doctors began speculating on a day in the near future when "people just won't die of pneumonia anymore."

Nowhere was the effect more electric than in the drug industry. There, Domagk's discovery seemed divine confirmation of Ehrlich's prophecy. A land rush ensued, with legions of chemists suddenly synthesizing and filing patents on thousands of new sulfa derivatives, then, as *Fortune* observed, "sending new substances every day to other investigators who force them down the throats of infected mice and rabbits and monkeys—and watch." Occasionally, in the race to exploit the public's hunger for miracles, the animal tests were simply overlooked, as when the S. E. Massingill Company of Bristol, Tennessee, began selling Elixir Sulfanilamide, a poison concoction that killed 108 people in the South—107 patients and the chemist, who killed himself. Recognizing the need for regulation in the new speculative climate and spurred by the Massingill case, the federal government responded by expediting passage of the Food and Drug Act, which strictly controlled the testing, development, and sale of new drugs.

Merck, still chiefly a chemical supplier, quickly began making large amounts of sulfanilamide under other companies' patents. It also tried to develop its own novel sulfas, placing Tishler in charge.

The effort, half-successful, was bitterly disappointing. Tishler's group made a molecule he hoped would combat malaria but that turned out to be too toxic for humans. However, it prevented coccidiosis, a poultry disease, and Merck helped revolutionize the broiler industry by introducing the antibiotics that would soon pave the way for factory farms.

But by then, the synthetics war between the U.S. and German chemical industries had become subsumed in the larger competition of World War II, a war in which emergent goal-oriented research was to be not just a bystander, but the decisive feature of the conflict.

The Carnegie Institution, ten blocks north of the White House, was founded in 1902 by steel magnate Andrew Carnegie "to secure for the United States leadership in the domain of discovery" and was governed in its early years by a genteel agenda ranging from stargazing to corn hybridization. A colonnaded mass of Federalist office and conference space girdling an imperial rotunda, it had become by the tropical summer of 1941 a citadel, the central marshaling point for science as the nation vacillated toward war. It was a novel idea—channeling research to achieve prescribed goals—and the Carnegie's normally clubby air was squelched in wartime officiousness and secrecy. Iron grilles covered the first-floor windows and armed secret service agents guarded the building around the clock.

The Carnegie was the impromptu headquarters of the Office of Scientific Research and Development (OSRD), the federal government's ad hoc commissariat for applied science and the brainchild of Carnegie's president, Vannevar Bush. As chief science advisor to President Franklin D. Roosevelt, Bush had pressed for and won extraordinary control over the nation's war-related research, most famously the Manhattan Project, which produced the first atomic bomb. Sharp-eyed and tireless, he also persuaded Roosevelt to allow the agency to direct the nation's wartime medical efforts and in May had engineered the choice of Alfred Newton Richards, a sixty-six-year-old pharmacologist from the University of Pennsylvania, as chairman of the federal government's nascent Committee for Medical Research (CMR). Richards, a staunch Republican and, like Bush, a minister's son, was best known as a kidney researcher and Solomonic man of science, eminent and wise. Bush trusted him to corral the merciless egos of the country's medical people just as he himself was having to restrain the physicists and engineers on the atomic bomb project.

Richards had another connection. When George Merck began steering his company into drug research in 1930, he had turned to Richards, who became the company's chief consultant and scientific architect. At a time when the drug industry was still suspect, Richards had given Merck and Company uncommon credibility, sacrificing his formal membership in the pharmacologists' fraternity to do so. He helped establish its labs, recruited its leaders, and allowed it to support some of his own work, explaining to contemptuous colleagues, "I saw in them no signs of the horns or tail." Now, assigned to the national interest, Richards pulled George Merck with him. Volunteering his company's services, Merck would soon be appointed head of the country's biological warfare program.

On August 7, 1941, the CMR held only its second meeting in the Carnegie's richly paneled library. Four months before Pearl Harbor, its goal was to assess the research already under way in the nation's labs and determine how best to apply it if and when the United States entered the war. Several vital areas came up—tropical and infectious diseases, nutrition, the blood supply—though the committee soon focused on one: aviation medicine. World War II was quickly becoming the first conflict in which air power was decisive. Between July 1940 and May 1941, Germany had dropped 54,420 tons of bombs on London during the Blitz. As bloody as the fighting on land had become, Germany's dominance of the skies over Europe was the most chilling augury of its claims of a thousand-year reich.

Germany, anticipating the primacy of air power, had begun exploring ways to sustain its pilots through the rigors of combat as early as 1934; Great Britain, Canada, and the United States had lagged, in that order. Now, the committee heard rumors that the Germans had isolated the active substance from the cortex of the adrenal gland—cortisone—and were giving it to their pilots, enabling them to fly at up to 40,000 feet. According to the reports, they were said to have cornered the Argentinian supply of calf adrenals and were importing it by U-boat.

Richards had little trouble believing the rumors. The development of performance-enhancing drugs—pharmacological warfare—had been a concern of his since World War I. And yet he also knew that efforts to isolate adrenal hormones had proven uniformly bewildering in the United States. Philip Hench, a biochemist at the Mayo Clinic, had isolated six such compounds but was unable to purify enough of them to identify their chemical structures, much less test them as drugs. He had turned in desperation to Merck, which for eight years had piled up its own failures in the area. If the Germans truly had identified the active molecule and devised a way of extracting it, the war could be over before Allied pilots would recover against drug-emboldened German fliers.

As Richards began formulating his own view of the nation's research goals, the necessity of a crash program in cortisone assumed instant priority. But Richards's day was far from over. Later, returning to Philadelphia by train, he was met by a British scientist, Howard Florey, who had once worked briefly in his lab. Florey and a colleague, Ernst Chain, were traveling the country to gather support for a promising new antibacterial agent that they believed might be better than the sulfas.

Now familiar, Florey's story was extraordinary. In 1928, a Scottish researcher named Alexander Fleming had been trying to identify the microbe that caused the 1918 flu epidemic when a green mold wafted through the open window of his lab in a London hospital and, landing in an open petri dish while he was on vacation, began to destroy one of his cultures. Growing more of the mold, Fleming noted that it was deadly to a variety of harmful microbes. Like Hench, Fleming couldn't find a chemist to produce enough of the pure substance, which he called penicillin, to test it in animals. But Florey and Chain, tantalized by penicillin's qualities, had painstakingly purified enough of the material to feed it to infected mice. By February they had obtained enough of it to give to a deathly ill London policeman. Within twenty-four hours after receiving his first dose, the man improved dramatically, but doctors soon exhausted their supply of the drug. Desperate, they began harvesting miniscule amounts of it from his urine and futilely reinjecting it. Still, by then penicillin's wondrous germ-fighting abilities were clear. With Britain's research infrastructure in ruins, Florey and Chain had come to the United States—and, ultimately, to Richards—to try to enlist American labs to develop the drug.

Richards was enthusiastic, though here, too, there were obstacles. U.S. laboratories had little experience with growing microbes and extracting their active ingredients, and penicillin was notorious. "The mold is as temperamental as an opera singer," a frustrated drugmaker would later declare. "The yields are low, the isolation is difficult, the extraction is murder, the purification invites disaster, and the assay is unsatisfactory." Too, there was a thicket of organizational issues. The birth of drug research in the 1930s had introduced a bristling new competitiveness as companies sought to protect their investments. Where patents were once reviled, they were now pursued ruthlessly. Squibb, which had one patent in 1920, had more than 200 by 1940. In 1937 alone, Merck had filed forty-six domestic and foreign patent applications. Any drug company willing to take on penicillin was unlikely to want to forfeit the exclusive right to manufacture and sell it, and even if it was, antitrust laws would bar it from collaborating with its compettitors.

On August 11, 1941, four days after the CMR meeting and his discussion with Florey, Richards wrote urgently to Hans Molitor, whom he'd been instrumental in recruiting to head Merck's fledgling Institute of Therapeutic Research: "I want very much to discuss with you . . . the whole question of how such a laboratory as yours can contribute to medical research contributing to national defense." He wrote a similar letter to George Merck. Responded Merck on September 10, "We are anxious to help you in any way we can."

Elevated swiftly to head Merck's penicillin project, Tishler drove his scientists blisteringly, himself even harder. He seemed fated to do so. Determined since he was a boy to fight ravenous infection, intent on vanquishing German chemical superiority and Hitler, spurred to prove Merck's primacy over academic and industrial labs alike, driven to make his mark in developmental research and overcome the failure with sulfas, he resolved to stop at nothing in developing the new drug. "The plant health department came out with a rule that you had to take a vacation," recalls Robert Denklewalter, who went to work for Tishler in 1943. "They decided we were risking our health by working all the time. Max's position was that he'd go along with it as long as it didn't interfere with getting things done."

Four companies in all—Merck, Pfizer, E. R. Squibb and Sons, and Lederle Labs—attended a secret meeting in Bush's Carnegie office in early October to discuss with government fermentation experts the prospects for growing the mold and isolating its active ingredient. From the beginning, none moved as determinedly or was as openly favored by Richards as Merck. The company vowed to make penicillin a priority and volunteered to share its methods and findings to the extent allowed by antitrust laws. The others, Richards observed, were "noncommital" and "less positive," a position he deplored and that, at least in one case, elicited his considerable wrath. "The imperfection in the Squibb material has given me much concern, not to say annoyance," he wrote scathingly in the spring of 1942 after a batch of the company's drug caused phlebitis in 100 percent of test patients with battle wounds at a Utah military hospital. "It is damn near criminal of them to have shortcut a process without finding out what the shortcut would do." Exacting, Tishler took no shortcuts. He was deadly serious, even reverential, about drugmaking, telling his chemists, "When you are working with those fifty to one hundred milligrams of penicillin"—about one-tenth the weight of a good breath of air—"you are working with a human life."

On March 14, 1942, barely five months after the October meeting, the CMR concluded that Merck had produced enough pure penicillin to begin testing it in humans. For four weeks, Anne Miller, the wife of Yale's athletic director, had been dying of an acute streptococcal infection, commonly called childbirth fever, in a New Haven hospital. Despite maximum doses of sulfa drugs, she was delirious with a temperature that had peaked at 106.5 degrees. At 3:30 Saturday afternoon, when she received her first shot of Merck's penicillin, her fever was 105 and she had "well over" fifty bacteria per cubic centimeter of blood. By 4 the following morning, her temperature was normal. By Monday, her blood was sterile. She was still alive in 1990 and living in Connecticut.

The world knew nothing of Mrs. Miller's wondrous salvation: It was a state secret. But as the companies and the government's laboratory in Peoria, Illinois, began producing enough penicillin to conduct widespread clinical trials, stories began to circulate about a new, unnamed miracle substance—superior to the sulfas and made, incredibly, from a mold. The CMR, tightly controlling its supply, released it sparingly to just a handful of leading infectious disease specialists, who gave it to their patients often without telling them what it was.

On November 28, 1942, as celebrants from the Boston College–Holy Cross football game poured through downtown Boston, a sixteen-year-old busboy lighting a match near an artificial palm tree in the city's oldest nightclub—the Coconut Grove—ignited a fire that killed 487 people. Suddenly, a major American city was plunged into a medical crisis that simulated the ravages of battle-torn Europe. In a grisly reprise of the flu epidemic that brought home World War I, Boston once again became a laboratory for the nation.

Moving swiftly to release all available penicillin to Boston, the CMR ordered Merck to step up production. For three days, several groups, including Tishler's, concentrated and purified all the crude penicillin broth in the company's fermenters. Fending off sleep, Tishler pushed his people around the clock in relays until they had enough of the drug. Finally, a thirty-two-liter steel container filled with injectable penicillin was packed into a car on the night of December 1. Picking up police escorts from four states as it moved slowly up the coast through steady rain, the "mercy vehicle" arrived at Mass General the next morning.

Far from providing aspirin to the dying, Tishler now delivered a drug that attacked not the symptoms of disease but the cause—a drug that worked. Eight months earlier, there had been enough penicillin in the United States for one patient; within fifteen months, by April 1944, all American military requirements would be met, and it would be the drug of choice for an array of infectious diseases. The federal government and the adolescent American drug industry, working cooperatively, had achieved a stunning success and a scientific landmark. And yet for Tishler, their triumph was just the beginning. For if science could identify and make this molecule, what else could it make? "The logic of survival in modern warfare," he would later write, "had dragged science from the periphery of our society right into the maelstrom of its center." It was there, after decades of boring inward, that Tishler and American biomedicine now prepared to punch through, to explode.

More than its own rewards, which were inestimable, penicillin proved that it was possible to find drug molecules of extraordinary power and precision in the world's simplest living things. Ever since Pasteur, biologists had known that every pinch of soil was a miniature Brooklyn, teeming with micromolecular competition. They knew that when an infected body dies and molders in the ground, the germs that killed it don't survive and assumed that other microbes destroyed them in self-defense. But they hadn't yet found any that could be ingested safely while having the same effect inside the body—until now. Penicillin bespoke a microbial Promised Land, and it did for microbiology what the sulfas had done for chemical synthesis. It exalted it.

In fact, the idea of harvesting "good bugs" to kill "bad bugs" had been developed around the same time that Fleming inadvertently discovered penicillin—and not by accident. René Dubos, an audacious twenty-six-year-old French microbiologist in his first job, at Rockefeller University in New York, started screening soil samples for disease-fighting agents as early as 1927. Like Fleming, he was looking for a substance that would destroy the deadly pneumococcus bacteria that had turned the 1918 flu outbreak into a cataclysm. By 1930, Dubos, who would later become better known as an environmentalist and Pulitzer Prize-winning author than as a researcher, found such an organism in a dirt sample from a New Jersey cranberry bog, not a drug but potent enough to cure mice infected with pneumonia. He quickly began expanding his experiments, exiling at least one early associate to a hospital roof to scour an "unpleasant brownish material . . . congealing into a sticky substance resembling uncouth ear wax."

Spurred by Dubos's success, other microbiologists were drawn to the search. Selman Waksman, a modest, bookish Ukrainian Jew barred from studying medicine in Russia, had come to the United States and through a series of detours had taken up identifying and classifying new strains of bacteria in soil at Rutgers University in New Jersey. Dubos had gotten his Ph.D. in Waksman's lab. In 1939, when Dubos announced he had finally isolated a bacteria-destroying molecule that was not the product of synthetic chemistry but of another microorganism, Waksman decided to launch the first large-scale screening effort for discovering other such compounds.

The idea of prospecting for what Waksman would soon come to call "antibiotics" failed to impress the university, which later tried to fire him, and the CMR, which turned down his proposal for funds. But it did interest another party—Merck. Desperate for money and a guarantee that if he found something, it would be developed for testing, Waksman agreed to grant the company exclusive rights to his discoveries.

From the beginning, Waksman was plagued by a nightmarish array of contingencies. Not only were there an infinite variety of organisms with numerous variations in each type, but the slightest change in diet, temperature, even the shapes of the flasks in which they grew, could alter them chemically. In its first year, Waksman's group found a promising germ killer, actinomycin, but it was so toxic that a single milligram could kill a five-pound chicken. Next, they discovered streptothricin, another potent antibacterial that first appeared safe enough for human trials. Feeding the drug to animals, Merck quickly discovered that it was fatal to kidney cells and canceled its development. By early 1943, Waksman decided to concentrate on finding an antibiotic to treat tuberculosis (TB), so-called captain of the men of death, which killed millions annually. Surely there were tubercle-killing substances in Waksman's petri dishes, but whether they would ever be safe enough to take and whether he could even find them remained increasingly in doubt.

Waksman persisted. Finally, in September, having cultured and tested thousands of strains of bacteria, he found what he'd been seeking. The organism—streptomycin—had come from the infected gizzard of a chicken that had died of TB and was apparently harmless to kidneys. Tishler and his chemists seized on the molecule. In the extraordinary time of four months—penicillin, by comparison, had taken more than a dozen years from discovery to clinic—enough of the substance was produced to begin animal tests. By October 1944, the first human trials were begun at the Mayo Clinic with a young woman hospitalized a year earlier with TB. Within six months, the lesions on her lungs disappeared. Eighteen months later, her sputum was free of bacilli. Released from the hospital in 1947, after four years, she later got married and had three children.

The implications of streptomycin were striking: It was the first drug deliberately discovered by screening natural products, found because it had been pursued; it had been developed in the United States by means of a technological paradigm that surpassed anything developed by the German chemical trust, which had spent the war producing not drugs but compounds to use in the Reich's gas chambers; Merck, by dint of its contract with Waksman, owned exclusive rights to a drug that might save the lives of millions of people.

Of all, only the last was troubling, as it raised the specter of one company profiting from a monopoly in an area where suffering was profound and there were no other treatments. Penicillin, with its joint development and diversified patents, ensured that the first of the "wonder drugs" would be distributed, at least initially, as a kind of public trust. But Waksman worried especially about the appropriateness of having turned over streptomycin to one company, however well intentioned, for "exploitation." Appealing to George Merck personally, he asked to be released from his contract. Merck willingly agreed. Tishler, who had watched his sister die of TB, was struck by Merck's magnanimity: "He used to say that if we discovered a cure for cancer, he'd not patent it," Tishler would say. "How can you keep it away from people? How can you charge a lot of money? What's the excuse? You can't do that."

Two years after World War II, the combined sales of penicillin and streptomycin equaled half that of all synthetic drugs, yet Merck, which had pioneered them both, was not the country's dominant maker of antibiotics (Pfizer was). But Merck had enthroned a new way of finding drugs: microbial screening. "Out of the earth shall come thy salvation," said Waksman, a self-taught talmudic scholar, paraphrasing Ecclesiastes in accepting the Nobel Prize in 1952. Asked later by reporters for the exact reference, he found a more precise translation with the aid of several rabbis: "The Lord created medicines out of the earth, and he that is wise shall not abhor them."

Hardly could Merck and the American drug industry abhor the lineaments of such a prosperous upheaval. Drugs from underfoot suddenly were making them richer and more respected than they'd ever dreamed, and they raced to discover the next great antibiotic like wildcatters in the neighborhood of an oil strike. Virtually every company began screening dirt samples, reaching literally to the ends of the earth to find new patentable molecules that their competitors might have missed. Squibb distributed vials to its employees and paid half their airfare if they returned from vacation with dirt samples. An Italian bacteriologist discovered cephalosporin, a broad-spectrum antibiotic, in the sewage outfall of the Sardinian city of Cagliari. Compost, humus, sewage, sludge (bogs, construction sites, cellars, sewage lagoons), wherever microbes swarmed, science now followed. Drug profits skyrocketed.

At Merck, Tishler was placed in charge of all drug development. Unmatched at synthesizing complex molecules, at knowing every detail of a problem, at pushing drugs through to market, he had become a dominant figure and one of just two rivals for managing Merck's rapidly expanding labs. His rise was a rebuke to the norms of science, which placed far more value on discovering new molecules than making them into drugs, but Tishler was overwhelming. "He knew everything," recalls Denklewalter. "If Max said something, it was almost as if it had come from God."

Idolized as omniscient, he also was ubiquitous, working on every project and at every level. No one knew what time he arrived in the morning or left at night because his car was always in the parking lot before and after everyone else's. He took his family to the Catskills every August, renting a cottage without a phone, but otherwise he never ceased driving himself.

In 1949, despite its success with antibiotics, Merck faced near certain failure with a far more complex and beguiling molecule. Virtually alone, the company had continued to pursue cortisone long after the initial intelligence reports of "hopped up" German superpilots proved false and national security was no longer an impetus. Merck's persistence paid off in 1944 when a twenty-seven-year-old chemist, Lew Sarrett, synthesized minute quantities of cortisone from ox bile. But Sarrett's synthesis—forty-two steps, with a final yield of one-hundredth of 1 percent—only reinforced the hopelessness of Merck's situation. Using his method, it was estimated, would take the slaughter of 14,600 cows to produce enough cortisone to treat one patient for a year. The cost per ounce was $4800, more than one hundred times the price of gold. With Merck sinking more and more money into the project, the company had produced by 1948 barely a total of ten grams of the substance.

No one knew what cortisone would do, but it was a spectacular trigger of a molecule, one of the body's master compounds for infiltrating and hot-wiring immune cells. It was of special interest to rheumatologists, who had no drugs at their disposal for fighting inflammation and were reduced to experimenting, without any particular rationale, with every new medicine that came along. In September 1948, Merck shipped six of its ten grams of cortisone to the Mayo Clinic for treatment of a twenty-nine-year-old woman so crippled with rheumatoid arthritis that she couldn't roll over in bed. The woman had already received massive doses of penicillin, streptomycin, gold salts, and sera with no improvement. Three days after her first injection, she was able to raise her hands above her head. Four days later she went shopping, declaring, "I have never felt better in my life."

If penicillin and streptomycin were miracle drugs, cortisone was Oz. Never before had a compound promised so much for such a panoply of chronic, incurable, and untreatable diseases. More, its wonders were preserved on film. Prefiguring *Awakenings*, the Oliver Sacks phenomenon, Mayo's doctors made a movie of the first fourteen cases in part to quiet disbelieving critics. It included footage of a woman who had been barely able to walk friskily running up and down steps and a man who had been in such pain he could not stand to be touched dancing a jig. In April 1949, Mayo's doctors would unveil the film triumphantly to the world, but they first brought it to Merck, where, trying to preserve their thunder, they insisted on a private showing with only the heads of research. Tishler exploded. He refused to let the film be shown unless those who had labored at the bench to make the molecule—some thirty-five to forty people—be allowed in. Reluctantly, the Mayo group agreed. "It was," one of Tishler's researchers would remember more than forty years later, "the most dramatic thing I'd ever seen."

Tishler now hurled himself into making Sarrett's synthesis viable. "I used to tell my teams, 'You worry about the first five steps,' 'You worry about the next five steps,' all the way through." Chain-smoking, gulping coffee by the thermosful, he seemed to be everywhere—in the labs, in the pilot plants. He was aflame, a pillar of fire. Once, a chemist dropped a highly valuable scarlet-colored intermediate compound on the floor. "That better be your blood!" Tishler thundered before ordering the precious liquid sopped up and its contents reisolated. He ultimately drove the synthesis down to a manageable—and potentially profitable—twenty-six steps, still the most complicated commercial process ever but adequate to begin large-scale production.

Scientists talk stoically about the *rate-limiting step* in their experiments, the one that ultimately determines its final yield. With cortisone, Tishler, in one fell swoop, had fractured the major rate-limiting step of all drug research—the inability to make complex organic molecules into synthetic drugs. "It is probably not too much to say," wrote Robert Burns Woodward of Harvard, arguably the century's greatest chemist, nominating Tishler to the National Academy of Science, "that Tishler's work in this field represents the most striking achievement in practical organic synthesis in the history of that art." Suddenly, chemists could conceive of making entire new classes of molecules—ones that were far more varied and sophisticated as drugs than anything they'd dared imagine. "If you took the kinds of molecules we're sythesizing today back to the 1930s, it would have the same effect as if you showed them your pocket TV," Boger would say forty years later. "That was Max."

Cortisone elevated Tishler and glorified Merck. Newspapers quickly filled with miraculous stories—cripples walking; a "statistically dead" eight-year-old girl still alive after being burned over two-thirds of her body; small children with severe eczema healed after nearly scratching themselves to death; a formerly gray seventy-four-year-old man made "bald as a billiard ball" by illness growing a complete head of dark new hair. "The number of diseases upon which it is reported to have at least some analgesic impact now approaches the galaxial," *The New Yorker* soon reported, naming some twenty-eight illnesses, ranging from asthma to ulcerative colitis, poison ivy to gout, as well as "shock, burns and fractures." By 1951 interest in the drug was so great that Merck, with Tishler now in charge of all research and development and again driving his production crews around the clock, was forced to take out full-page newspaper ads blaming shortages on the huge surge in demand. Even as cortisone's extraordinarily violent and far-reaching side effects—deterioration of joints, gross obesity, moonface, hypertension, diabetes, softening of the bones, nausea, headaches, skin eruptions, and, occasionally, madness—also began to surface, the company was exalted for its scientific leadership and progressivism. In August 1952, George Merck was pictured on the cover of *Time*, above the caption "Medicine is for people, not for profits."

Yet even as he was being lionized publicly for his altruism, Merck was being forced to move quietly to pay for it. Despite leading the world in the production of cortisone, Merck and Company's sales *dropped* from 1951 to 1952, leading it to merge the following year with Sharp and Dohme, a Philadelphia drug firm best know for its aggressive sales force and over-the-counter throat lozenges, Sucrets. Tishler resented the merger and feared its consequences. Recalls Denklewalter, "Max had an almost militaristic respect for authority, but we had this new CEO [Henry Gadsden] who'd come up through marketing at Sharp and Dohme. We had a meeting in research where he said to us, 'There are more well people than sick people. We should make products for people who are well.' He gave us three examples. Man-Tan was big at the time and he wanted us to develop a quick-tanning formula. He also thought we should have a morning-after pill and that we could do well with something for straightening kinky hair that would sell to blacks. I remember wanting to vomit when he was talking, but Max was silent. He would never say a word. On the other hand, we never heard of any of those projects again."

Now firmly in control at Rahway, Tishler continued as he always had, pushing to find important new drugs. With the apparent rout of infectious disease, the company—and its competitors—began focusing on the next major matrix of killers: cancer, heart disease, and stroke. Meanwhile, Merck began screening 50,000 new microbes a year at a plant it had built in Spain. Tishler had become convinced that the near infinite variety of active organic molecules in soil might produce drugs other than antibiotics. The key was in being able to test them against the right targets. With chemistry fulfilled, biology now became the rate-limiting step in devising new drugs, and Tishler turned Merck's science against it forcefully.

Merck continued to grow, but haltingly. In mid-1957, Tishler was named president of Merck, Sharp and Dohme Labs and placed in charge of 1600 reseachers at Rahway and at Sharp and Dohme's former campus in West Point, Pennsylvania. In November, George Merck died at home of a cerebral hemorrhage. Vannevar Bush, recruited by Alfred Newton Richards, became the company's new chairman. Though Bush worried about Tishler's insistence on doing everything himself—"He keeps all the threads in his own hands," Bush complained to Richards—the two worked closely. Bush also distrusted the "opportunism of the sales organization" and saw Tishler's vaunted labs as his chief bulwark against it.

Facing mandatory retirement, Tishler left Merck in 1970. His name was on 109 patents, including ten of the top-selling drugs of all time. He had set in place a scientific legacy that would soon pay off mightily with a succession of billion-dollar drugs and that would make George Merck's company, ironically, not only Wall Street's favorite drug stock but its favorite stock, period—a company with the fourth highest market capitalization in the country. He was not sorry to go. Tishler was not retiring. He was merely redirecting his energies, resuming the career he had left off in 1937 as an undergraduate chemistry professor.

Arriving at Wesleyan University within the same year—a year when the seismic outfall of Cambodia and Kent State suffused all cross-generational contact with a near-toxic edginess and distrust—Tishler and Boger found each other with surpassing speed.

Boger was nineteen when they met, obviously brilliant, superior, and more in need of a mentor than he would like to admit. His parents had begun to fight bitterly throughout his high school years, drawing sides and forcing him and his younger brother to choose between them (both boys favored their mother), and the young freshman chemistry teacher whom he admired, Peter Leermakers, had died that summer after flipping a Jeep onto himself at his California ranch. Though he'd gotten the highest grade in his freshman physics class, Boger had quarreled adolescently with his professor. "I found out he was a devout Roman Catholic, so for my extra independent project I wrote a computer program to launch a missile from Middletown and bomb the Vatican," he recalls. "Josh didn't suffer fools," says a friend from his CRAW days. "And he didn't suffer stupid science."

Tishler, sixty-four and released at last from the pressures of industry, had always been keenly interested in young scientists but had been too consumed by his work not to feel often violently impatient with them. Now teaching Boger as a sophomore, he saw at once the inquisitiveness, intellectual rigor, capacity for hard work, and raw insistence on knowing everything and being right that he believed all great scientists shared. For his part, Boger saw in the stooped, white-haired Tishler an archetype, someone for whom science was not a dry exercise, but a powerful tool for changing the world.

Tishler taught Boger to be a chemist. "I have a distinct visual snapshot of me holding a burette and having Max's hand over the top of my hand to show me how to grab it one-handed and open and close down the stopcock," Boger says. Nurturing Boger's pedigree, Tishler sent him to interview Selman Waksman in a New Haven nursing home just weeks before Waksman died. Moreover, he instilled in Boger the ethos he had cultivated at Merck. For an extra-credit question on a final in medicinal chemistry, Tishler offered $50,000 to any student who could develop a synthetic step that was cheaper than using microbes for making vitamin C. Like everyone else, Boger couldn't find one—the drug industry had been toying with the question unsuccessfully for almost forty years—but the problem itself qualified all Boger's assumptions about what science ought to be and what he wanted to do with it.

At Harvard, Boger continued to go his own way. He completed a postdoctorate with future Nobel Prize–winner Jean Marie Lehn, visiting from Strasbourg, while finishing all his course requirements in one semester and doing exceptional work on enzymes under Jeremy Knowles. His hair, grown long at Wesleyan, grew longer. He had a ninety-pound black Labrador retriever named Isaac that accompanied him everywhere on campus. Knowles, fresh from Oxford, told Boger he didn't allow pets in his laboratory, to which Boger asked if Knowles had ever had a dog in his lab before. When Knowles was forced to concede that he hadn't, Boger responded, "Well, Jeremy, don't you think you ought to do the experiment?" Knowles did, and the dog stayed. On the few weekends Boger managed to get away from Cambridge, he drove to North Carolina—stopping off in Middletown to visit Tishler—with Isaac sitting upright in the passenger seat of his VW.

By now Tishler saw in Boger the makings of a great drug maker and promising scientific leader, and he helped him along. Merck had a hiring freeze the year Boger finished at Harvard, but Tishler called Ralph Hirshmann and told him to find him a job. Boger was an instant anomaly at Merck. Most of its chemists had been trained like Tishler, if not by Tishler and his minions, in synthetic organic chemistry: They were expert at making molecules. But Boger was interested in how proteins, and particularly enzymes, worked as drug targets. His thinking was inverted from theirs. He was more interested in what drugs needed to do—what spaces they had to fill, what contacts to make—than in their actual configurations. Of locks and keys, he saw locks as primary.

Boger's insurgent attitude—"angular," Knowles described it—won him few friends. He was not above gloating. Dismissing the first project proposed to him as "snooze time," Boger began designing molecules he believed would reduce hypertension. His target, a molecule in the bloodstream called renin, was obscure, one of those proteins that, like HIV protease, snips apart other proteins with atomic scissors. Within eighteen months, working with one assistant and using the X-ray structure of a related enzyme to produce crude models, he made a molecule 1000 times more potent than the company's previous best inhibitor. Quickly, Boger's star began to rise. His results were published in *Nature*, attracting international attention and, of equal importance, the attention of Merck's publication-sensitive management. Merck had another hypertension drug, Vasotec, which would eventually make more than $1 billion a year, causing Boger's accomplishment eventually to fade. But by that time he was being quickly pulled up through the ranks. By 1987 he was not only in charge of Merck's rational drug design but head of a task force of another hundred researchers on a project in immunology. Two years earlier, he had had two assistants.

Tishler witnessed Boger's ascent proudly but with diminishing influence. In 1984 at age seventy-eight, he got pneumonia. The disease, though curable now, immobilized him, and he was forced to slow down. It had remained his habit at Wesleyan to arrive at work before 7 A.M., but now he could barely shuffle a dozen steps without sitting down on one of the small wooden chairs he had placed about the halls and in the labs. A lifelong smoker, he also had emphysema and had to retreat to his office twice a day to breathe through an oxygen machine. Merck helicopters still arrived on campus when his successors needed to consult with him, and he never missed a faculty meeting, but Tishler was increasingly ravaged by his condition. Retirement seemed not to cross his mind.

Boger stayed in close touch, writing warmly on Tishler's eightieth birthday that Tishler had taught him the value of becoming "engulfed in all facets of a problem" and that "only important problems are interesting." Yet as he mastered the system of science that Tishler had bequeathed, he also found it increasingly inadequate for his ambitions. "What you need in this business," he would say, "is more information than the other guy. Not more smarts. Not more intuition. Just more information. I began to realize that Merck wasn't set up to generate and use the kind of information I was going to need."

In mid-1988 Boger received a phone call at home from Kevin Kinsella, a San Diego venture capitalist. Once before Boger had been approached to start his own company and had gone so far as to write a business plan before opting to stay at Merck. However, Kinsella is a force of nature. The son of a Broadway actor and a fashion model, his résumé reveals a breakneck ambition and fathomless drive: Eagle Scout; B.A. in electrical engineering from MIT; M.A. in economics from Johns Hopkins; founder, at age forty-four, of seventeen companies, including his San Diego venture firm, Avalon, a name that he picked because A names sit atop alphabetical lists. Six feet three inches and built like a ski racer, Kinsella stands over most people in conversation the way Lyndon Johnson did, enveloping them, blocking their sky, as irresistible, consummate, and overachieving an entrepreneur as Johnson was a politician. Unambiguously motivated by wealth and power, he once posed for a business magazine cover with his hands steepling a stack of $100 bills; if he couldn't be a venture capitalist, he has said, he'd like to be president.

"Kevin can leave a trail of people wondering what's hit them," says the president of one of his companies. "It's like being exposed to a vacuum cleaner; he Hoovers out every piece of information." Now, Kinsella wanted to launch a structure-based drug company and had decided, of one hundred names he'd collected, that he had to have Boger to lead it. Boger was intrigued enough to listen. During the next few months, the two met several times in person and Kinsella began calling nightly, turning up the heat. But Boger wanted more control than Kinsella was offering and dismissed Kinsella's West Coast SAB as "suboptimal . . . not enough horsepower." Simultaneously, Kinsella's financing began to unravel. "I'm on the phone in the Reno airport," recalls Kinsella, who'd gone skiing one weekend at Lake Tahoe believing everything was sewn up, "and this fucking deal has turned to ashes."

Kinsella immediately flew to Boston, arriving at Harvard unannounced and not leaving until he had recruited the entire SAB that Boger had insisted upon. Six weeks later he flew to New Jersey with a new proposal, taking Boger and his wife, Amy, to dinner and unloading full force. Kinsella had once recruited an entire section of the National Cancer Institute—fourteen scientists—moving them from Bethesda, Maryland, to Seattle to launch another company. He was not about to be refused now.

"I want to ask you a question," Kinsella said, after laying out terms that effectively gave Boger complete control over the new company. "Do you remember who the person at Apple was who developed the personal computer?"

"Steve Jobs," Boger replied.

"Now who's the largest PC maker in the world?" he asked rhetorically.

Again Boger answered effortlessly, "IBM."

"OK. Do you want to be the guy at IBM or the guy at Apple? Do you want to end up with a good watch? What kind of credit do you want for your work?"

Boger's departure from Merck was swift, determined; ten days after submitting his resignation—ten days, fittingly, in which Navia's paper on the structure of HIV protease, announcing the first concrete inroad of the new religion of structure-based drug design in AIDS, was being vouchsafed by *Nature*—he was gone. It was the final week of 1988, and Tishler was ailing gravely. "Why didn't Merck manage to keep Josh?" he asked Knowles feebly. "What's wrong?" Merck sent Ralph Hirshmann, now retired from the company and a professor at the University of Pennsylvania, by helicopter to visit Tishler in the hospital, and Scolnick wrote him a letter detailing the company's efforts to keep Boger. Neither assurance was satisfactory. Tishler died in mid-March, just as Boger was flying around the country scrounging start-up capital with Kinsella, who carried a camera so that he could have his picture taken in each moneyed suite they visited. At Tishler's funeral, Betty Tishler, his widow, refused even to acknowledge Boger.

"Joshua," she says icily, "is a great disappointment to us."

CHAPTER EIGHT

Back from Japan, Boger was rhapsodic. Touring the palatial headquarters of the Japanese drug firms, whizzing on the bullet train out and back to their spectacular but oddly underutilized labs, he'd been able to measure the gulf between their expectations and practice and it elated him. The Japanese drug industry in 1990 was a reflection of America's in mid-1941: second tier, but with a small number of companies poised to move up. Just as Tishler and Merck had done by synthesizing vitamins in the 1930s, the companies had prepared themselves by importing technology and becoming virtuosos within the existing sphere: FK-506, a triumph of screening, was the first potential blockbuster drug developed wholly by a Japanese firm. They were supported by an activist government that in 1980 made developing pharmaceuticals for export a national priority. Having telescoped in less than forty-five years the entire evolution of industrial research from coal tar dyes to microprocessors, Japanese science, recapitulating the U.S. model, was now chasing the world's most complex, powerful, and profitable molecules. All it needed was direction. Boger, MacArthuresque before, couldn't but see in their courtship of him a reverence that was not entirely undeserved.

Unlike the big American drugmakers, all of which—except Merck—Boger had visited and had told him to come back when he had a promising lead, even the largest Japanese firms welcomed him and Aldrich avidly. They staged elaborate banquets for them, plying them with abalone and puffer fish, which the Americans dreaded. And yet it was the pharmaceutical wannabes, those giant noodle and steel and tobacco companies now dabbling in drug research, that enticed them most. Pulling up to the gleaming research center of Nissin, the largest Japanese noodlemaker, Boger and Aldrich were nearly toppled by three scientists in yellow pants, white shirts, and white shoes. The lobby was a tomb of black marble. "I feel like I'm going to see Dr. No," Aldrich muttered as the scientists ushered them into an immaculate chemistry lab in which a lone technician staged experiments with yellow and pink liquids. Of all the sources of money Boger could imagine, these companies—with billions of dollars in cash and all but devoid of scientific sophistication—looked to be the easiest mark. "They wouldn't intrude, the check would clear, and there would be zero synergy," he said. He began thinking he might be able to sell not only immunophilins to Chugai but some other, unspecified project to another Japanese buyer as well—a prospect that had him, three days after his return to Vertex, huddling with an accountant and the company's lawyer, his brother Ken.

"We're looking at three to four million in 'excess revenue,' " he told them proudly, "which would put us in a tax situation in our second year. If we take Nissin's money, too"—he already took for granted Chugai's—"we're going to have to figure out how to avoid running a profit."

Aldrich, contrastingly, was burnt out after the trip. A foot-long chain of Post-its hung from his computer screen, and his desk, normally as prim as a clerk's, was suffocating under an avalanche of laminated folders. For several days after his return, he played "purposeful phone tag" with Glaxo, exploiting the time difference with London to evade Leslie Hudson even as he tried to reach him. An inveterate planner, he couldn't plan until he heard from Chugai, and it grated on him. In by 8 A.M., he didn't leave until 7:30 P.M., then went to a gym to work out "so I don't feel like a total slug." He downed a vodka every night before collapsing into bed. He daydreamed about taking off a weekend to do his taxes. "What kind of life is that?" he asked, distraught, knowing too well.

Aldrich had more experience than Boger with the Japanese and was sullen about the prospects of an easy deal. For most of the past ten years he had tried futilely to put together agreements the size of the one he and Boger were trying to engineer. Each time it had been a drawn-out, treacherous, heartbreaking failure. At Biogen he had coordinated the development of a promising anticancer drug: After spending $80 million, the company was close to licensing the molecule when the results of its first clinical trial came in; of 350 patients, none showed any improvement. He had two deals collapse at the last minute at his next firm, Integrated Genetics (IG): one when a Japanese partner backed down after a U.S. patent was issued to another company for the same drug; the other with Merck, which had been eager to license a genetically engineered antithrombolytic until such drugs proved no more effective than conventional molecules one-thirtieth the cost and, in a few well-publicized cases, caused bleeding on the brain. "We were in big-time negotiations—money, proposals changing hands, a $50-million-type deal. By the time things had taken their course, we couldn't get people to pick up their phones." As a practical matter, Aldrich valued Boger's pose of inevitable triumph, finding it useful in negotiations and for morale. But he knew the odds were against one quick deal, much less two, and fretted privately about Boger's effect on the Japanese. "They have an expression, 'The nail that sticks out gets hammered down,' " he said. "I think a lot of them think Josh is cocky. They think he'll get put in his place."

In fact, because he was both storyteller and subject, Boger's role in any negotiation was to be at once agreeable and insolent, personable and larger than life, and he fulfilled it in part by playing to other people's expectations of him as a golden boy. This left Aldrich for ballast, part tactician, part monitor, glowering when Boger soared too high. ("Josh is the accelerator," he would say; "I'm the brake.") Up until now the two of them had worked capably together, but as more detailed legal issues evolved, Boger began to rely more and more on his brother Ken. Five years older, Ken Boger was Boger's height, but broader, more manly looking, with a kinder, less angular face. A partner in the white-shoe Boston law firm of Warner and Stackpole, he was more introspective than Boger, his quiet stature and firewall demeanor reminiscent of Robert Duvall's *consigliere*, Tom Harken, in *The Godfather*.

On March 12, 1990, with no word yet from Chugai, both Bogers and Aldrich met with Harvard patent officials to discuss the situation with Schreiber. By the pendulum of Harvard's internal debate over faculty business interests, it was a perishingly bad time. That morning, the *Wall Street Journal* reported that all major universities were now starting to rein in their biomedical researchers and that Harvard had proposed the strictest controls. Scheduled for a vote later in the month, the plan would bar faculty members from owning stock in companies supporting their research—in the case of Vertex and Schreiber, a direct interdiction. The patent office still wanted a deal, but otherwise Harvard was riven, inhospitable.

Boger managed to elicit a tentative agreement, in part by appearing generous; he offered to let Ken do the drafting. Harvard still insisted on owning all Schreiber's discoveries, but Vertex now won enough guarantees to block other companies from licensing them; if Vertex couldn't control Schreiber, neither could anyone else. It was hardly what Boger, who believed Vertex had as least as much right to Schreiber's work as Harvard did, would consider just, but it would have to do.

Thus armed, Boger now prepared to face Glaxo. Calling twice a day, the company was urging a meeting in London by the end of March. "Their tone is, 'My God, what were we thinking, of course our strategies are compatible. In fact our noncompatible/compatible strategies are the best possible approach,' " Boger said. "I think they got the message: eleven days in Japan."

There was no clear route through the shifting negotiations. But two phone calls on the morning of March 16 affirmed Boger's exultation, even as they gave Aldrich more to fret. Finally getting through to Aldrich by phone, Leslie Hudson's secretary invited him and Boger for two days of meetings with Glaxo's senior scientists in London on March 27 and 28. Meanwhile, Benno Schmidt called to say that Chugai appeared to accept Vertex's terms but that its board of directors needed to approve them at its next meeting in early April. Without misleading either, Boger and Aldrich decided to go hard at both companies. It was twenty-one business days since the initial meeting with Nagayama in Schmidt's office and barely five months since the meat market at the Vista. Even Aldrich in his deflationary state began to revive. Boger beamed like a TV tower.

Chugai's assurances weren't firm enough to keep Boger and Aldrich from traveling to London, but they fortified them while they were there. Straight from the plane, they arrived at Glaxo with a preemptive gloat. Boger wanted support for twenty-five to thirty researchers. At $200,000 per scientist per year for five years—a rule of thumb that included compensation, equipment, and supplies—that was $25 to $30 million, or considerably less than the $40 million Schmidt had proposed to Chugai. If they produced a drug, Boger proposed splitting the revenues in half. Facing a group that included Hudson, Hammill, and several other ranking scientists and executives, Aldrich assumed his "practiced M.B.A.-proposal scowl" in presenting Vertex's terms.

Glaxo, unsurprisingly, had its own ideas. The company had never completely believed that Vertex would develop a drug, only that it would help Glaxo's scientists find their own. Nonetheless, it made an unsolicited proposal: $25 million for world-wide rights to any new immunosuppressant that Vertex produced. Suddenly, what Boger had intended as a rout turned into a scraping negotiation, with Glaxo setting performance milestones and royalty schedules. Boger was incensed and dismayed. As he and Aldrich returned to their hotel, he was determined not to go back for the second day. Aldrich, however, preferred to call Glaxo's bluff. Declining its offer, he countered by saying that it might be more consistent with Glaxo's thinking simply to give Vertex the $25 million for biology, since Glaxo didn't seem to value Vertex's chemistry or biophysics or believe that it would find a drug. "We decided," Boger said afterward, "to come in deeply provocative."

Impatiently, the same Glaxo group listened to Aldrich's restructured proposal the following morning, then sent him and Boger out of the room for forty-five minutes. They were supposed to meet through lunch, but "when we came back," Boger recalls, "they said, 'We've sent for your car.'

"It was great. It was the perfect negotiation for us. It showed we were better able to structure and think through a deal than they were.

"I began to think we really should be doing something together. We have similar arrogance levels. Arrogance doesn't bother or impress us. We understand arrogance."

If Glaxo's unceremonious dismissal gave cause for worry, neither Boger nor Aldrich seemed troubled. "Glaxo can't afford to be beaten in this area," Boger said in early April. "It's too front page. They'll be back." Harvard, however, was another matter. Its arrogance, real or imputed, was world famous, a whipping boy for U.S. presidents. The university didn't need Vertex, as Boger believed Glaxo did, and Vertex had no fallback, such as Chugai. The delayed faculty vote on the new conflict of interest guidelines aside, Schreiber's status remained as fraught as ever. Seduced by Chugai, mired in brinksmanship with Glaxo, muddling scientifically as his staff pressed him increasingly to do something about Schreiber, Boger now resolved to bring the situation to a head.

For six hours on April 4—"six hours of trading insults," as he would recall it—Boger, Ken, and Aldrich met in Vertex's conference room with Harvard's licensing team. They knew they couldn't prevent Schreiber from working with FK-506 but were relying on Harvard's desire to avoid what Ken Boger called "a very messy situation" with respect to one of its distinguished professors.

Joyce Brinton, director of Harvard's licensing office, also seemed eager to eliminate the problems arising from Schreiber's overlapping research deal with Roche. However, she was now joined by a representative of Harvard's sponsored research office, who'd been invited to address how and when any discoveries resulting from a Vertex-Schreiber collaboration would be publicized and who allowed with some satisfaction that Harvard was now considering investigating the entire matter.

Ken Boger, who seldom uses a litigious tone, leaned forward in the manner of a courtly but calculating southern lawyer. "We would certainly be very interested in learning what conceivable basis Harvard has for an investigation," he said. Joshua, impassive, said nothing although he dreaded such a probe, not for what it might reveal, but how it would look. "It would be your basic congressional investigation," he said later, "covered by the *Crimson*, which is the only student newspaper in America read in the executive suite of every major drug company. We can't afford that. We can't afford to be scapegoated."

Nettled by the escalating ironies of making concessions to Harvard, which he considered irrelevant, to protect his relationship with Schreiber, whom he distrusted, in order to have access to science he had little use for other than to impress companies whose research he disdained, Boger now grew impertinent. "I bet your college education still isn't paid for," he snapped at the sponsored research man. "My research grants not only paid for my education, but also supported the English Department. You should be thanking me." When the man said, "Harvard doesn't accept the notion of confidentiality. Harvard exists for the dissemination of information," Boger spat, "Is that why you charge $30,000 a pop for FKBP?"

Boger was ready to kill. In four weeks, as outside Vertex's ramparts Cambridge slouched morosely toward spring, he had fallen from conquering Japan to berating academic functionaries.

And he still had no deal.

If negotiating with Glaxo and Harvard resembled a stag match in rutting season, with weeks of posturing punctuated by the crack of heads, contact with Chugai was more subtle, a caress with a mysterious figure slipping in and out of a mist. After its board met in early April, the company faxed Boger a three-paragraph letter stating its intention "to move forward with the project." To Boger, self-schooled in Japanese culture, the letter represented a carefully worded statement of principle—a bond—which it would be a grave dishonor to break. Aldrich, who'd been misled before, considered it annoyingly vague and felt manipulated.

"Americans think that what people say publicly is not the real truth; the real truth is what people tell each other in private," noted David Livingston, who knew nothing about the communication at the time—none of the scientists did—but who had negotiated with Japanese companies alongside Aldrich at Integrated Genetics. "In negotiating with Japanese firms the opposite is true. What businessmen say to one another privately is thought to mean very little. They'll say anything. What matters—what lasts—are more public forms of communication, like letters."

Intrinsically, Boger agreed. He began assessing Vertex's chances of doing a deal with Chugai at 95 percent. Again he told Aldrich to stall Glaxo, which, as Boger predicted, recovered from its pique and was now pressing for a more detailed proposal.

Besides Vertex's dire need of money and benediction, Boger's and Aldrich's own ambitions also now came into play. Barely a year old, with only twenty-five employees and still very little science to speak of, the company was more of an experiment than a corporation. How big it grew and who would lead it through the wrenching expansion that seemed to lay ahead depended largely upon its ability to raise enough money to incinerate in the labs *and* satisfy its investors. The company's board of directors—Kinsella, Schmidt, and several other venture capitalists—was delighted with Boger's performance so far. But in every emerging business there lurked the specter of the fallen, overreaching founder. Steven Jobs, whose example Kinsella had used to extract Boger from Merck, eventually lost control of Apple Computer when its board sought a more seasoned executive to manage its sprawling operations. The sooner Boger and Aldrich could claim a major deal, the quicker the board would put aside discussion about the company needing what Aldrich called "white hair"—bona fide executives who would take over and run Vertex as it became a drug company. Conversely, if they lost Chugai and/or Glaxo after coming this far, the board might be dismayed and want to replace them soon.

The Chugai negotiations were particularly sensitive in this regard. Boger and Aldrich had done all the legwork with Glaxo themselves, but acting on their own, in the fall, they'd twice been unable to arrange even an introductory meeting with Chugai's licensing office. They would hardly be impatiently exchanging provocative faxes, much less contemplating the transfer of tens of millions of dollars, without Schmidt's bearhugging of "Sam" Nagayama, the company's young deputy president. Chugai was Benno Schmidt's deal. As desperately as Boger and Aldrich wanted it, a Vertex/Chugai alliance would do more to enhance Schmidt's legend than Boger's story and would require them to prove themselves in yet another negotiation.

On April 26, 1990, four days before Boger and Aldrich were to meet with Nagayama in Schmidt's office—and a week and a half after Glaxo's deadline for a revised proposal had passed—Chugai wrote that it was no longer interested in the type of deal the two companies had been discussing. The pullback seemed to derive from Chugai's growth strategy, which is to buy heavily into those U.S. companies whose products it licensed. As a second-level Japanese firm with global ambitions, Chugai was determined not only to barter its way into the U.S. market, but gain access to its newest technologies. The previous fall, it had paid $100 million for Gen-Probe, a California diagnostics firm, eliciting shrieks of protectionist horror from those who feared a raid on U.S. biomedical "seed" much like Japan's infiltrations of other leading industries. Just as American dominance in biomedicine, one of the country's last unchallenged technological strongholds, was coming full circle fifty years after its triumph over the Germans, Chugai was pushing hard to be a leader among the new class of upstart Japanese contenders. It knew precisely what it wanted from Vertex, which included enough stock to warrant a seat on the company's board. It was displeased with Boger's reluctance to include more than token equity in any agreement.

Characteristically, Boger and Aldrich interpreted Chugai's sudden reversal differently. "They played us perfectly," Aldrich fumed. "They kidnapped us. They kept us from talking to anyone else for six to eight weeks. It may have really fucked us up." Boger, alternatively, was cavalier. "It's dropped back to a 45 percent deal ironically because they've become more interested in us," he said. "Are they really going to walk away from this because we disagree on the price of the entry fee?

"If they want to buy 30 percent of the company for $100 million, I'll think about it. I won't say I'll do it, but I'll think about it."

Whatever their disagreements on the meaning of Chugai's hesitancy and their varying stakes personally, Boger, Aldrich, and Schmidt all now favored a firm response. They agreed that Chugai wanted a deal badly but was exercising the Japanese custom, as Boger put it, of "beating you up at the end on price." With Aldrich straining to revive the discussions with Glaxo, they resolved to force Chugai into a decision—a task that fell to Schmidt during the meeting on April 30.

"We've had a lot of other inquiries about this program," Schmidt told Nagayama and four other senior company officials in his office. "Joshua," he turned, "it's time to start returning some of those phone calls." He then set a two-week deadline for Chugai to respond. "That shouldn't be a problem for Chugai since I don't think we can close another deal by then," Schmidt said, again eyeing Boger. Immediately after the meeting, Nagayama and Schmidt flew together in Schmidt's private jet to Washington, where they lobbied FDA officials on EPO, the promising antianemia drug that Chugai had licensed from GI and that was still held up by patent problems. Whatever Nagayama's views of Vertex, it was unlikely they were unaffected by being squired around Washington by Schmidt, whose connections there were still strong.

Said Boger, "That trip may be the difference between Chugai being a $1-billion-a-year company and a $2-billion-a-year company. The key now is how much rope the board gave Nagayama."

"This is high-stakes bluff poker," Boger said. "It's an objectively tough sale. I mean we turned down $21 million. Turned it down. We're having to reeducate everybody. People don't do that. We may make a deal in the twenties rather than the thirties, but if we do that, we're going to take back chunks of the world."

Chugai's fax, dated Friday, May 11, was as explicit as its previous ones were vague. The $40-million deal Schmidt and Boger were proposing was too rich. Chugai countered by halving the offer. Boger, predictably, was confident an agreement could still be struck. He saw Chugai's refusal as a ploy. Aldrich, equally, was somber, morose. Because Nagayama had left New York favoring a deal, Aldrich believed that only a bloody battle on the board of directors could have subverted it. If that was true, there were now ominously bad feelings in Japan that would prevent Chugai from agreeing at a lower price. "I guess we played hardball too hard," he said sullenly. "They couldn't handle the money."

Schmidt's original proposal called for splitting the U.S. market with the rest of the world going to Chugai, which already had Asian and European sales forces "on the ground." Now, working through the weekend, Aldrich drafted a counterproposal to reduce Vertex's asking price. Facing the central dilemma of all small biomedical firms—how to avoid giving up, as the price for developing a new drug, all but a shred of its value—Aldrich suggested taking back Europe. Vertex could decide to sell a drug there itself, or it could license it to another partner, perhaps netting another $50 million in a few years. In effect, Vertex would now capture—for the reduced price of something less than $30 million—half the world market for a molecule it was nowhere near yet designing. Not as breathtaking as Schmidt's original gambit, the idea was audacious nonetheless and Aldrich wasn't counting on it. Reviving his counterproposal to Glaxo, he prepared to send it the minute that Chugai turned Vertex down.

Schmidt called Nagayama with the new proposal. It was still his negotiation, although his faith in it, conveniently for Boger and Aldrich, now appeared to waver slightly. Soon they would begin to say that Schmidt had interpreted Chugai's last fax as irrevocable and had given up hope of making a deal. Schmidt disagreed. "Maybe they were more disheartened by the letter than I was, but I never thought the deal was dead," he said several months later. "I do think it was fortuitous, though. I could ask for things I would not have asked for before, in a context of responding to [Chugai's] wishes.

"If Chugai got a big drug out of it, what we were asking them for was chicken feed. Survival for us, but chicken feed for them."

Predicted Boger, ultimately caring less than Aldrich whose deal it was as long as it succeeded: "I think we'll be eating sushi real soon."

Boger's equable demeanor, his humor, his arrogance, his confidence in negotiations, all derive from the central theme of his own competence. Believing he's right and that others—even nature itself—will eventually come around gives him a consistently placid air. Seldom does he show strain or even annoyance without also joking about it. Once, while lightly describing a tortuous and costly three-month campaign to recruit a key scientist away from Merck, he calmly tore a Pepsi can in half, but otherwise few difficulties raise in him more than a hollow chortle or a quick verbal jab.

Schreiber is the major, outsized exception. Two days after Schmidt conveyed Vertex's final terms to Chugai, as Boger waited anxiously but unemotionally for a reply, Schreiber sent him a preprint of an article he had submitted on the binding activity of FKBP. Boger was livid: "This is Dingell Committee stuff!" he exploded unironically. He was referring to Michigan Congressman John Dingell, whose staff had recently launched several highly publicized investigations of scientific fraud. The article itself made no large point. It seemed intended, as with much of Schreiber's lesser work, to "plant the flag": keep his name associated with the protein while obtaining publication credits for his students and identifying new areas of inquiry that he could claim to have suggested first in print if they became important later on. In it, however, he cited—with far from thorough attribution—data that Boger knew had first been developed at Vertex and Merck. Calling the paper "a gross abuse of ethics," Boger concluded more determinedly than ever that Schreiber cared more about his own reputation than doing legitimate science and that he couldn't be trusted—a view, he would say, that Schreiber helped confirm by announcing around that time at Stanford that "Manuel Navia at Vertex" had crystals of FKBP.

"At this point, I'm ready to go to Glaxo and lay every single dirty piece of laundry on the table," Boger fumed. "I'll tell them, 'You don't want this guy.' I'll stipulate that we would consider it an unacceptable breach of security to have Stuart anywhere near this project."

"He's a loose cannon," Aldrich agreed, staking out the alternative position that Boger, in his rage, had suddenly abandoned, "but if he's on our deck, we can chain him down. If he's on someone else's deck and he's pointed in our direction, we can't deter him."

"And if he's an exploding cannon, who's deck do you want him on?" Boger asked.

Even if Boger was ready to break with Schreiber, his question raised the potential of an even more damaging scenario, that Schreiber would "drag the rest of the SAB with me," as he had recently threatened to Matt Harding.

"It would be incredibly embarrassing to have an SAB blow up at this point," Aldrich said. "Apocalyptic."

Boger grinned for the first time in the conversation. "There are certain advantages in life you wish only on your enemies."

John Thomson limped into the lunchroom, put up a pot of coffee, and announced he wasn't leaving the building until he'd concentrated a slug of FKBP sufficient to grow crystals that were large enough to solve its structure. The night before, he'd gone drinking with Mason Yamashita, Vertex's young second crystallographer, to celebrate his one-year anniversary at the company. The two had recently become close friends. Working the same hours, driven by similar compulsions, it was their work—Thomson's with supplying protein and Yamashita's with divining its atomic structure—that Boger was relying on chiefly to advance Vertex into the arena of designing drugs. Thomson was drained—"Every little thing that happens now breaks me," he said—and Yamashita, under Navia's direction, was next in line to inherit the protein and its attendant burdens. Thomson was determined to bequeath him both.

"The moment I see spots on that screen," Thomson had said, referring to the radarlike color monitor atop Yamashita's X-ray machine that indicated if a crystal was diffracting and thus ordered enough to reveal a protein's structure, "I'm on my bike and out of here."

"Where?"

"West."

"How far west?"

"West."

Thomson's frustrations had only mounted since he first isolated FKBP in mid-February. To grow protein crystals—ideal, unnatural forms in which huge numbers of floppy and unstable molecules spontaneously arrange themselves in a lattice, then repeat the process billions of times outwardly in all directions—one must first create an infinitesimal Nirvana. The individual molecules must be bathed in a state of calm or they begin chewing each other up or themselves up or become agitated and bolt from solution rather than submit to a rigidification they abhor. In other words, they must be doped. To keep proteins happy and serene once they've been wrenched from their native environment, biochemists suspend them in *mother liquors*, chemical buffers that have the effect of amniotic fluid laced with heroin. The problem is, as chemical entities, all proteins are different, and thus identifying the right mother liquor is hit or miss. Worse, one must sacrifice scarce protein to do so and may still find that no perfect elixir exists. Only a fraction of the tens of thousands of known proteins has ever been crystallized, largely because the right conditions for growing them are so exasperatingly hard to find.

Torn from the thymus juice and placed in its own solution, FKBP was jumpy, excitable, a resistant hostage. For several weeks in February and March, Thomson and Matt Fitzgibbon had pushed batch after batch of the material to final purification, only to watch it "crash"—precipitate out of solution. They eventually delivered Navia and Yamashita a total of five milligrams—500 times as much as Harding and Merck had isolated initially in discovering the protein—but it was unstable. Worse for Thomson, the crystallographers' goals and priorities were the opposite of his. Though he had struggled mightily to keep the protein in solution, they now, with considerably less care, tried to coax it out. Navia, though competent at growing crystals, was traveling a lot and seemed distracted by other projects. He had little of Thomson's devotion to benchwork, leaving most of it to Yamashita, who had never worked on such a scarce or important protein before. Though it had been all he had labored for and though he liked Yamashita, Thomson now found it excruciating to pass the protein to other researchers whose qualifications for handling it he couldn't help but distrust.

Navia quickly dispelled Thomson's grief by growing fulsome crystals of what he believed was pure, native protein, but they were too small to identify; he needed more material. To Thomson, still optimizing his procedures, the smallness was also an invitation to push the protein to sharply higher concentrations, which if it worked would help the molecules pack together more, or if it didn't, could make them crash out before the crystallographers ever saw them. "We'll all have FKBP earrings before long," he boasted to Navia and Yamashita on April 5.

That was six weeks ago. Now Thomson began another siege that would last, with few interruptions, until the end of May. And unlike his previous attempts to isolate the protein, what he feared most wasn't what he didn't know, but what he did.

For his final concentration, Thomson used a pump-loaded microfiltration system that pushed the protein through a narrow, eight-inch column packed with ultrafine silica, a sandlike substance. First he saturated the column with a salt concentrate, then extruded the protein through with a thin stream of distilled water. Seeking refuge from the caustic salt, the protein molecules jumped to the outside of the column, then "caught the wave" as the water flushed them, drop by drop, into a rack of test tubes. Only by watching a computer readout could Thomson tell whether this molecular surfing was successful or, as had happened with his previous batch, the protein stuck to the silica and was "gone forever."

Late in the afternoon on May 28, Thomson loaded the column with what he calculated to be almost 200 milligrams of pure FKBP, the result of more than a month's preparation. It was a significant portion of the world's supply, half as much as Schreiber's group had produced recombinantly in six months. Thomson had been in the lab for five days, his chin a mass of stubble, his eyes red and raw. Tapping his foot incessantly, he stared unblinking at his equipment.

"Damn," he said suddenly, "it's going off scale. It's too concentrated." Tiny white particles now bubbled at the surface of his sample. The protein was metastable—at the very margin of stability, like a volatile liquid about to explode—and Thomson quickly added a buffer to stanch the precipitation. "I'm scared that getting it ready to put back on that column may have pushed it over the limit. Shit.

"You've missed the fun and games, Matt," he told Fitzgibbon, his assistant, who had drifted over with several others. "We may have lost the lot."

"I hate this protein," Navia said. He put his hand on Thomson's shoulder: "Take it easy, big guy."

"The big guy is having a coronary here," said Saunders.

Sucking in air, exhaling deeply, Thomson riveted his eyes on a liquid crystal display readout. "Get up!" he said. "Get up!" It was several minutes before the test tubes containing the FKBP solution were full and he could remove them from their carriage. Holding them to the light, he flicked each one three or four times with his thumb. The liquid was clear. There was no change in density from his earlier samples. Thomson heaved a voluminous sigh.

Next morning in the lunchroom, he quietly handed Navia two small nose-cone-shaped locking plastic tubes, each smaller than a thimble, containing 130 milligrams of high-concentration FKBP.

He had wanted to conduct the exchange in private, but Saunders, whom Thomson had still not forgiven over the loss of his girlfriend, noted what was happening and quickly gathered witnesses from the library. The room erupted in applause. In the highest scientific tradition, Mark Murcko, a computer design expert recently arrived from Merck, proposed a new coinage: the Thomson Unit—one hundred milligrams of ultrapure enzyme.

With 1.3 Thomson Units of FKBP, Navia and Yamashita quickly began sketching a barrage of new experiments to increase the size of their crystals. It still wasn't clear whether the protein was the target Vertex was looking for or even whether the earlier crystals were of FKBP, but science proceeds mostly from an abundance of materials, not understanding. Having at last experimental quantities of protein gave Vertex one of the two vital reagents—the other being money—it most desperately needed. "Either we've already been beaten," Navia said, "or we're going to win."

Two days later, Thomson, his vitality inching back, went to his doctor for a three-hour physical—a requirement for working with human tissue. As he had stated back at the first SAB meeting, he was now planning to begin cutting up spleens.

"I have a rare ailment called thymophilia," he reported to Laura Engle, one of Boger's assistants. "The only known cure is six weeks in the Bahamas, followed by a liquid diet administered in the south of Spain. Four or five months and they said I should feel like new."

"What's up?"

It was midafternoon, late the following Thursday. Navia had just returned from an NIH site visit and observed Yamashita typing quickly but desultorily in the X-ray lab. A plate-glass window between the lab and the hall forbade any privacy.

"Not much."

"Anything doing."

"No."

Yamashita's face was a mask, dark and rigid. He looked like a young kick boxer or medical student, slight, with square shoulders, a broad face and glistening black hair parted unceremoniously in the middle. He wore an oversized Big Dog T-shirt, black polyester pants, running shoes, and, around his neck, lightweight headphones attached to an ever present Sony Discman. At twenty-seven, he had gotten his Ph.D. a year earlier at UCLA, then moved east to work for Vertex, buying a duplex that he seldom saw around the corner from the labs. Though Boger had hired him as a full scientist, his absence of any personal life, endless days in the lab, and all-around slavishness shouted postdoc.

Yamashita was lying: A lot was doing, all bad. Immediately upon getting the protein from Thomson, he and Navia had begun trying to grow FKBP crystals under a variety of conditions. By the time Navia had left two nights earlier, they each had incipient crystals, and Yamashita had been ecstatic. Growing crystals is the trickiest, most critical task in finding a protein's structure, and the part Yamashita was least skilled at. At the same time, FKBP was now fast becoming one of the biggest prizes in biophysics. A recent conference of the American Crystallographic Association had ranked it among the most important protein structures to be solved, ratcheting the competition. Wafted by the opportunity to lunge ahead on a career-making problem and perhaps impress Navia in his absence, Yamashita had tried the night before to see if the crystals diffracted.

X rays, like light, deflect off some objects and pass through others. Because their wavelengths are about 1000 times shorter than visible light, they slice through organic molecules, which is why they permeate flesh and tissue like wind through a fence and stop only at dense, mineralized bone. Train an X-ray beam on a large, well-ordered crystal, however, and it scatters against the electrons: thus the rationale for X-ray crystallography. It's like firing a laser at a crystal chandelier, then trying to determine from the millions of spots on the ceiling, walls, and floor the precise placement of each lozenge of glass.

Crystallographers nearly always solve a protein's structure once they have a large enough crystal. But they need to know if they've crystallized protein and not some other constituent in the mother liquor. This is what Yamashita had attempted the previous night. Inserting a crystal into a sealed capillary tube, he had mounted the tube, like a slender drill bit, in a rotating chuck atop Vertex's X-ray generator. The machine, filling a third of the lab, is the size and shape of an industrial freezer and is jacketed with computers. It buzzes constantly. Yamashita, perched on a library stool and hunched over the superstructure that sends the beam through the crystal and catches the diffracted X rays, set the crystal precisely, punched in several commands at a keyboard, then waited for a pattern to emerge on a monitor. He was listening to Sinead O'Connor's "Nothing Compares 2 U" loud, on his headphones.

At the sight of the first spot, Yamashita began jumping up and down. But after several minutes, a few more appeared. Proteins, with hundreds of atoms and thousands of electrons, produce galaxies of spots, but this was something else. Two or three atoms at the most. A salt molecule maybe. Yamashita was bereft: "It was throat-cutting time," Thomson recalled. Miserably, they left the lab and went to a bar.

Like Thomson, Yamashita abhors failure. He, too, wanted desperately to prove that Vertex could design drugs and saw his own work as vital to the company. And yet now in his first real job, he was also ambivalent.

He had come to Vertex not, as Boger had, along a fast stretch of straight track, but circuitously and alone. Raised on army bases, he had alternated in adolescence between being an amateur Pentecostal and a rabid existentialist. Ever since, he had drifted in isolation, seeking a moral way to live. He had few friends and was reeling from the break-up of a lopsided romance. The main constant in his life was science—in particular, computational chemistry, through which he believed one could know the truth about the world's most basic events.

That crystallography, in which he had "extreme trust," confronted him now with an undesired outcome pained Yamashita deeply. Normally polite to a fault, he spent the morning cursing. He still didn't know what he had crystallized. Meanwhile, Navia began examining his own experiments. He, too, was dismayed. Opening the well on a particularly promising crystal, he instantly noticed the faint odor of compost. "Bacteria shit," he scowled. Microbes had fouled his sample.

Hearing from Yamashita about his failure the night before, Navia suggested poking the suspect crystals with tweezers. Protein crystals are delicate, crumbly, like Halva: The individual molecules are attached by few atoms relative to their mass and are easily separated. Salt crystals, on the other hand, are brittle. They snap. Yamashita was horrified by the thought of deliberately destroying a crystal, but he eventually gave in. When a few moments later Navia cracked one of his crystals sharply like a sliver of mica, Yamashita burst out of the room.

"I don't want to be here," he hissed, though clearly he also didn't want to be around Navia, with whom he compared himself. It was an inapt comparison: Navia was seventeen years older, highly successful, a well-regarded investigator. He knew it sometimes took years to grow decent crystals and that once grown, they might never grow again. Navia, more equable about his own disappointments, thought he understood Yamashita's: "His point is, my career is established. I can take a few hits. But he can't. He's got to do everything right the first time."

Yamashita found Thomson at his bench and offered to drive him to Stoughton, in the suburbs, to pick up Thomson's new motorcycle. It was a used Honda V 1000 R, a 440-pound racer with more horsepower than an Accord. A hellish machine, it wasn't as fast as the one Thomson had totaled in the fall, but still what transplant surgeons call a "donorcycle." Thomson had planned to take it away for a week but now, hearing of the setback in crystallography, decided to stay. Sympathetic, he was also put out: "How do you think I feel, their growing bugs and precipitating salts in the world's most sought after protein?"

That night, Yamashita slouched in the blackened conference room, his feet up on a chair, listening morosely to the girl group, the Bangles, on his Discman. Navia sat studiously in the library, his tie still straight, reading and taking notes until well after 9. No one else was in the building, but they didn't seek each other out. "A fraternity of grief," Mark Murcko once called science. A fraternity particularly unforgiving of failure.

"I've got to talk to him," Navia said, although plainly he wished he didn't.

CHAPTER NINE

A burst of hiring in May and June brought Vertex by the first week in July to more than forty people. In the labs, benches were divided, hoods shared, egos disjoined as new scientists, a disproportionate number of them Ph.D.'s, arrived weekly. Rivalries flared as some groups grew faster than others. There still were no permanent desks for the scientists, although some, like Navia, asserted squatter's rights and were alternately deferred to and reviled for it. The squeamish Massachusetts economy doomed several proposals for more permanent space, forcing Boger to lease a former airplane-parts factory on Putnam Avenue, a block away. Twice as large as the original building, it wouldn't be ready until mid-August. Then, the scientists would be divided between the two sites, kindling other jealousies. Now, Boger knew he was tightening the lid on a pressure cooker. Exalting turmoil, he thought it a good thing.

As reagents in Boger's social experiment, the new scientists catalyzed several critical reactions. Synergy, the combined action of disparate forces and a favorite concept in drug design and of Boger's, increased sharply. So did entropy, the tendency in all things to pull apart and degrade. If once there had been a singular Vertex identity modeled upon Boger's—male, aggressive, entrepreneurial, brazen—that character was no longer so simple or viable.

Many of the new hires were women and many came straight from graduate schools and postdocs. They had different sensibilities, different missions. The rough and tumble of business thrilled them less than Boger's promise, taken at his word, of doing academic-grade science in attention-getting areas. Vertex's core remained its ex-Merck contingent, but with Boger declaring all but a moratorium on hiring any more Merck scientists, its preeminence began to wane. Vertex was now filling with people who'd never heard of Max Tishler, much less deemed themselves his heirs. Many of them had been impressed by Boger at their interviews but had few direct dealings with him since. As with the scaleup of any chemical process, Vertex's new size bred uncertainty, unfamiliarity, irreproducibility, diminished yield.

Boger's office door remained open, with the scientists wandering in to talk while he churned through Niagaras of electronic paperwork. He never stopped working during these discussions, but benignly incorporated them into the flow of what he was doing, like a river absorbing a new tributary. His eyes remained locked on the big Macintosh IIci on his desk and his fingers tapped intermittently at his keyboard, giving the conversations a one-way, confessional air. Not that he wasn't engaged—his mind could still move like a laser when he wanted to make a point, and he was relaxed and personable. But as they slumped in one of two chairs, so close to his desk that their knees knocked, the scientists often sensed that they were receiving only that percentage of his attention that his hypothalamus, the brain's autonomic regulatory center, had allotted.

When there had still been little to discuss and the number of scientists was small, Boger had used these occasions to talk up the company's business. Now he was more guarded. "Deals can be killed simply because word leaks that one is in progress," he told the scientists via electronic mail (E-mail). If anyone asked, he counseled, "Just smile." He, of course, smiled constantly.

A year earlier Vertex's entire scientific staff could fit in Boger's Toyota Camry; now it overflowed the lunchroom. Within the building, Boger communicated primarily by E-mail, preferring it to large staff meetings. Still, on July 7, he called a companywide meeting at noon. Atypically, he gave no reason, inviting rumors and, among the new people, curiosity. A few of them had never seen him lead a large group.

His mother's son, Boger enjoys drama. He calculates flourishes carefully for maximum effect. Now, wordlessly, he switched on an overhead projector, revealing in the semidarkness a document covered except for two signatures at the bottom, his and Nagayama's. He smiled broadly. The scientists didn't know what they were looking at. He then revealed the rest of the page, a letter of intent between Vertex and Chugai. Again, many of the scientists didn't understand what he was showing them.

Boger quickly put up a second transparency. Entitled "The Standard," it was a description of the research deal between Cytel and Sandoz, the Swiss pharmaceutical giant and developer of cyclosporin. When it was announced the previous year, the deal, which brought Cytel $30 million but guaranteed Sandoz all rights to a drug and 30 percent of Cytel's stock, was considered a record for small companies working at the juncture of immunology and chemistry.

"Pitiful," Boger murmured.

The scientists muttered among themselves, then quieted down as Boger now put up a third transparency: "The New Standard." It referred to the tentative agreement, signed by fax the previous night by Boger and Nagayama. The deal gave Vertex $30.25 million—"Not an accidental figure," Boger said—and a 50 percent share of the world market. It would cost the company only 5 percent of its stock, a token amount.

"We get more money and give up half, less than half, of what Cytel did," Boger said. "This is the world's best deal by a factor of two."

The room erupted. More than the money, which was crucial, the announcement rang with deliverance. Boger had promised everyone he hired that Vertex would be better than other companies: They were the original buyers of his story. And yet the constant dire need for money had discouraged many of them. They worried privately whether they had made the right choice and whether Boger was to be believed. Now, it dawned on them—some for the first time—that Boger's optimism, his aura of glowing success, was justified. Suddenly, they were much closer to being a drug company and getting rich than their own work indicated. Corporately, if not scientifically, the company had legitimized itself much sooner than even Boger had expected.

Boger lavishly credited Aldrich as the deal's architect and savior. "Benno helped enormously," he said, "but ultimately even Benno thought it was a lost cause. None of the board members thought it could be done." Aldrich was more modest. Taking no credit himself, he hoisted it all back onto Boger. A few of the scientists still distrusted Aldrich and, despite their euphoria, clucked disapprovingly under their breath.

In fact, in the month since Chugai had first responded positively to Schmidt's ultimatum, Aldrich had been engaged in one continuous negotiation and at least once had kept it from collapsing entirely. "I have to restrain myself," Boger had said upon reading Chugai's initial response in June, "but there are no issues in this." Worse perhaps than issues, however, were the fine points, nuances, understandings, and subtleties of language—"hidden grenades," Boger called them—that Aldrich now addressed in a series of daily, multipage memos back and forth to Japan. From the time he came in each morning until he left after dark, he worked on sharpening Vertex's demands, then slipped them into its fax machine before going to the gym. When they rolled out of a similar machine in Tokyo minutes later, Chugai's lawyers and licensing people attacked them as Aldrich worked out and tried fitfully to sleep at home; they shipped their own end-of-the-day fax just as they were leaving work that night and Aldrich was arriving at Vertex the following day. Segue upon segue, over thirteen time zones, it rolled on this way, a textbook negotiation, hurtling twice as fast as it would with a company next door, and proceeded thus until two days earlier, when a glitch arose.

A midlevel functionary in the Chugai licensing office suddenly challenged certain aspects of the deal that had already been agreed upon. Aldrich fired back angrily that if Chugai changed its position, the deal was off. Aldrich held his breath. He knew the potential price for his adamance. It was not inconceivable that the deal would unravel and that it would be perceived as his fault. "Rich got a bit grayer," Boger recalled. Fifteen hours later a three-paragraph fax arrived with Chugai's answer. "I apologize for raising and addressing issues in a rather slovenly way," the author of the previous fax wrote, impaling himself in a way that no American manager ever would. Several hours later, with the debris finally cleared and the deal back on track, Boger and Nagayama signed the letter of intent, paving the way for the scene in the lunchroom.

Aldrich, recalling his experience at Biogen and IG, cautioned the scientists not to count the deal as done until the final papers were signed and Chugai's check cleared at the Bank of Boston. Much could still go wrong.

But for Boger the full impact of the deal was already clear: It was his deal; the mantle of glory, Schmidt notwithstanding, was all his. It was Boger's vision, his scientists, his system of science, that Chugai had been willing to pay for at such a premium and on such apparently unfavorable terms. And it was he who would benefit. He now had a powerful ally, someone who could take Vertex's molecules to market and yet would leave him in charge. He had money, enough to afford Vertex time to meet his most ambitious goals. And he had validation.

Especially this last. Fourteen months earlier, at the first meeting of the board of directors, he'd said he'd be happy to do a $500,000 deal in the company's first two years. Now it had done a deal sixty times that size—120, if one accepted Boger's analysis comparing it with Cytel's. If there had ever been any doubt about Boger's promise or his competence to run a company, it was now shattered. Schmidt, who had always admired Boger's business acumen, began singing his praises loudly to the rest of the board. "Joshua is the best fella in this field I've seen," he drawled, "and he's as good as anybody starting a new company in any field I know of. He and I work together in a negotiation like two musicians that have been playing the same duet all their lives." Occasionally going so far accidentally as to call Boger Jock, after his friend and mentor Jock Whitney, he smothered any further suggestion about bringing in seasoned management.

Boger now had a secure mandate to run Vertex's business and science as long as he pleased. He was in total, unmitigated control—god mode.

"I'm going to have to resist the temptation to rent a sedan chair for the board meeting next Tuesday," he said.

• • •

Tom Starzl strode through the mobbed liver transplant unit on the fifth floor of the Falk Clinic. The suite was small, semimodern—four examining rooms, a sparsely furnished office that doubled as a conference room, a cramped waiting room with a TV dulcetly delivering Joan Rivers from high atop a Formica shelf. In the waiting room, hollow-eyed patients, some with walkers, huddled like refugees on a steamer. Along the walls stood relatives, friends, transplant candidates, case workers from donor banks. Inside, surgeons—many still dressed in blood-spattered scrubs and running shoes after operating all night—raced between rooms followed by nurses, case managers, residents, and interpreters and demanded updates, prognoses, indications, charts, reports. At up to fifteen transplants a week—more, until recently, than the rest of the country combined—the clinic joined the magnificent aspirations of Lourdes with the grit and commotion of an inner-city emergency room.

Starzl is king here; FK-506 is king. After the conference in Barcelona eight months earlier, he was besieged with requests for the drug. He tried to keep up, but the demands on his energies soon became homicidal. There was too much to do. He was running the transplant service, pushing ahead with multiorgan grafts, directing new experiments with autoimmune diseases and cellular and intestinal transplants, and administering several burgeoning clinical trials that compared FK-506 with cyclosporine for patients receiving new livers, kidneys, and, soon, hearts. His Monday night FK-506 meeting ballooned. Starting at 7:00 P.M., when Starzl entered crisply in his customary windbreaker and slacks and look of eternal stamina, it routinely lasted past midnight.

If Starzl had once insisted on directing every aspect of the drug's testing, he now had no choice. Fujisawa, working with the FDA and European regulatory agencies, had been scheduled to begin offering FK-506 at sixteen other sites in April but had failed to develop acceptable protocols. It was now July, nearly a year after the patient revolt in Pittsburgh, and still no one else had the drug. With the world's transplanters clamoring for FK-506 and unable to get it and exhorting Starzl for more data to support his claims, his work with FK-506 grew more controversial than ever. Once again, his credibility was at stake. "Starzl's a giant; I'll be surprised if he doesn't win a Nobel Prize," said Dr. Ronald Busuttil, chief of liver transplantation at UCLA. "But the joke has already gone around: 'FK-506? It's a unique drug. It only works in Pittsburgh.' "

Characteristically, Starzl responded by immolating himself in work. He pushed himself as hard as he always had, but he was now sixty-four and the pace took its toll. He looked strained and gaunt, with liver spots freckling his worn face. He had always been a notoriously irresponsible eater: flying out of the operating room in his scrubs and blood-spattered disposable booties to gorge on donuts or heaps of french fries smothered with melted cheese at the Original, a hot dog place and student hangout a block from the medical center. That hadn't changed, though he was also now eating a lot of pizza; his office, a third-floor walkup, was above a Pizza Hut. A former three-pack-a-day smoker, he had quit ten years earlier after suffering chest pains but had maintained the nervous energy and compulsiveness of the reformed addict. A competitor once observed that Starzl's energy quotient is "so far off the scale of most humans that it is almost unbelievable," but now there was also a measure of desperation to his drive as if he was racing not only against himself and the world, but against time. Working furiously, he was bogged down increasingly on the phone, in meetings, and with paperwork. He came to the liver clinic not to see his own patients—he seldom had time to operate anymore—but as a general whose few relaxations from command included visiting the front. It was here, and only here, that the full measure of FK-506 could be gauged.

"You look outstanding," he gushed to an energetic woman in her mid-forties. She was sitting on an examining table, amid an entourage that spread to the clogged doorway and beyond. Her exposed torso bore the signature scar of the liver-graft recipient: a cross-hatched, upside-down T running from her sternum to below her navel and from hip to hip. Two months after her transplant, the scar had begun to darken, a rusty color now, no longer pink, and she had stitch abscesses. She complained of soreness, but said she couldn't have the abscesses drained that afternoon; she had to take her daughter to the doctor. The woman's hair was thinning slightly, and she had some tingling in her hands. Otherwise, she said she felt fine, an assessment that seemed to be borne out by her appearance. She looked no more haggard than any other middle-aged parent of teenagers. Starzl ordered her off acyclovir, a powerful antiviral drug, and bactrim, an antibiotic. He had already taken her off prednisone. Because of the generalized effects of suppressing the immune system and the toxicity of cyclosporine, most transplant patients take up to a dozen other drugs, many of them toxic in their own right, for as long as they live. With no signs of rejection, the woman would now take only FK-506. Starzl was elated. "That's our objective," he told her. "To liberate you from us."

Passing through the throng, Starzl swooped in next on a burly man with a thick black beard and volunteer fire department cap who was being examined by another surgeon. Three and a half months ago the man had been rejecting his new liver. He was jaundiced, his kidneys were failing, and he shook uncontrollably, Starzl believed, from the neurotoxic effects of cyclosporine, which is known to affect the brain. He was also taking heavy doses of prednisone. Now the man was off both drugs, his liver and kidneys had revived, and the tremors were nearly gone. Starzl wanted them gone completely.

"We're going to drop your FK dose to three milligrams twice a day," he said. "It's what we give kids."

In the hall Starzl was buttonholed by a young visiting oncologist. She asked if he would be willing to recommend FK-506 for a patient dying of leukemia. "Sounds interesting," he said. "Maybe we can get a compassionate." Because FK-506 was still approved only for certain transplant patients, all other requests had to be sanctioned individually by the FDA, which depended heavily on Starzl's opinion. Some he approved pro forma: For instance, the drug had already been shown to work dramatically in extreme cases of psoriasis, clearing up within weeks rampant, open sores that resisted all other treatment. Starzl was determined to test FK-506 against every disease for which there was a clear scientific rationale, especially those like juvenile-onset diabetes, which had shown a response to cyclosporine. It was well established that cyclosporine, if given soon enough, could stop type A diabetes from developing in children. The problem was, the drug was too toxic to be given over time. The relief, tragically, was only temporary. Starzl was convinced that FK-506, because it was less toxic, could cure the disease permanently and was pushing hard to begin a clinical trial in Pittsburgh. He was not indiscriminate, however. He told the oncologist about a Michigan doctor who asked for FK-506 to treat a rare tumor. Starzl had declined. "Sounds like a shark," he said.

The more patients Starzl saw, the more he was convinced of FK-506's singular potential, even when its side effects mimicked cyclosporine's. In the next room, a distraught black woman, thirtyish and wearing a red dress, told him the drug had caused her to have hallucinations; FK-506, like cyclosporine, seemed to be hitting an unspecified receptor in the brain. Terrified, the woman had asked for her dosage to be cut back, but Starzl observed from her most recent biopsy the aggregation of white cells that indicated she was beginning to reject her graft.

"We're seeing some nibbling," he told her, his manner clipped but not unfeeling. "We're heading for disaster unless we do something about it. With this drug, in a few weeks, you can go off everything else, and I think that'll help. But you're being undertreated." Reluctantly, the woman agreed to a higher dose.

Viewing complaints about side effects from the extreme vantage point of life and death, Starzl, as an experimental transplanter seeing the most desperate patients, has always had a higher threshold for considering toxicity a problem than many of those he treats. Certainly, he is willing to overlook many side effects more readily than most other doctors—a disparity that now had critical implications for FK-506 and for Vertex. Hallucinations and kidney poisoning were one thing if you were going to die; they were another if you had an itchy scalp or were an eight-year-old diabetic. In other words, as Starzl was finding—and as Boger had predicted ten months earlier—FK-506 was a remarkably benign drug compared with the multiple traumas of transplant surgery and cyclosporine, but it still might be too toxic for most autoimmune diseases. Second-generation molecules would undoubtedly be needed.

The similar toxicity profiles were important for another reason. The range of side effects Starzl was seeing in patients with FK-506, though lesser in degree, was almost identical now to those with cyclosporine. In a way, their duplication was even more intriguing than the fact that the drugs worked alike. Whatever their targets, however they behaved on a molecular level, the two drugs were so similar that they were likely to be interfering with the same chemical pathway. It was extraordinary. To understand how they worked, what they did, would perhaps unlock one of the most basic secrets of how the body defends itself. It even might help explain how cells—all cells—transmit information internally, one of the premier mysteries of biomedicine.

Starzl felt himself reaching across a gulf that no one had ever bridged: a surgeon drawn across the whole expanse of medicine to plumb the most basic questions of molecular immunology, of life. He had spent his career groping in the dark, feeling his way along the walls of a tunnel, going farther than anyone else, and now he was near some transfixing light that seemed to indicate its source. More than a drug, FK-506 was a beacon, a probe, and Starzl was determined to track it, no matter where it led. "Transplantation," he began saying suggestively, "may just be a footnote to this entire story."

If Starzl was insatiable before, this new larger mission drove him to flights of religious ferver. He seemed mesmerized by FK-506, enchanted, consumed. After two hours, nearing the end of the clinic and still pumping, he marched past the reception area and beckoned, like an evangelist who can't stop pleading for souls after an all-night revival, the one or two patients still in the waiting room. "Come on back, somebody," he waved. "I'll steal a room."

And yet he was also exhausted and prey to an unfamiliar lassitude. He had to push himself harder and harder just to keep up. In June, he took his first vacation in ten years, traveling with his wife, Joy, to Hawaii. He then went on to Japan to give a series of talks on FK-506, which, ironically, because of Buddhist strictures against violating corpses and an absence of laws governing brain death, is barely used in that country. Kidney transplants from live donors are sanctioned, but otherwise patients like those that Starzl routinely restored to a normal life with the aid of Japan's first great drug were, as he saw it, cruelly and needlessly relegated to die for want of legal organs.

Returning from overflow lectures in Osaka and Tokyo, Starzl drove to his office the morning of Saturday, July 11—four days after Boger's announcement of the Chugai deal—to plow through paperwork that had piled up while he was away. Halfway up to the second floor above the Pizza Hut, he collapsed. "The slightest movement," he later wrote, "caused a cylinder of fire beneath the breast-bone which erupted like a volcano into [my] neck." He dragged himself, inch by inch, to the second-floor landing, where he lay sweating and panting for an hour. He then did the same to get to the third floor. He eventually pulled himself into an upright position at his desk. For the next twelve hours, speaking breathlessly into two dictaphones, he answered three weeks of mail before stumbling downstairs and driving home.

The next day doctors discovered a 99 percent blockage in Starzl's dominant right coronary artery and told him he risked a heart attack unless he had a bypass operation at once. He refused. He and his team were now rushing to finish more than forty papers in time for the international meeting in San Francisco in mid-August of the Transplantation Society, the same organization that had sponsored the conference in Barcelona. It was to be the most rigorous hearing to date on FK-506, made all the more pressing—and critical for Starzl—by Fujisawa's failure to expand its clinical trials. Surgery was out of the question, Starzl said. He agreed to an angioplasty—expansion of the artery with a balloon-tipped catheter—but so reluctantly that he dislocated his shoulder resisting the straps on the operating table. "I needed," he explained, "to be in San Francisco."

Starzl returned to work two days later, but his stamina was diminished. All summer long, his conditioned worsened. By the time he reached San Franscisco, he was in constant pain, yet he dragged himself through the conference, winning adherents with his frank appraisal of the new drug. He flew home immediately afterward and was operated on the next day.

"Between the angioplasty and the San Francisco meeting it was kind of touch and go," he conceded, returning to work the following week. "But it was worth the flip of the coin, I thought."

If Starzl's collapse had any effect on Vertex, it could only be construed positively: Chasing FK-506, the company stood to gain from any delay in its clinical progress, however small or ill derived.

Even without knowing FKBP's structure, Vertex's chemists had begun making small molecules that bound to it almost as tightly as FK-506. Such binding is the first, primitive measure of a molecule's effectiveness and in no way indicates whether it will become a drug: In the evolution of safe, active pharmaceuticals, good binders have the same relationship to good drugs as troglodytes to brain surgeons. But Boger knew that science is an incremental progression, won and lost by inches, and that every gain is worth optimizing. "This class of compounds would be so easy to manufacture," he said about Vertex's tightest binders. "I'd be happy to have an FK-506 look-alike eighteen months behind Fujisawa. Theirs is going to cost $1000 a gram; ours is going to cost $1000 a tankload. Even if our drug has the exact same therapeutic profile, I can wipe them out."

His prediction, of course, precluded the use of structure-based design. Still without protein crystals, the company was several months away at least from having the kind of detailed information about FKBP that would influence what molecules it made; more—perhaps much more—time would be involved if FKBP turned out not to be the relevant target and the process had to be restarted with another protein. Though it certainly was conceivable, given Starzl's resistance to comparison testing with cyclosporine and Fujisawa's mishandling of the regulatory agenda, that Vertex would have a drug on the market less than two years after FK-506, it was unlikely that such a molecule would be designed in a way that proved Boger's theories or brought him closer to his scientific goals.

Indeed, as a demonstration project for designing drugs, FK-506, with its devilish receptor and uncertain biology, was beginning to seem a much poorer choice than the company's other prospect, HIV. As its name implies, structure-based design requires a reasonable certainty about how molecules are shaped and how they fit together. Researchers need to know the exact protein target, correctly configured, and the precise correlation between the way it "talks" with other molecules and the biological activity they hope to affect. With FKBP, in the summer of 1990, science had none of this information; with HIV protease, it had it all. There was no proof that if you inhibited the protein folding action of FKBP, you automatically had an FK-506-like drug. However, it was now all but certain that if you placed a well-designed molecule into the active site of HIV protease, you could cripple the virus's reproductive ability and thus slow the spread of AIDS. You might not destroy the virus, but you would wound it gravely.

The problem with designing protease inhibitors was not how to make them but how to make them into drugs. Practically all of the research to date—"thousands of man-years," observed Boger—suggested that the best way to shut down enzymes that cut and slice proteins is with peptides, small chainlike molecules made from amino acids. Peptides can be made to infiltrate and disarm aspartyl proteases, like HIV's, with remarkable precision. But peptides are useless as drugs. They're fragile, easily broken down in the gut. They can work ideally in a test tube, as almost every major drug company has shown, and yet if you ate grams of them, continuously, almost none would make it to their destination within cells, leaving science with a tantalizing challenge: to make molecules that look and act like peptides but are structurally different. Called peptide mimetics, such molecules are the grail of that considerable portion of AIDS research focused on inhibiting HIV protease.

In terms of pure science, there were many at Vertex who believed HIV protease was custom-made as a test project for the new company, an opportunity for them to succeed where screeners and medicinal chemists were doomed to fail. It had everything: known biology, a solved protein structure, established chemistry, well-developed model systems employing similar enzymes. More, Vertex seemed uniquely positioned to exploit it. Navia's solution of the structure at Merck; Boger's leading work in inhibiting aspartyl proteases; the company's high-tech approaches, theories about small molecules, and integrated labs—all suggested a powerful edge, if not an outright franchise.

Yet Boger himself was less than convinced. From a business standpoint, AIDS—with its muscular competition between industry giants, each with scores of able researchers and a multiyear head start—was still no place for a small start-up company, he believed. Nor was it at all clear that nonpeptidal protease inhibitors worked as drugs—no company had ever made one—or that one of those behemoths, particularly Merck, wasn't infinitely better suited than Vertex to produce the first such molecule. Boger proceeded cautiously. He authorized Roger Tung, a former Merck chemist who was eager to begin working on his own project, to make the Janssen compounds that had prompted Navia's "hallucination" in March. He asked Dave Livingston to develop an enzyme assay so that they could be tested. But he stopped short of a full-scale call for crystallographic quantities of the protein. Notoriously hard to produce, in part because, by definition, it begins cleaving itself almost as soon as it's made and because unlike FKBP it's unavailable through tissue extraction, HIV protease can only be obtained through large-scale fermentation or by Herculean chemical synthesis. Vertex, Boger believed, hadn't the personnel or facilities to spare for either.

Navia, understandably dismayed, pressured Boger for more enzyme, but he was chronically stretched by the demands of his career and the variety of his interests and was no longer the project's main champion.

Mark Murcko was. Five feet seven inches, sandy-haired, stocky, with a thick mustache, Murcko, thirty-one, was the last scientist Boger had hired from Merck and the one he'd gone to the most trouble to get. He was also the most pivotal. A molecular modeler and computational chemist, he stood between crystallography, which churned out huge amounts of data with up-to-the-minute technology and computers, and chemistry, which made molecules with almost no technology more advanced than an espresso machine. If, as Boger said, Vertex was set up to generate and use more information in drug design than anyone else, Murcko stood astride that river of data at its widest, wildest point. The colossalness suited him. Fast-talking, glib, bristling with intelligence and enthusiasm, he had all the attributes (low-to-the-ground construction, broad-beamed confidence, gravelly appearance, hectoring banter, high schmooze quotient) of a catcher in baseball.

If Vertex was to design drugs, Murcko and people like him would design them. On his first day at the company, in May, he sat down at one of a bank of Silicon Graphics workstations that Boger had ordered in anticipation of his arrival and worked, his hands flying over the keyboard, until 3 A.M.—putting himself instantly in league with Thomson and Yamashita, with whom he became fast friends.

He had kept up a similar pace ever since, programming, simulating molecular activity, setting up calculations that could "pin" Vertex's computer network for days on end and would easily pin it forever if other researchers hadn't also been clamoring to use it. Murcko's phrase for this was "speculative science." It's impossible to know, of course, how molecules, which vibrate at billions of times per second, actually connect with one another. But with big, fast computers and three-dimensional computer graphics, it's at least possible to venture educated guesses.

For instance, it's understood that molecules, adhering to the laws of nature, favor those conformations that require the least energy. Thus there are "good" interactions—those that conserve energy—and "bad" ones—those that tax it. Similarly, individual atoms, which Murcko thinks of as "like Nerf balls, a little squishy," smush or repel one another on the basis of an amalgam of almost unknowably small, yet measurable, forces—heat, from the formation and destruction of individual atomic bonds; gravity, exerted by subatomic particles; electric charge. In imagining how molecules bind, Murcko couldn't predict—at least not without a crystal structure showing it—how a drug molecule and a protein might fit together, but he could speculate mightily on how, based on certain proclivities, they might want to.

Murcko is an expert at calculating the activities of atoms. He came to biochemistry, not as almost everyone else at Vertex had, through an interest in life science, but through a passion for computers and a curiosity about the physical aspects of chemical reactions—the "underlying reasons, the mechanisms, the properties of things, the fundamental forces." The computers were first. As a twelve-year-old growing up in Fairfield, Connecticut, a Bridgeport suburb, in the early 1970s, he took a field trip to a science center that had the force, judging from the loving detail with which he retells it, of a revelation. By ninth grade, he was spending twenty hours a week after school and at night writing programs in the computer lab at Fairfield University, where he eventually went to college and majored in chemistry. He then went to graduate school at Yale. There, midway through his dissertation, he became interested in the molecular life of drugs.

"If you believe you understand something fundamental about the way molecules interact and you want to find out if you do understand it," he says, "now, pick a difficult system: Imagine a complex system where your confidence can be tested." Practically by default, proteins, the most complicated of all molecules, became that system. In the fall of 1985, Murcko, who knew nothing about protein structures or drug design, began applying to drug companies, which not surprisingly were mystified by his interest in them.

"I would go into an interview with a representative of a Pfizer or a Glaxo or a Lilly and they would look at my résumé and say, 'Well, you've obviously done interesting things but you don't seem to have the kind of background that would lead somebody into the pharmaceutical industry.' And then the representative from Merck came, and it was Joshua. And he starts telling me about the modeling that he himself had done. . . . This was incredible to me. He was the only one who seemed to appreciate that you didn't have to be a medicinal chemist to contribute to this work."

As a pioneer in the use of molecular modeling, Merck recruited Murcko avidly, though in fact Boger had already persuaded him that there was no other place to be. Joining the company in the spring of 1987, just as Boger was beginning to organize Merck's rational drug design effort, he was assigned to its West Point labs. With Boger now in Rahway, the two never worked together directly, but Boger continued to track Murcko's career, and shortly after he left to found Vertex, he called him. To that point, Boger had failed to win only one scientist whom he'd set his sights on, but Murcko was not easily converted. After persuading him to go to Merck, Boger was now trying to pull him away, and Murcko, his catcher's crouch set, dug in. It took three months and a hefty draft of founder's stock—three months in which Vertex's senior scientists were regularly platooned to call Murcko every other day—to extract him. For Boger, it was a tour de force, a major victory against Merck's best hands, a personal triumph, although one that would also leave him troubled that he had perhaps ruffled Mother Merck one too many times. Seeing the battle for Murcko in Machiavellian terms, he worried about the cost of such a prize.

Says Murcko: "Joshua has the unfortunate situation of having a psychiatrist and two lawyers as brothers, and that, combined with his own scheming, complex, twisted intellect, causes him to see plots and patterns where there probably are none. I can't swear that there's nothing to that notion, but it strikes me as a little paranoid." On the other hand, Merck was noticeably put out by Murcko's decision. Unlike previous defectors to Vertex, who were given up to a month to leave—and perhaps reconsider—Murcko was told initially to be out in four days. Not wanting to waste several hours driving, he flew to Boston the day he finished at Merck, leaving his wife, Kathy, a teacher, to complete the school year back in New Jersey.

Murcko had expected to begin work in immunophilins. "No one ever said to me in an unequivocal way, 'Yes, we have crystals and we're, oh, fractions of an inch from getting a refined structure,' " he says, "but the clear impression was that Vertex was very far along on that path, and that turned out not to be true."

And yet he was far more dismayed by the alternative. For the previous six months, he'd been working eighty to one hundred hours a week designing HIV protease inhibitors at Merck. With major-project status, HIV was discussed in rare detail at regular Monday morning meetings, and Murcko had attended those sessions. He was privy to all Merck's strategies, leads, data, talking points—and legally sworn not to disclose them. Assuming, as Boger himself had, that Vertex would never work in AIDS, he had comforted himself that he would be free of any conflicts in his new job. Yet now, asking Navia on his first day what other areas the company was considering besides FKBP, he was stunned to learn that it had a feeler project in HIV.

"I thought I was going to have a heart attack," he says. "I don't know what a heart attack feels like, but this was a heart attack. Here I was, fresh out of Merck, which I hadn't exactly left on the best of terms, and this icy, steel ball just knots up in my stomach, and I start twitching and drooling as it turns out they want me to start modeling right away on HIV. I couldn't do it. I felt as if anything I did had the potential for being a legal problem.

"I felt really uncomfortable. But I talked to Joshua and other people here and what I ended up deciding was that it was a situation where I had to use my best judgment on a case-by-case basis. It was unfair to me to simply say, 'Well, since you've worked on HIV somewhere else, for the rest of your life you can't work on it.' "

Storming Vertex's computers—and still with no structure of FKBP to work with—Murcko hurled himself into modeling HIV-protease inhibitors that were determinedly unlike any he'd designed, or seen, at Merck. He placed himself in a kind of deep denial, "lobotomizing" himself to forget what he knew, a tortured bit of self-surgery. "At one point, one of the chemists showed me an idea for a compound, and I knew before he was done talking exactly how to synthesize it and how it would perform in several assays. I put my hands over my ears and walked out. It hurt. That really hurt." (Later, Tung concluded that Murcko's repeated self-disqualifications cost the company up to six months.) Meanwhile, by mid-July he was also immersed in a series of modeling experiments using the Janssen compounds and another potent antipsychotic, Haldol, that was being promoted as a powerful inhibitor of the enzyme.

From the outset, the work proceeded slowly. Now that he had committed himself, Murcko began haranguing Boger—"Joshua the indeflectable," he called him—about the need for protein. He reminded him incessantly that without real structural data on either project, he couldn't begin the actual process of designing drugs, and that Vertex, already behind on HIV, could only catch up by exploiting such information. Meantime, he ran endless simulations, devouring masses of computer time. Asked during this period how much computating power he considered optimal, he deadpanned: "Infinite."

His chief ally was Yamashita, with whom he bore the same ambivalent relationship as Yamashita had earlier had with Thomson, when Thomson was still trying to isolate FKBP. Bound by a common belief in structure-based design, brought together by long, late-night theoretical conversations that often stretched until morning, depended upon equally by Boger, they became close confidants in Murcko's "fraternity of grief." And yet Murcko was stalled in his own work because of delays in crystallography. He didn't want to pressure Yamashita—Mason pressured himself enough—but like the last runner in a relay, he was forced to watch in exasperation each time crystallography stumbled and fell behind.

On a Thursday night in late July, Yamashita mentioned to Murcko that he and Navia had grown several new crystals with Thomson's protein and that he was planning a third diffraction attempt early the next day before the labs filled with onlookers. At home later that night, Yamashita logged onto Vertex's computer at about 1 A.M. to monitor some calculations. There was a memo from Murcko in his directory. It concluded with a snatch of dialogue from *Star Wars*, which Murcko had seen a dozen times and could quote at length:

> OBIWAN: Vader was seduced by the dark side of the Force.
> LUKE: The Force?
> OBIWAN: An energy field produced by all living things. It surrounds us, penetrates us, it binds the galaxy together.
> (much later)
> HAN SOLO: I been all around the galaxy, kid, and I seen all kinds of strange things, but I've never seen anything to make me believe in some all-powerful "force" that controls everything.

Publicly, Murcko thought it best to humor Yamashita, though privately he was less sure. "If that crystal doesn't diffract," he said earlier that night, "a lot of people around here are going to be crushed.

"I would watch Mason, follow him around, take his car keys."

Thomson could work in a forty-degree cold room all morning in a T-shirt, but Yamashita, having spent the last fifteen years in Hawaii and Los Angeles, hated cold. He bundled up in a stadium coat. Even before Murcko's inadvertent reminder, the steel-walled meat locker where he and Navia grew many of their crystals had always made him think of Darth Vader, its unseen compressor exhaling nightmarishly like a cancer patient dying into a microphone. At 7 A.M. Friday, midsummer, Yamashita stood blowing into his hands amid its steel racks and Styrofoam coolers and boxes of supplies from the Spectrum corporation—"laboratory products for the third millenium"—preparing to mount, for the third time, presumptive crystals of FKBP.

It was painstaking work, and Yamashita was wise to ward off any sudden shivers. The crystals, still tiny though substantially larger than those he and Navia had first grown in April, floated luminously in droplets of mother liquor: Under the microscope, they looked like coffin-shaped diamonds suspended in water. A protein skin on the surface of each drop needed to be peeled back without disturbing the crystals. A slip could easily set him back weeks, a reversal he preferred not to consider.

The unintelligible but hypnotic lyrics of Ireland's Cocteau Twins blared through his headset as Yamashita hovered over the double-barreled eyepiece of his microscope, steady as a gem cutter. He clasped in his right hand a small plunger that was attached at one end to a glass capillary tube, 0.5 millimeter in diameter. Slowly retracting the molecular skin with the end of the tube, he then poked it into a clear area of the liquid. Finding a spot near a well-formed, isolated crystal, he subtly withdrew the stopcock with his thumb and forefinger until the crystal rose halfway inside the tube. Sealing one end of the tube with hot wax, he inserted into the other a hair-thin filament wick, which he dipped in mother liquor before sealing that end as well. He then repeated the process two more times with two other crystals that had been grown in different conditions and were more lozenge shaped. Together, the three toothpick-sized tubes looked like tiny glass barbells with tear-shaped orange wads of gum on the ends, a barely visible silverfish floating amidships in each.

"There," he said, referring to the crystals. "That ought to keep them happy."

Stung by the pain and embarrassment of his last diffraction attempt, Yamashita had worked hard in recent weeks to control his feelings, both high and low. He affected a seasoned and, given how desperate he was to succeed, anomalous midrange view, an I'm-OK-you're-OK emotional competence that fit him like an oversized suit. Still, pleased by the success with mounting the crystals, Yamashita was buoyant as he entered the X-ray lab. Murcko had come in early to join him, and Navia was once again out of town: at the very least, if the crystals didn't diffract, he wouldn't have to repeat June's excruciating psychodrama.

"That's very good," Yamashita said, observing several bright spots on the monitor, "I'm impressed. Manuel grows very good crystals." It was just the signal he was looking for, the one that had evaded him both times previously, irrefutable proof that the crystal was protein. But there was bad news as well: If the number of spots indicated that he had gotten a protein lattice to diffract with, their distance from an established reference point at the left of the screen also showed that the grid was of a poor quality. Yamashita might be able to solve the molecular structure of the protein with such a crystal, but his data would be only marginally reliable. Would it be the correct structure? He wouldn't know. Could Vertex depend on it as a template for drug design. Probably not. "This is actually the toughest result we could have had," he confessed to Murcko. "It means we're going to have to work on it, but it's going to be very hard to work on."

If the outright failure of his previous diffraction attempts had left Yamashita distraught, the ambiguity of this one was far kinder. It was a start. He knew that. And though it was a disappointing one, he also knew it meant the structure could be solved. He had heard through the grapevine that Merck, too, had crystals but no structure, which meant that though he and Navia might be behind, they were still in the race. Despite his concern about their quality, the pinhead-sized crystals he had mounted on the beam that morning contained information he now knew to be worth keeping, and he began instantly collecting data. He was calm, assured, relatively composed, if privately worried about what lay ahead.

Boger, arriving by 8:30 and immediately visiting the X-ray lab, was ecstatic by comparison. "Great," he said, hearing the result. He paraphrased Archimedes. "Give me a place to stand," he said, "and I'll move the world."

With the discovery of information he considered most vital for designing a better drug than FK-506—Vertex's long-sought-after "place to stand"—now being only a matter of time, Boger felt he *could* move the world. Unlike Yamashita, he wasn't primarily interested in who got the information first and received credit for it, though that clearly mattered to him. What mattered most was that it could be gotten. It had taken Navia less than three months from the time Merck crystallized HIV protease to solve its structure, and he had vowed that he and Yamashita would take even less than that with FKBP. By Thanksgiving, perhaps, they would have the native structure of the enzyme. After that, they would begin feeding Murcko a succession of structures, first FK-506 bound to active site, then Vertex's own proprietary compounds—structures that would show how changes in Murcko's "fundamental forces" affected the biological activity of the molecules.

It was all laid out before them now. Thomson's hellish struggle with the protein, the difficulties in getting crystals, Murcko's impatience—in the weeks ahead, Boger watched them dissolve in the face of what was from here on, a vastly more predictable, if not always orderly, process. Getting diffraction-grade crystals was the prime rate-limiting step in crystallography. Now that Vertex had them, its scientists *would* solve the structure. They *would* make better molecules. Indeed, they already were. By mid-August, while both Boger and Aldrich were in Japan working out the final details of the Chugai deal, the chemists received word from Harding that one of their molecules was now one hundredfold less potent than cyclosporine in switching off T cells. Like the binding and enzymatic assays that showed Vertex's compounds behaving much like FK-506, regulating cellular activity in test tubes was still a primitive measure. And Vertex's molecule was weak—too weak to become a drug. But making a molecule that was active at all was another major step. Vertex now had compounds that appeared to be patentable and that had a desired biological effect—a drug in form, if not content. And Vertex was gathering information about its presumed target, information that would help improve it.

Boger, returning from another trip to Tokyo, was as confident as the scientists had ever seen him. "If in the next six to nine months we can improve by a factor of thirty in cells, six months after that we have a preclinical candidate," he said. "A year after that we have a clinical candidate. That's three years from start to finish—at least two years ahead of where I thought, in my most wildly optimistic projections, we could be."

It had always been implicit in Boger's story that time is money in the drug industry and that structure-based design, in addition to all its other virtues, would vastly increase the speed (thus reducing the cost) of discovering new drugs and bringing them to market. Says Murcko: "Molecular modeling is largely a game of making mistakes faster than your competitor can"—screening ideas, quickly and efficiently, at a keyboard, rather than laboriously scouring dirt samples by sifting through foul-smelling, labor-intensive fractionation broths. Satisfied that Vertex was setting a record pace in all departments, Boger departed on August 15 with a foot-high stack of publications under one arm and gripping a MacIntosh laptop with the other and started out on his first vacation since he and Kinsella began plotting the company. He and Amy took the three boys to North Myrtle Beach, slowing down to make the trip in two and a half days, for their first time off together in more than twenty months.

There was nothing to stop him now. In less than a year, Boger had accomplished all that he had set out to and more. He had known that before Vertex could make a drug it had to make a deal, and he and Aldrich and Schmidt had now brought in the deal of the decade, or so he could argue. The company had protein by the Thomson Unit, crystals that diffracted, and patent applications pending on novel, tight-binding molecules that showed discrete biological activity. It had the germ of a second project, and companies were interested in that, too. Perhaps most significant, it had a powerful ally, Chugai, contributing money, expertise, and credibility—"benedicting," to use a word of Boger's, the company. "Together we now have the biggest research program in the world in this area," he told the scientists. "We had the best, but not the biggest. Now we're bigger than Merck, bigger than anyone." With the exception of permanent labs, Boger had delivered on every one of his major promises, no matter how grandiose or improbable.

It was time to assert the prerogatives of his new station: "the night of the long knives," Aldrich called it. On September 26, 1990, within hours of receiving the last signed legal documents from Chugai, Aldrich sent two letters to Harvard by bonded courier, backed up by certified mail. The first was to Martin Karplus, a founding member of the SAB, who was attempting to solve the structure of FKBP with protein from Schreiber's lab. The second was to Schreiber. They were termination notices. Aldrich wrote officiously, without explanation, that Vertex now considered its relationship with both of them untenable and that as of December 31, 1990, they would no longer be associated with the firm. Vertex, he wrote, would buy out their unvested stock, although they could keep what they already owned—75,000 shares apiece. Though the letter didn't say so, the stock was the price of Vertex's freedom, all but guaranteeing that Schreiber and Karplus wouldn't complain too loudly or counter with lawsuits.

For Boger, the worst part of an otherwise charmed year, his yawning distrust of Schreiber and the need to debase himself with Harvard, the one matter that had persistently eluded his control and provoked his anger, was over. The negotiations with the Harvard patent office were moot, discontinued, as what Schreiber did with his compounds and protein were no longer Vertex's affair. If Schreiber was a loose cannon, he was no longer on the company's deck. Where he pointed, as with all things Boger couldn't control, ceased instantly to matter to him, or so he would make it seem.

CHAPTER TEN

The formal signing of a business agreement is a paradoxical event. Like a wedding, it marks the end of the breathless seduction that spawned it and the start of the frank, purposeful relationship that is supposed to follow. In Judaism, couples are traditionally wed under an arborlike *chupah*, a metaphorical house, which shelters them momentarily from these and other contradictions. On October 3, 1990, a day in which a sudden north wind ushered fall into Massachusetts like a bugle blast, Vertex became such a house, or, rather, a stage set of one.

Boger had prepared for the day to be a circus. By the end of it he would rachet his assessment to "Felliniesque."

Photographs of Mount Fuji in all four seasons, a prenuptial gift from Chugai, lined the walls in the conference room, and an architect's rendering of a new four-story, postmodern company headquarters—the third, and most recent, proposal for new labs to have fallen through—hung impressively in the lobby. In the lunchroom, several graffiti-laced New Yorker cartoons and two dated Merck I.D. pictures of Boger and Navia that had been taped on the refrigerator were replaced with NASA decals commemorating shuttle launches that had borne Navia's experiments—so far all unsuccessful—into space.

Boger had spent most of the previous day's staff meeting issuing last-minute directions. On props, he said: "Absolutely every instrument should be on. If it's got a display, make it do something—in color preferably"; on staging, "I want all the Putnam people [those working in the new Putnam Avenue labs] here for the morning. It'll make things look more exciting"; on costumes, "Be reasonably accurate; don't go out and buy a new suit, but if you wake up tomorrow and you have a choice to reach for the grungy pants you usually come to work in and the nice ones you sometimes wear, pick the ones you sometimes wear."

Boger himself arrived that morning looking as if he'd just stepped from the shower, sartorially correct in his dark blue "power money suit," monogrammed shirt, and paisley tie, his hair and beard neatly trimmed and combed. Leading Chugai's uniformly black-suited delegation on a tour of the labs, taller than each of them by a foot, he stood out like a giraffe among pandas. He beamed cleverly, refreshed by the view.

Chugai was paying for Vertex's scientists and science, but the laboratories were an indication of the company's wealth and breeding—a secondary asset, like a trousseau—and Boger liked to show them off to advantage. He steered the Chugai men briskly through the wet labs, past the X-ray and protein facilities, and finally and inevitably into the darkened modeling room, where they crowded around a workstation at which Navia demonstrated how molecules bind. On the screen, stick diagrams of hundreds of connected atoms in brilliant reds, purples, and blues rotated gently, like hair-thin Tinkertoys, in a fathomless black sea. Navia handed around 3-D glasses, which all but Boger accepted and which thrust them, at the push of a button, inside the molecular cosmos.

"I'm afraid I need translation," said Sam Nagayama, Chugai's young deputy president, adjusting his glasses. Navia rubbed his fists back and forth, mimicking the interdigitation of atoms as they talk to each other.

"These simulation show identical thing that happens in body?" Nagayama asked, to which Navia protested gently that they weren't simulations but "experiments."

"It's all Greek to me," Nagayama smiled, adding brightly, "You make me feel like a fool."

As hosts, Boger and Aldrich had taken exceptional care to satisfy the needs of their guests, to make them feel, despite Vertex's boasting, that the agreement was equitable and that Chugai was getting as good as it gave. Had Nagayama's remark seemed less admiring or more than a too literal translation, they might have been concerned, but he seemed to imply no unpleasantness. The night before, repaying the ritual feasts Boger and the others had politely endured in Tokyo, Vertex had held a banquet at the plush Four Seasons Hotel, across from Boston's famed Public Gardens. It had consisted entirely of New England cuisine—roast quail, shelled lobster, medallions of sweet potato, pumpkin soup in whole pumpkins—and though not as inflationary as the $1000-a-head dinners provided by Chugai in Tokyo (where, as Boger observed, a whole melon "with a stem and a leaf" cost $140), the Chugai group was impressed nonetheless. Now, as they swept into the conference room for the formal signing, Nagayama looked like a man every bit confident that he'd made the right purchase, even if he didn't entirely understand what he'd bought.

The signing was Chugai's idea. "The Japanese like ceremonies," Boger told the staff, "so we're having a ceremony." Boger had planned for it to be low-key and private: signatures, a champagne toast, just the principals and the photographers from their respective public relations firms. However, just as it was about to start, Kinsella, who'd flown east to New York instead of Boston so he could engineer an invitation from Schmidt to finish the trip aboard Schmidt's private Gulfstream, burst in and stood beside Boger. Privately, he was angry at Boger about the loss of Schreiber, whom he saw as a hugely bankable asset and whose name he hated to see dropped from the marquee of his most promising company, and Boger might have wondered if the interruption wasn't hostile had he not known that Kinsella found such moments irresistible. Leaving right after the pictures were snapped, Kinsella marched into the lunchroom, buttonholed a reporter, leaned into him like a figurehead on a victory ship, and instantly began promoting one of his latest business ideas: the first privately owned potato chip factory in non-Communist Poland. As ever, he was on to the next thing.

In that, Kinsella and Boger had always been same. Kinsella made his money and his reputation hatching new companies and staying with them only until they went public and he could sell his founder's stock at a windfall. He liked to put things together and walk away. But Boger had no other temptations or affections. For him, the Chugai deal was leverage for one thing and one thing only: Vertex's future. In a value-added world, a deal's ultimate measure was not the money, but the larger catch it could induce. Thirty million dollars was barely one-seventh of what Vertex might need to make a drug. But the increased financial security it represented had probably doubled the company's value overnight. Anticipating the day, presumably still years in the future, when Vertex would sell shares beyond its current limited circle of investors, Boger had never stopped trying to inflate Vertex's worth or play to potential buyers. As always, his next thing was to bring in money, much more money, enough money to continue to, as Aldrich said, "feed the beast." More than anything, the deal gave Boger his first significant opportunity to sell Vertex's story to a wider audience, an opportunity he intended to flog at the day's next and featured event, a press conference.

What Boger had been telling other drug companies and small, indifferent groups of investors like those at the Vista, he could now tell readers of the *Wall Street Journal*, *Harvard Business Review*, *Boston Globe*, *Scrip*, and the four or five other journals that had survived the blandishments of Vertex's PR firm and sent reporters anyway. Like all journalists, they had seen too many staged new events and press conferences to look more than congenitally bored as they tried to jumpstart themselves with coffee and backgrounders. Boger was confident he could win them, as he had won others, with his story. But first he had to yield to Nagayama, who had his own story to tell and had earned, as a paying guest, the right to tell it first.

"Chugai," Nagayama said (pronouncing it chu-GAI), standing at a lectern borrowed from the nearby Hyatt Regency Hotel, "recently decided to make investment in very interesting company called Vertex. We have been very impressed with very rational approach to designing drugs." He was stating the obvious, and the reporters understandably found little in his remarks to copy down, much less write about. And yet Nagayama's story, like Boger's, brimmed with subtexts, subtexts that gave away not only much about the company's ambitions and motives but about Japan's.

Like most industries just prior to becoming part of Japan's export juggernaut, its leading drug companies—Chugai included—were now all stagnating, all for the same reason: They had succeeded too well at home. They had grown up in an overheated domestic market that, by American standards, was almost unimaginably forgiving. Unlike in the United States, for instance, doctors in Japan are allowed to own pharmacies. More than 60 percent of total drug sales are from medications that doctors prescribe, then sell to their patients at prices set by the government. Breathtaking in its potential abuses, this system has had two overarching results: Japan's people take more prescription drugs and live longer than anyone else on earth. Yet now, as Japan was getting older, it was also saturated with drugmakers and drugs, and the government, under pressure to bring down rising health costs, had begun by slashing prices.

The shrinking Japanese market was the main reason behind Chugai's global expansion and its interest in Vertex. It had spurred a blistering new competition among Japanese drugmakers that had forced them to revert to a traditional strategy, borrowed from the Chinese, of forming alliances with remote partners in order to defeat those nearest to them. Americans tend to think of Japanese companies as, paradoxically, both xenophobic and having their sights set irreducibly on worldwide dominion. Yet the simpler fact often is that they are scraping to protect slender market shares at home. This was what Nagayama had flown halfway around the world to say. The journalists seemed notably uninterested. They barely reacted, even when he injected a note of internationalism. "We don't see any longer the borders in business and science," he said. "Our major mission is to help patients suffering around the world."

If the reporters were underwhelmed by a Japanese businessman telling them he was appropriating some of America's most advanced technology in order to beat up other Japanese companies that would soon, if they hadn't already, be doing the same, and that he dismissed the protectionist angst then gathering in the United States, Aldrich was stricken. The one thing he and Boger feared all along in doing a deal with Chugai was anti-Japanese backlash. Like Nagayama, they also saw the transfer of biomedical innovation between the United States and Japan as inevitable and, in its likely result of generating more new drugs, desirable. Nevertheless, they had been extremely careful about structuring those parts of the deal involving Chugai's access to Vertex's technology. Chugai had wanted, for instance, to send three of its best young scientists to train at Vertex for a year; Boger refused. The contract allows for frequent visitation but no outright siting of trainees. The first visiting researchers wouldn't arrive for more than a month, but Boger had already begun security measures to restrict their presence. Gazing at the uncomprehending faces of the reporters, Aldrich drew a sigh of relief.

Free of the need for damage control, Boger now took the podium. Chugai deal or none, his objective as always was to position Vertex as the forerunner of the coming revolution in drug research. For lay audiences, this usually meant starting with a discussion of microbial screening and scrolling logically, inexorably ahead.

"There's nothing wrong with screening if it works," he told the reporters. "But it rarely works. It's trial and error. And when it fails its a very frustrating process because *you can't do anything about it.* We're not happy with statistically successful probabilities. We want to solve *problems.* We don't want to set up ten screening programs and hope that one of them pays out."

He went on: "We like to go into projects where we're very sure of the biochemistry. We believe we have that with FK-506. But FK-506 is very difficult to change chemically. The optimum drug is one that fits into a receptor with very little to spare, but you can't engineer the bad parts of the molecule out by simply knowing the structure of the drug molecule. You need both parts of the picture. You need to see every atom."

Boger's last slide was one he had made just for the occasion. It showed comparative time lines for drug development. On top—the traditional approach—was a colored bar extending four to six years, the time it generally takes to begin testing a promising molecule in people. Nearly half the bar was titled "Discovery." On the bottom was Vertex's approach: "Discovery" was cut by about a third, shortening the entire bar.

"The question is," he said, "What does that get us? Since if there's no payoff on the bottom line, there's no reason to do this. It gets us this: We're in control of the process. It's an information-based process, not a random process. It means we can get drugs faster to market and that they'll be better drugs."

The reporters, who hadn't heard such a story before, perked up.

"Will you be manufacturing a drug or will Chugai?" one asked.

"We haven't decided yet. The deal calls for a straight fifty–fifty share of the responsibilities and rewards."

"How will you divide the project?"

"We're working together."

"Could you be more specific?"

"The discovery will be done here. But that doesn't mean Chugai won't have a role to play. This isn't press-a-button-and-get-a-drug-out. It's an interactive process."

One appealing—and forgiving—aspect of representing a research-based company in public is the presumption of secrecy. Boger didn't mention the nagging questions about the biological relevance of FKBP or the difficulties in solving its structure, nor was he obliged to. He didn't explain that no small company could screen as effectively as a large one and that Vertex's novel approach was not only a matter of choice but of survival. Speaking with the press, he could maintain as pristine and uncomplicated a version of Vertex's story as the one he presented at the Vista, almost a year earlier, when the company had no science to contradict it. In that respect, he performed impeccably. The reporters were satisfied. Unlike many such sessions, this one gave them something to write about.

Business done, Boger now led the assemblage to the nearby Hyatt for an outdoor lunch and reception. For most, it meant trudging across the weed-choked rail yard and down an alley that runs between two warehouses-cum-laboratories before spilling into an industrial backstreet a block from the Charles River. Sparing the Japanese and the board members, Boger sent them ahead in cars. The Hyatt, at a point just above where the Charles widens into an artificial basin broad enough for sailing, is a hollowed-out, fifteen-story brick and glass pyramid crowned by a revolving bar. With splendid views of the looping river and of downtown Boston, it is an essential stop on the recruitment tours not only of Vertex but of many other Cambridge companies and the universities as well. What the X-rated Combat Zone once was to conventioneers, the

Hyatt is to researchers visiting the new Cambridge: a site of multiple seductions. Boger had worked his wiles there often.

In a private courtyard, a banquet table groaned with posh food—medallions of beef, scallops wrapped in smoked salmon, tortellini Alfredo, melted brie, an antelope dish. Waiters floated by with champagne glasses on silver trays. A bar was set up in a gazebo, though the scientists, not knowing whether they were supposed to get drunk or return to work, abstained at first. It was chilly. Small knots of people warmed themselves in the sunny corners of the yard or huddled in their jackets around cocktail tables, mingling like dancers during a break.

The signing and press conference had been restricted, but now they were all here: Boger and Nagayama, receiving congratulations and trading confidences; Aldrich, finally unprepossessed, and Dr. Hiroyuki Ohta, Chugai's chief of U.S. operations, sharp-eyed seconds and now, in a way, brothers-in-law; Schmidt, the rich, back-slapping uncle, and his limo driver, waiting to shepherd him out early and back to the moneyed canyons of New York; Kinsella, the eternal bachelor, effervescing about his next conquest (using bee pollen to deliver drugs to the lungs); the SAB (though not Schreiber and Karplus, who were explicitly barred); the other board members; Chugai's research sachems; the press; the scientists. Seventy or eighty people in all. Only Thomson, in Faustian protest over what he considered a violation of the purity of science, stayed away. "I called over and told him that nobody wants him anyway, but he still didn't come," said Laura Engle, whom Thomson, emerging from his isolation, had quietly begun dating.

In an atmosphere so warmly suffused with self-congratulation, the talk was uniformly bright and confident. There were no expressions of misgiving, though perhaps, as Thomson's absence suggested, there might have been. The marriage of biomedicine and money brought together, as the assemblage showed, a bizarre assortment of bedfellows. It was not hard to imagine the discomfort that a George Merck—much less a Max Tishler—might have found here. Or the kind of sharks, attracted by the opportunities for wealth and glory, that were now circling in nearby waters. There were compromises—great compromises—inherent in this mix, and they were now made stranger and more glaring with inclusion of the Japanese.

"I told these boys to go over to Japan and bring home the bacon," said Frank Bonsal, a venture capitalist and board member from Maryland, elbowing his way into a conversation between Nagayama and Boger's brother Ken.

"Bacon?" Nagayama said.

"Money."

Just as Boger had foreseen that before Vertex could do any real science it had to have a story, he had also known that before the company made a drug it must make a deal. Now he had such a deal and was already looking beyond it to the next stage. Part of that—the easiest part—was widening his audience, which was helped considerably by the next day's *Wall Street Journal.* "The agreement promises to catapult Vertex, a closely held start-up company formed last year, into a leading role among companies employing 'rational drug design' to develop new medicines," the paper wrote. As Boger knew, being anointed by the world's leading business journal as a leader in innovation, even if it had no way to measure or support such a claim, was tantamount to leading in fact. In business, perception leads reality, and Vertex was now beginning to be perceived—to the extent that it was known at all—as the best company in its class, just, of course, as Boger had always said and believed. It was a major coup for so young a firm.

And yet it also pointed up a larger stress, one that even Boger, to use his own term, couldn't benedict away: the contradiction between business and science. As Thomson had hoped to register with his protest, they were obverse systems based on fundamentally antagonistic beliefs, and there were grave perils in allowing the gap between them to become too great. If perception led reality in business—and even Aldrich now called Vertex's business development efforts, only half-jokingly, "blue smoke and mirrors"—in science the exact opposite was true. Science without verification—proof—was nothing, a shell. It needed facts, data, evidence, rigor. Vertex's business was science, but no amount of profile building would produce a drug.

Boger had come to this place of uncommon success almost entirely on his business acumen, his story. But to design drugs he needed information, answers to basic questions, and those he still didn't have. He had never turned away from science, not entirely, but the ultimate test of his foresight, his intelligence, the veracity of his aims and the potency of his ambitions, was in the labs. And there, unlike the undefined heap Vertex was hurtling to the top of in the *Journal*, he faced a much stiffer, and potentially ruinous, competition. Merck, his claims of superiority aside, still towered before him. What was more, he now had to contend with a freshly antagonized and equally ambitious Schreiber, whose own contradictions concerning business Boger himself had dramatically simplified. In that jurisdiction, Boger had yet to prove himself, and he now pointed himself headlong toward it, reasserting his leadership over Vertex's science and plunging into those questions he'd been forced to neglect while he had been consumed with making a deal. This, he knew, not deal making, is what it would take to build a better molecule.

Boger had not come this far to falter now.

PART TWO

THE CHASE

CHAPTER ELEVEN

I, Stuart Schreiber says, sotto voce, "am ecstatic in my life."

It is the least academic of settings. The office is large, irregularly shaped, orderly, serene, and sleek as a $300-an-hour Miami lawyer's. Bold expressionist paintings, suffused by recessed lighting and the warm halo of a black torchère, broadcast a taste for art and for expensive design trends. In that part of the room where an earlier generation of chemists would have anchored an ancient seminar table seared with cigarette burns and coffee stains, gleams an immaculate, low-slung glass and steel cocktail table encircled by a perfectly equilibrated constellation of plush chairs. Anchoring here is an imported olive green sofa, leather soft as vellum.

Schreiber himself is tall and supple, not wide, but no longer lanky either. In his mid-thirties, he still brims with a kind of boy-wonder enthusiasm, as if stunned by his own cleverness. His speech is smooth and precise, though sometimes, when he thinks he has gone too far or revealed too much, it trails off in an Annie Hallish vacuum of awkward self-censorship. With a cleft chin swathed in a semipermanent three-day beard ("so that when you sit on airplanes, little kids and old ladies don't want to talk to you," he says), chestnut cheeks, goggle eyes, and a receding cap of stylishly cropped prematurely gray hair, he coils in a sidechair with the poise and equanimity of an ostrich, just as Boger, at his desk, resembles a crane.

His aestheticism, his sense of perfection, extends beyond himself and his office to his labs. Most university laboratories are grim, utilitarian places, scruffy and metallic, like the waiting rooms of welfare offices. Schreiber's gleam strenuously. The hoods are a brilliant tomato red, the cabinets blond. In the cold room, a floor-to-ceiling picture window ensures that his students won't have to suffer unpleasant isolation while conducting their experiments. When Schreiber arrived from Yale in 1988, Harvard assigned him two wings in contiguous buildings joined only by a hallway. At his insistence, it built a lounge to connect them—a move that required filling in all the floors above and below. "I'm sure this is the most expensive lounge Harvard ever built," Schreiber says, "but it's worth it."

That nurturing Schreiber's "personal equilibrium" is "worth it" few Harvard administrators would dispute. Like all research institutions, Harvard is a business. Its products are ideas and scholars, and its relationship to its senior professors, particularly in the sciences, largely that of a central bank. From that standpoint, Schreiber's fusing of synthetic chemistry, which he is trained in, with cell biology, which he's not, promises to be a bonanza, and Harvard has supported him avidly. Having saved itself—and lost a fortune—with the first round of biotech companies spawned by its professors, Harvard has been eager not to neglect the "major industry" in small molecules that Schreiber believes he is now inventing in its labs.

"The day I realized I was going to move" from Yale, he recalls, "I compiled a list. I thought it was fairly responsible, but I was worried that they would be upset by it. I know the Yale administration would have gulped. Suffice to say I put my figure down and they immediately came back with a counterproposal that was larger than that figure.

"That's when I realized I'm not dealing with a nickle-and-dime group. These are people who are deadly serious about the sciences."

So much euphoria, such great good fortune, might be expected in so young a researcher to unleash a ranch-sized ego, and Schreiber is far from modest about his status in the new Harvard plutocracy. "This is where I had to be," he muses. "I knew I would end up at Harvard. I knew it! Whatever it would take I would achieve that. I wouldn't stop working until I got there. If I had to give up everything else, I would be willing to do that." On the other hand, his self-awe seems largely to devolve from and be measured against Harvard's singular history. "This is the Mecca for organic chemistry. It always has been. I quickly came to the conclusion that, yes, I was good at certain problems in synthetic chemistry. I had published manuscripts that basically caused people to say, 'Gee, he's a clever fellow.' But being here, you realize that's not good enough. You've got to create something new."

This was Schreiber's charge to himself in late 1988 and early 1989 as he assembled his labs: to do something new in chemistry, something creative, beyond mere cleverness. Indeed, it was greater than that. Equating his call to Harvard at age thirty-two with a thirty- to thirty-five-year mandate to advance the frontiers of organic chemistry, he set out to redirect and enlarge the field. Like previous generations of synthetic chemists, he would concentrate on making biologically active molecules. But he wouldn't stop there. He would use those molecules as probes, to uncover a wider world. Exploring how small molecules affect intracellular processes, he would push chemistry to the very heart of the revolution in biology.

It was a potentially historic manifesto. For decades, cell biology had been moving to a more molecular understanding. At the same time, its principal tools and methods came from biology, particularly the recombining of genetic material. And yet biotechnology had its limits. A wide array of protein receptors on the surfaces of cells had been identified, for instance, using genetically engineered probes, spawning a new generation of drugs and drug research. But such molecules were too big to penetrate the cell membrane.

Here was Schreiber's vector, his opportunity: small molecules. He resolved to use smaller, synthetic compounds to breach the cellular barrier. He would make molecules that could be used to identify proteins in the cytoplasm, extract them, purify them, elucidate their architecture, identify their partners. He, a chemist, would penetrate and explain the most fundamental biological events—the chemical interfingering of molecules inside cells. He even had a model: FK-506. Schreiber determined to use the molecule and variants of it, like 506BD, to attack the most hermetically encoded secrets of the cell: how proteins, which are made of lifeless atoms, talk to each other, how they travel purposefully within the nexus of the cell. He would answer questions that biologists usually ask, and he would do it better, faster, more brilliantly, by using new molecules that only he—and those Harvard licensed them to—had.

"Biochemists," runs an old definition, "are people who talk about chemistry to biologists, about biology to chemists, and about women among themselves." For more than a generation, the leading biochemists had mostly been molecular biologists, people who looked at life through the prism of DNA. To understand biology at its most basic level, they said, one must start with the gene—the act of creation. Schreiber took a more situational view. He would start with the trigger of all biochemical events: the binding of two molecules. Changing how molecules bind, he would revise their structures, affect their activity, and demonstrate the physical relationship between the two. He would use his knowledge of chemistry to explain biology and perhaps, in so doing, alter it.

This was Schreiber's mission, his trajectory.

Of course, he wasn't alone. That Boger, another Harvard-trained nonbiologist whom Schreiber then knew only slightly and who indeed thought of him as just such a "clever fellow," should just then have embarked on the same path with the same molecule was surely the most startling and epic coincidence in either of their lives.

Once at Harvard, Schreiber moved quickly to consolidate his charter. He's prone to epiphanies, and when he's had one, it prompts in him a dauntless confidence. He places himself, as Mark Murcko puts it, "above the noise." Schreiber had catapulted himself to a central place in immunophilins research by collaborating with Harding on the discovery of FKBP. He had furthered his priority by having his group isolate the gene that encoded the information for making the protein, then, borrowing recombinant technology, cloning it and churning out new protein by the Thomson Unit. At each stage, his luck had been enormous. He had never discovered a protein before, never cloned a gene, never overexpressed an enzyme. Yet he and his network of students and collaborators had routinely tied or won against some of the best labs in the world. Calling FKBP a "blessed molecule," which for Schreiber and his students it surely was, he now sought to use it to make his mark on all science.

That Schreiber should be doing all this at Harvard was easily as surprising as the successes themselves. Certainly nothing could have been further from his own view of his future when, as a profoundly indifferent high school student, he enrolled in a work study program so he could avoid taking classes, then skipped those few that were required. "I was not academically oriented whatsoever," he recalls. "I never thought about going to college. I thought about being a carpenter. I was thinking about whether I wanted to do flooring or roofing. That was the level of resolution I was considering. I really didn't think I needed college; those people were so uninteresting to me."

What interested Schreiber, growing up in semisuburban Virginia in the 1960s and early 1970s, were dirt bikes, sports, carousing, and girls. His father, a retired army colonel and ballistics consultant, was a strict disciplinarian; his mother, a doting homemaker, uncritical of her youngest son. Yet both his parents resolved to let their children live their own lives. Working at a pizza shop, Schreiber became a virtual stranger at school. "I never had a book in high school," he recalls. "They gave me books at the beginning of the year—I never thought this was odd—and a locker and a combination. I took the books, put them in the locker, then at the end of the year, they asked for the books back, so I had to go back and ask for the combination. I never once went to the locker during the year." Schreiber was a shop rat. He took Shop 1, 2, and 3, electronics, automotive maintenance and repair, and a course he remembers as "bachelor wardrobe planning"—the boys' equivalent to home economics. It was taught, he recalls, by "a thick redneck guy who told us about his escapades, which involved either beating up people or chasing women."

"I never heard of chemistry until my last year," he says. "They brought us into the auditorium and we watched a Walt Disney movie. That was our introduction to chemistry. My impression was that chemistry had some analogy to the planets rotating around the sun."

Despite his lack of motivation, Schreiber seemed to have an uncanny ability to absorb certain information, particularly at test time. He had a flair for abstract concepts and a kinship with shapes. "I thought there was something strange about me. I was able to sit down and figure out a geometry exam right on the spot. And yet I hadn't been to the classes. Everyone else was complaining about how difficult it was, but it just seemed so logical to me." A guidance counselor suggested he take the SATs. Racing through the six-hour exam after a night of "raising hell," Schreiber scored among the highest in his class. "I remember people saying, '*Sheeee.* Schreiber! How the hell did he do so well?' "

On a lark, he decided to apply to the University of Virginia (U.Va.) and Virginia Tech. "I didn't really care if I got in or not—in the back of my head I was still thinking about roofing or flooring. But I did get in, which was another big surprise."

In most hagiographies there comes a period of bitter wandering before the advent of a profound vision. Such is how Schreiber describes his arrival at U.Va. He was miserable. He didn't want to be there. The students were nothing like his buddies back home at the pizza shop. Thinking he might want a career where he could work outside, he contemplated biology and forestry, but was told he first had to take chemistry, a notorious "flunk-out course." Schreiber considered the fact that he'd never studied before, thought about Walt Disney and his revolving orbs, and signed up for all liberal arts courses instead. "I absolutely hated it," he recalls. "In one course we had to read Sartre's *No Exit.* There were a lot of Northerners, and they took the book so seriously. It made me want to vomit.

"I quit going to classes. I spent my first three weeks absolutely convinced that I was leaving the university, which was fine with me. Then I realized that I could take advantage of my street sense and have a lot of fun. I realized there were a lot of young women I could spend my time with. So that's what I did.

"After the third week I called my sister and told her I was going to quit. I explained why, and she said, 'Well, you should do what you want to do. You should take the chemistry course if that's what you want.' The minute she said it, it seemed perfectly reasonable: *Well, of course, take the chemistry course.* I went and saw the instructor. He told me, 'You missed the first three weeks. The first exam is on Friday'—I think this was Monday. He said, 'I'll let you into the course, but you will not be excused from taking the first exam.' " Schreiber laughs. "I thought, 'What difference does that make to me?' I didn't care about failing an exam. I obviously wasn't going to make it, so fine, great with me."

He continues: "Now that Monday, that was a very important day. I went into the big lecture hall. That was the first time I'd been in a real bona fide lecture. I remember my first impression. I walked into this room, I looked around, and *everybody* had notebooks, and they were taking notes. And I wondered, 'How did everybody know to do this?' So I asked someone. I said, 'How did you know to get a notebook? Did somebody tell you? Did I not get something?'

"I sat in on that lecture. The instructor—his name was Russell Grimes—went to the board. And he started drawing something that I had no idea about. It turned out he was drawing atomic orbitals. In fact, he was discussing a set of five *d*-orbitals. These are sort of geometric shapes. They have large lobes. One of them has two large lobes and a donut ring around the intersection of the two lobes . . . He used colored chalk.

"I looked at that and thought, 'My *God!* This is chemistry? I thought chemistry was like the planets revolving around the sun. This seemed like geometry. It was spectacularly beautiful, with great aesthetic appeal in what he was drawing. I had no idea what it was. But the colored chalk. The different shaped orbitals. I thought, 'This looks really interesting.'

"I went down to the bookstore and bought the book. I took it back to my room. I decided, 'OK, here's where I run into my dead end.' I had heard, around the dorm, a lot of the students complaining about how difficult this was and how they didn't understand it. So I read the first chapter, carefully, sentence by sentence, waiting to be told something that was uninterpretable *and it just never happened.* It all flowed so wonderfully. It seemed so clear and logical."

Cramming, Schreiber got an eighty-eight on the exam, missing just three questions—"the last three problems I missed for the rest of the year," he recalls. "After that, I started going to every lecture. I couldn't get enough of this. It was really exciting. It was absolutely clear to me that I was really good at this, and I really loved it."

Schreiber was now on the road to Damascus. He discovered that the next course was organic chemistry, which he found "orders of magnitude more interesting" than general chemistry. Insatiable, he bought the second-year text and pored over it throughout the summer. "I pretty much mapped out my career at that stage," he says. "I remember going to the chairman of the department. I'd read through the graduate brochure and knew what everyone was doing. I sat down and said, 'I've given a lot of thought to what I've studied and I want do synthetic organic chemistry in my career. I want to be a synthetic organic chemist. I want to be a professor at a major university, probably on the East Coast. And furthermore, I want to work in your laboratory.' "

He still hadn't entered his first lab or run his first reaction, but Schreiber, at age nineteen, now devoted himself to a life of making complex molecules. "I went from two incredible extremes—no academic orientation to this tremendous passion for organic chemistry. Then I learned synthetic chemistry, and that was *truly* exciting. I realized that it was much like studying architecture. You have this complex target molecule and a large collection of reaction processes. Then you have to analyze logically a sequence of reactions that will take simple materials and convert them to complicated materials, just like building a new building. There are an infinite number of solutions, but some are clearly characterized by elegance and efficiency. There is an aesthetic appeal to the way you assemble a certain number of these smaller fragments and they just snap together. You know you've got it. There are other ways of doing it, but they're not as interesting."

Schreiber fancied himself a master synthetic chemist from his first moment at the bench, although now he concedes that the pace of his research was "pitifully slow." It hardly mattered. He was a phenom, a natural. By the end of his sophomore year, he was devouring chemistry texts in rapid succession and compiling an academic profile so hyperattenuated to one discipline that the university didn't know what kind of degree to grant him. Of 120 total undergraduate credits, he amassed 105 in the sciences, 85 of them in chemistry. He took every graduate course in the department and got straight A-pluses. Totally dedicated, he abandoned all other pursuits, considering them hopelessly dull. By the time he graduated, he recalls, "I had *tremendous* confidence in myself. I don't think I was an arrogant person. I wasn't hard to get along with. But I *knew* I was very good at this."

Schreiber had entered college not knowing how students knew to take notes and left it gaining automatic acceptance to Harvard, the best organic chemistry department in the world. What's more, he had come, by the supple ease with which he managed everything associated with his new path, to expect no less. "Harvard acceptance? Of course! They had to take me!" he says. He had begun an uneducated naif and ended a wunderkind, bionic. There was little question what was next for him. Arriving in Cambridge in the fall of 1977, Schreiber quickly approached Robert Burns Woodward, by almost every estimate the greatest synthetic organic chemist of the twentieth century, and asked to study in his lab. A towering, romantic figure already well on his way to beatification, Woodward indifferently agreed.

Four years after his introduction to science—to books—Stuart Schreiber had entered the holy city. Immoderately, typically, he chose to set himself against perhaps its greatest living icon.

Claiming six U.S. presidents, thirty-three Nobel laureates, and twenty-five Pulitzer Prize–winners, Harvard competes only with itself in the manufacture of academic legends: Institutional narcissism is the weather in which Harvard egos move and grow. Yet few Cambridge personalities have ever been as encompassing, as jealously revered, as R. B. "Bob" Woodward. "What was Woodward like?" shrugs a colleague of more than forty years. "He was a genius." Says Schreiber: "He absolutely, completely overshadowed everyone else in the field. If Woodward walked into the lab and said, 'Cut off your arm,' you'd ask, 'Which one?'"

Woodward knew he was great: He'd always known. Arriving at MIT in 1933 at age sixteen, he already had taught himself more chemistry than the university expected of departmental majors. By his sophomore year, the faculty voted to give him his own laboratory and stipend and excuse him from attending classes. He responded by taking fifteen courses in a single semester and finishing his Ph.D. at age twenty. From there he went to Harvard, entering as a postdoc in the fall of 1937 and taking over Tishler's former lab on the third floor of Converse Hall. (The scene of Tishler's dramatic fire and rescue, it is now part of Schreiber's biology wing.)

"None of us thought he was really that great," Tishler would recall. "He had such great press. He was also sort of irritating." But Tishler and Woodward soon became close mutual admirers, united by an obsessive devotion to making molecules. Tishler wanted to prove that any compound that could be made could be made commercially; Woodward, that any molecule found in nature, no matter how complex, could be synthesized in the lab. They were kindred spirits whose informal synergy came to dominate synthetic organic chemistry for the next forty years.

"Max," observes Peter Jacobi, chairman of the chemistry department at Wesleyan, who worked with them both, "thought Woodward was the best chemist who ever lived." Surely he was among the most Promethean. In 1943, at the height of World War II and at age twenty-six, he synthesized quinine, helping to break Japan's stranglehold over natural supplies of the drug. Four years later, he again stunned the world by stringing amino acids together into proteinlike chains—"silk that doesn't come from worms, wool that doesn't come from sheep, and fur that doesn't come from any furbearing animal." Though he claimed not to emulate God—asked once if he hoped to synthesize life, he said, "No, I am quite happy with the way it is done now"—he deemed to match himself against nature at a higher level than any chemist before him.

In 1949, under a contract from Merck arranged by Tishler, Woodward undertook the total synthesis of cortisone. For five years, the company had struggled to produce the drug from ox bile with prohibitive results; it took the slaughter of forty head of cattle to treat one patient for one day by a process with forty-two chemical steps. Tishler ultimately would condense the synthesis to a more manageable—and profitable—twenty-six steps in what an admiring Woodward called the greatest feat in the history of commercial chemistry. But Tishler believed—and Woodward agreed—that if cortisone was to become universally available, it had to be made from a starting material more abundant and less gruesome to obtain than the precious fluids of bovines. As the *New York Times* reported, the competition to produce the drug from another source had become "the greatest international race in modern chemistry in which the world's top chemists have participated."

Woodward, still in his early thirties and released the previous year by Harvard from teaching so he could work on the effort full-time, was relentless. Chemists had learned to make steroids, large multiring compounds, from other steroids, but no one had yet made one from scratch, the way nature does, by assembling them from individual atoms. Using a derivative of coal tar, Woodward learned how to direct groups of atoms to reshuffle, like dancers at a reel, into the precise shapes he was seeking. The result wasn't cortisone, but another steroid equal to Merck's ox bile derivative for staging cortisone's final synthesis. At twenty chemical steps and made from cheap, clean, abundant materials, Woodward's steroid represented a breakthrough of stunning scientific and commercial appeal, auguring the availability not only of cortisone but of the entire family of human hormones.

The announcement of Woodward's discovery in April 1951, at the height of Merck's crisis in supplying the drug to a cortisone-starved world, was incandescent. "The achievement was hailed . . . as 'one of the greatest in the history of chemistry' . . . of 'incalculable importance' to millions of sufferers from rheumatoid arthritis, rheumatic fever, burns, blindness-causing eye diseases, and a host of other chronic ills, as well as for the future welfare of humanity," the *Times* reported on page one. Woodward himself was less breathless. "We have not yet synthesized cortisone," he said. "I don't know how many operations will be required before we get cortisone nor how long it will take. It may even be impossible."

In fact, Woodward was right; the discovery was less epochal than it appeared. Synthetic chemists were enthralled, but they no longer defined what was important in the life sciences. The far bigger story that spring belonged to Linus Pauling, a biochemist at the California Institute of Technology who had, after fifteen years of work, published a series of papers that would affect the future of organic chemistry substantially more than Woodward's synthesis. Pauling, a puckish, brilliant experimentalist who would go on to become one of only three people ever to win two Nobel Prizes, one in chemistry and one for his disarmament work, had determined the essential rules by which proteins folded.

Pauling's discovery, like Watson and Crick's a year later of the structure of DNA, inverted the hierarchy of science. For seventy-five years, reseachers had been baffled about whether proteins conformed to discrete shapes and whether those shapes determined their activity. Pauling not only answered with a resounding yes, but detailed all the major motifs by which they were formed. Instantly, as chronicler Horace Freeland Judson has observed, the *structure* of molecules—not their chemical composition—became "the central and most productive question of modern chemistry." In structure lay function, and what a molecule did determined its importance. Synthetic chemistry, which didn't explain molecular behavior but only reproduced it, fell in stature.

More disappointments were to follow. Tishler had hoped that the Merck–Woodward collaboration on cortisone would cement a long-term alliance, but though Tishler and Woodward valued their connection, others were less comfortable. Woodward eventually arranged to consult with Pfizer, one of Merck's chief competitors. "This broke my heart," Tishler would say almost forty years later. "We had a problem. There were one or two people who wouldn't tolerate bringing him in because they thought he'd take over." Worse, in July, less than three months after Woodward's announcement, Syntex, a little-known drug company with laboratories in Mexico City, disclosed that it had succeeded in making cortisone from a wild, inedible Mexican yam. The process was cheaper than Woodward's and Tishler's combined syntheses, and Syntex quickly went on to become the world's largest maker of cortisone and other hormones despite Merck's earlier lead. Salting the wound, the average age of Syntex's chemists on the project was twenty-seven. Woodward, like many prodigies, had exulted in his youth, saying that most synthetic chemists were through by age thirty-five. Now he was thirty-four—two years older than Schreiber would be when Schreiber was recruited to Harvard—and had so increased expectations about what molecules could be made that other chemists were beginning to surpass him.

Woodward continued to make more and more complex molecules—chorophyll, lysergic acid, strychnine. ("If we don't make strychnine," he vowed characteristically, "we'll *take* strychnine!") In 1965, he won the Nobel Prize in chemistry, then in 1972 went on to synthesize vitamin B_{12}, the most complicated synthesis up until that time and a feat so extraordinary that many chemists believed that he, like Pauling, would eventually win a second Nobel Prize for it. That he didn't only seemed to drive him harder to expunge the stigma of having won just one.

Woodward labored toweringly. He seldom slept more than a few hours a night. He reviled wholesomeness, which he blamed for most of the problems of others, and smoked and drank with ferocious abandon. "We generated three axioms about Woodward," a former postdoc would write. "He never got drunk, he never got tired, and he never perspired." To his students, he was a demigod, and they treated him so by bearing him half-solemnly to the podium for lectures in a blue sedan chair emblazoned with his initials.

But Woodward in a sense performed too well. He had shown that nearly any organic molecule could be made in the lab and had pioneered the methods for making them. Yet after that, each succeeding synthesis, however bold technically, could only make a narrower point. Science, like galaxies, grows fastest at its fringes, and two areas where divergent fields overlapped and shaded into each other—biophysics and molecular biology, the fields pioneered by Pauling, Watson and Crick, and others—had begun by the early 1950s to offer far more powerful means for understanding how molecules behave and why. Woodward had staked his frontier, conquered it, and stayed there, bringing synthetic chemistry to the center of the scientific world. Yet by the late 1970s, when Schreiber arrived in his lab, the greatest advances were coming almost entirely from someplace else.

At first, Schreiber didn't notice the change. His newfound belief in himself and in making molecules was so exhilarating, his exaltation of Harvard, as its Mecca, so profound, that he dismissed everything else. "I felt sorry for anyone who wasn't doing synthetic organic chemistry because they were missing out on what was so obviously the most important science," he says. He was particularly contemptuous of biology.

Woodward by now had withdrawn almost entirely from his students, leaving them to direct their own projects. He still held late-night poker games, but Schreiber wasn't interested in playing and so got little insight into his mentor. What he got, he explains, was an invisible hand instilling in him the confidence to take on any problem and the princely feeling of being part of an elect. "You felt very good about yourself in the Woodward group," he says. "You felt it was rubbing off on you. People would do anything—anything—to achieve something that would capture his attention."

Schreiber had joined Woodward in the fall of 1977, just as Boger was finishing his doctorate across the courtyard in Jeremy Knowles's enzymology lab. Twenty-two months later, midway through Schreiber's Ph.D., Woodward, then sixty-two, died suddenly of a heart attack at his home. Schreiber, perhaps understandably, was less remorseful about Woodward, whom he barely knew, than about himself. He had been progressing boldly, inexorably, and now his future was clouded. In fact, Woodward's death could hardly have been more fortunate for him. It freed Schreiber to establish his own credentials as a chemist. Invited by another professor to finish his dissertation in his group, Schreiber ended up publishing, while still a graduate student, two prestigious papers on which he was sole author. Not even Woodward had escaped the necessity of a postdoc, yet Schreiber, finishing his Ph.D. in just three and a half years, was fully credentialed. Pursued hotly by other universities, he began plotting his next move.

"There were some discussions about [my] staying *here*," he says. "I was invited at any point in time I wanted to, to interview, which would mean have a small group discuss my research proposals, go out to lunch, and make a decision. But everyone's advice was 'It's best for you to experience a different environment, and maybe you'll come back.' " As Schreiber understood, there was a subtext to this friendly encouragement. In the 120-year history of the department, only Woodward had started as an assistant professor and gotten tenure. As in Tishler's time, Harvard maintained a maddeningly self-important, albeit unwritten, policy of granting full professorships to a vanishingly small percentage of its junior faculty. The other members of the department may have felt they were doing Schreiber a favor, and perhaps accelerating his return, by exiling him.

Schreiber, full of himself, dismissed their concern. "I never once thought about tenure. It was absolutely inevitable that I would take care of that quickly." Nonetheless, he says, "Their advice turned out to be right on the mark. I went to Yale and experienced something I hadn't anticipated, that I know I couldn't have experienced here. Here I was a graduate student [who] had done very well and I would have made a transition to assistant professor. . . . Instead, I went to Yale, and I sensed that everyone there felt about me that I was the future. I was instantly a member of the faculty."

Like Woodward when he came to Harvard from MIT four decades earlier, Schreiber was preceded in the move by annoyingly good press and a reputation for egotism; his career was in near vertical ascent and he knew it. And like Woodward, he was looking for molecules on which he could advance science and his career simultaneously. The most significant of these turned out to be a compound called periplanone-B, a synthetic aphrodisiac for cockroaches. For billions of years, virgin female roaches have been sending males into a benighted sexual frenzy by emitting chemicals called pheromones. Recognizing the possibility of using such substances to lure the insects into traps treated with insecticide, scientists had been trying for decades to isolate enough of one of the compounds to test it. In one famous attempt, a Dutch professor raised and dissected 75,000 virgin females over a seven-year period, netting just 200 micrograms—200 millionths of a gram—of active pheromone. Clearly, only by synthesis could the material—and a potential breakthrough in the multibillion-dollar pesticide business—be tested.

Schreiber, as ever, was totally dedicated to the task. For two and a half years he and an assistant worked eighteen-hour days to build a synthetic pheromone—a large molecule with a characteristically daunting ten-membered ring. The project produced endless sneering within the neo-Gothic corridors of Yale's Sterling Laboratory, although Schreiber's wife, Mimi, regarded it seriously enough to insist that he wash his hands carefully so as not to bring home a trail of cockroaches. Finally, on Christmas Eve 1983, the work was completed. Schreiber's compound was so potent that several femtograms—several quadrillionths of a gram—were enough to send a half dozen male cockroaches into an orgy of sexual self-devastation. The insects immediately stood on their back legs and started flapping their wings frantically. Fifteen seconds later, they had broken antennae, gnawed legs, and tattered wings—and apparently no further appetite for arousal. "It is easy to see that they are suffering from severe sexual fatigue," Schreiber observed dryly.

Coming during the darkest days of World War II, Woodward's first great synthesis—quinine—was accorded high moral and national purpose, though Woodward, characteristically, cared more about the science. Schreiber's announcement of periplanone-B, despite its potential for protecting foodstocks in cockroach-infested regions of the Third World, was made during a far more commercial era and in a year—Orwell's 1984—in which scientific motives were particularly suspect. Schreiber was ridiculed. Along with John DeLorean, Louis Farrakhan, Bob Guccione and Michael Jackson, he was singled out by *Esquire* in its annual Dubious Achievement awards for creating "a dating service for cockroaches." "Sex and Roach at Yale" quipped the *Times* editorial page.

But Schreiber emerged from the episode knowing now just what he wanted and where he was going. Periplanone-B attracted male cockroaches by targeting a receptor within their nervous systems. It had tremendous biological activity, which Schreiber had witnessed as the male roaches writhed frantically, flapping their wings like stuck chickens. No longer satisfied simply to make molecules, Schreiber now determined to study their interaction with protein receptors. He had ignored biology assiduously. Now he would study the biological consequences of chemical events. Woodward's dominance of synthetic organic chemistry had created, in the end, an untenable situation for his students: They could never hope to match his example. Schreiber, infused with ambition, resolved to fulfill the terms of his own rise by launching himself anew.

Crossing into deeper scientific waters, Schreiber was emulating perhaps the noblest fraternity in science: those apostates who have departed, and exceeded, their host disciplines by plumbing what French microbiologist Louis Pasteur called "the mysteries of life and death." Ehrlich, Pauling—both had been chemists before the lure of biology drew them to cross the double-yellow separating academic fields. Pasteur himself was a chemist and avid crystal grower when, as a consultant to the French wine industry in Strasbourg in the 1850s, his studies in fermentation led him to discover microbes: He later posed—and proved—the germ theory of disease, ushering medicine into the modern age. "Fortune," he said famously, "favors the prepared mind."

Boger, several years older than Schreiber and recently chartered by Merck to head his own drug design group, was making a similar transition. His renin work had catapulted him upward within Merck and attracted the attention of other chemists. Yet if renin had proved anything, it was the value of using a structural basis for designing drugs and the necessity of including protein chemistry and biology in an overall strategy that Boger himself could control. Ever since his years with Knowles, whose work with enzymes bridged several disciplines, Boger had aspired to do swarming, multidisciplinary research. But he'd never been able to act on it until now.

Between 1985 and 1987, he was given charge not only of immunologists and biologists but of protein chemists and X-ray crystallographers. Like Schreiber, he, too, was suddenly in a position to move beyond making molecules to a more open, tumultuous, and visible scientific realm. He, too, considered biology "too mushy." ("I mean, what are the basic concepts of biology and how sure are we of them?" he would say. "Well, there aren't any, hardly. It isn't that the people are stupid, it's that the data isn't there.") Consequently, he intended to bring it new rigor through resurgent chemistry. Though his goal was to discover drugs and Schreiber's to use synthetic molecules as biological reagents, they were now on the same path, chasing the same objective, the same role.

Cyclosporine simultaneously focused them and brought them together. Boger saw the molecule as a jumping-off point for a program in immunosuppression, an area in which Merck was weak and which he'd inherited. "The immunology and biology groups I worked with at Rahway had all kinds of exploratory biology going. I said, 'That's all very interesting, but there's nothing for me to do here. There's something for me to do here in cyclosporine. It's a drug.' " Schreiber, typically, was intrigued for another reason: "It was interesting to me from the point of view of molecular recognition . . . I wasn't interested in the fact that this was a very useful drug." With Merck collaborating with Yale, Boger and Schreiber—who'd known of each other since they were both at Harvard and had met during Boger's recruiting trips to New Haven—now began seeking each other out. Not as personally fond of each other as Woodward and Tishler, they were united by something just as potent: immense, unrelenting self-interest.

Fusing Boger and Schreiber even more tightly than cyclosporine were its frustrations. Neither of them was set up to accomplish what he hoped. Boger didn't have his own molecular biologists and so depended on Yale for protein. "We were trying to do it all by collaboration," he says. "It wasn't going fast enough, but I knew what I had to do." Schreiber, meanwhile, made some molecules based on the structure of cyclosporine, but they were "doomed to failure. . . . The geometric shapes we were aiming for were all inappropriate. We needed protein chemistry." Throughout 1986 and early 1987 Boger and Schreiber independently began not only to face the same problems, but to draw the same conclusions. Each foresaw a self-contained, interdisciplinary, structure-based research effort—a kind of project-derived institute—with himself in complete control.

Scientists, curiously, talk a lot about luck. As murderously as they work, as dedicated as they are to rigor, as much as they may believe in their own perfection, they concede that great scientific careers are almost always favored by something else: great timing or an unseen hand connecting the observer and the observed. Pasteur's oft-used remark about fortune encapsulates the view, almost universally shared among scientists, especially in the drug industry, that they'd rather be lucky than good. The inevitable counterpoint is that its best to be both lucky and good.

Boger and Schreiber have been not only lucky but charmed throughout their careers. Much of what accounts for scientific good fortune is looking at the right problem, and both have a knack for positioning themselves in highly productive areas. Part of that is simple, competitive intelligence gathering, a desire to push and stay ahead. Schreiber, for instance, subscribes to a service that reviews all European patent applications, which issue ahead of those in the United States and thus are a frequent first indication of new molecules. Boger complements his own "rabid" reading by foraging widely among an array of data banks and computerized searches. Information junkies, both of them are always looking for anything to give themselves a fresh edge. Now, in mid-1987, both suddenly saw something obscure on the horizon that hit them equally with the force of a revelation: FK-506.

Like college roommates who fall in love with the same woman on the same day, each insists he noticed the molecule first—Schreiber among his European patent searches, Boger in a copy of Ochiai's slides from the August 1986 meeting in Helsinki. Far more significant, besides such niggling, is how they understood it. They each saw, ahead of most others, a potential bonanza almost solely on the basis of the molecule's shape. Boger: "I knew what I wasn't looking for. I wasn't looking for a tricycle, heterocycle ring, flat molecule with five nitrogens in it—that's just ugly. FK-506 was a beautiful molecule. It fit everything I was looking for." "This is too good a story to leave alone," Schreiber told himself after noting the similarity to rapamycin. "This is too remarkable."

Each leapt without hesitation. One obvious first question—What did FK-506 bind to?—led to the quasi-tie in discovering FKBP. Even more critical for Boger and Schreiber, given their previous frustrations, were organizational issues: personnel, lab space, reagents, logistics, deployment. Now that each knew exactly how he wanted to approach a major new molecule and had such a molecule in hand, speed became everything. Drug companies have long argued they can marshall resources quickly that academic scientists must assemble painstakingly through onerous, often messy, collaborations. Merck's narrow victory with FKBP seemed to prove that. Yet as Boger and Schreiber began building their respective programs, the opposite seemed true: Schreiber's incorporation of multiple disciplines into his lab, combined with Merck's huge size and lumbering organization, seemed to favor Schreiber. "I immediately began to think how I was going to wind down cyclosporine and wind this up without having to crash cyclosporine," recalls Boger, who for the first time began to doubt whether Merck was "set up to generate the kind of information I was going to need."

Boger and Schreiber each now had the molecule he'd been seeking. Each knew exactly what he wanted to do and had the support of a great institution. But each still lacked what he most wanted and needed: complete control. As long as Boger remained at Merck, he would have to fight to get the people and reagents he needed when he needed them. At Yale, Schreiber would never have a broad enough group to dispense with most collaborations. In the end, the result was much the same: in Boger's phrase, "not enough horsepower."

Within eighteen months, each of them was gone. Schreiber leapt first. Called to Harvard by Derek Bok one morning while shaving, he quit Yale and returned to Cambridge in the fall of 1988. Four months later, Boger, enticed by Kinsella, established Vertex. Like exiled princes suddenly in power, they quickly allied themselves. Kinsella, at Boger's insistence, recruited Schreiber for Vertex's SAB. Schreiber, for his part, was enthralled: "I thought [Boger] was a spectacular choice. I was impressed that Kinsella was willing to aim that high." By mid-1989, the two were aggressively playing badminton together at a backyard barbeque at Boger's new outsized suburban Colonial and toasting an alliance that anyone familiar with the field could only find fearsome.

Drawn across the terrain of science to a common meeting place, they now stood so close that their shadows intertwined. And yet they, and seemingly everyone around them, overlooked the obvious. Boger and Schreiber had engineered a storybook collaboration—Magic Johnson and Michael Jordan in the same backcourt—but the competition between them was innate, overarching, inevitable. They had wanted the same thing and, having gotten it, soon realized that there was room for only one of them on the solitary plateau where they were heading next.

They had created—less than two miles from each other—competing programs as much like one another as they were unique from anyone else's. Indeed, to hear Schreiber talk about his new labs—the "Schreiber Institute," Vertex scientists called it—was eerily like listening to Boger discuss Vertex. "We can do synthetic chemistry, make molecules," Schreiber says. "We can use those to purify receptors with the protein biochemistry effort, use molecular biology to clone the genes, overexpress the proteins. We have the genes themselves available for transfection into mammalian cells. . . .

"It's really," he says, "completely circular now."

Completely circular, and a mirror image of Vertex except for one notable omission: X-ray crystallography. It was the one piece of armamentaria that Schreiber, expecting to work with Navia, had failed to include—the one imperative, Boger would say, he still hadn't stolen.

Schreiber moved aggressively to bridge the gap. Having decided to pursue FKBP, he knew that he couldn't concede the crystal structure to others, least of all to Vertex, which, despite what he thought of it, had still fired him. A high resolution X-ray structure was a necessity for anyone aspiring to broad dominance in the field. On the other hand, since Schreiber had been spurned, it wouldn't hurt for him to have a less visible collaboration, one that wouldn't attract the attention of the Cambridge rumor mill.

Within days of receiving Boger's "letter bomb" discontinuing his contract, he called Jon Clardy, a highly regarded crystallographer at Cornell. He and Clardy had spoken before about the enzyme, and Clardy, who had never solved the structure of a protein before, had wanted to work on this one.

Although Schreiber's contract with Vertex extended through the end of 1990, Schreiber and Boger were now also formally rivals. Boger, of course, knew nothing of Schreiber's calculations. He could only guess how Schreiber would act. Indeed he had hoped that the stern language in Schreiber's contract combined with the fact that Schreiber was still a major shareholder in Vertex would cause him to hesitate. But to think he would hold back now defied everything Boger knew about him.

Their split was total, inexorable, and resonated with more history than either of them knew. Forty years earlier Tishler and Woodward, the best academic chemist and the best drugmaker of their generation, epoch-making Harvard men, teamed up to make the most spectacular molecule of their time: cortisone. Encapsulating the most productive era in the history of drug research, they opened up new avenues not only in chemistry and drug development, but throughout medicine. Ironically, bitterly, their collaboration signaled its own demise and that of their field.

Boger's and Schreiber's own collaboration on FKBP had promised multiple redemptions. As heirs of Tishler's and Woodward's legacy, they would fulfill it by improving on perhaps the most tantalizing biological molecule since cortisone, FK-506. More, they would do it by catapulting research to its next—and perhaps ultimate—stage while reenthroning chemistry at its core. Yet this was a different age. Just as cortisone and the other triumphs of the 1940s and early 1950s produced a high-water mark for collaborative research, they also set in motion—ironically, by providing the world with powerful new molecules—an entrepreneurial era in science that made collaboration more and more difficult. Competition now ruled and overruled. Ideas of fraternal vengeance, of sons atoning for fathers, of bringing history full circle, were subsumed by the exigencies of winning.

Now that their collaboration was dead, Schreiber had less trouble than Boger dissecting their shadows. His prominence in immunophilins, plus the natural tendency—and need—for academics to be recognized, had always given him a higher profile. Starzl, for instance, with whom Schreiber also now began collaborating, thought Vertex was "Schreiber's company"—a not uncommon understanding. Schreiber had clearly never needed Vertex as it had needed him, and though Boger would say the company's interest was purely in Schreiber's marquee value, Schreiber believed it was Boger's wounded ego that forced their estrangement. "With young start-up companies," Schreiber would say, "it's very important that they establish their own identities." He dismissed the suggestion that what Boger and the scientists considered his inability to keep quiet about the company's secrets had anything to do with it.

Freed in his mind, if not contractually, from any further obligation to the company, Schreiber swiftly rededicated himself. "Science is all that matters," he said cooly. "They got themselves in this situation by getting into in a very competitive area.

"I just hope they didn't think that anyone would ever slow down for them."

CHAPTER TWELVE

Boger needed no lectures, least of all from Schreiber. For most of the past year he had concentrated on finding a corporate partner. He had sold Vertex's story in a down market by projecting confidently what the company could and would do. Now, the goals were scientific and more stringent by nature. Immunophilins research had reached a steady boil. Besides Merck, Glaxo and Sandoz, the obvious front-runners, nearly every major U.S. and European drug company now had a program in the area, and dozens of top academic scientists from every field were stampeding either to collaborate or compete with Schreiber, Starzl, and the other early leaders. With a perspicacious Chugai closely marking the performance of its newest U.S. asset, and show-me pharmaceutical and financial critics fervently waiting for Boger to deliver—or better yet, as some at Merck hoped, fail—he needed to do competitive science. Vertex needed to make a drug.

All of the company's senior scientists including Boger had worked on promising leads before, but not one of the molecules had, strictly speaking, become a drug. "A compound may be brilliantly designed—everything absolutely rational," explains a former Merck vice president, showing something of the company's bias toward pills, "but until that compound has been shown not only to do clinically what you want it to do, but to be safe, to be active orally, to stay around in the body, and not to give you nightmares, it's not a drug." New drugs are exceedingly rare; novel ones still rarer. Of the hundreds of thousands of new compounds tested each year by the pharmaceutical industry, only about thirty are eventually approved by the FDA. Only three or four of those, like cyclosporine, actually do something new by working through a new pathway or (like FK-506, which was still not approved) suggest new uses because of their potency. Most others are variants of existing molecules. Like sperm cells, which are launched with great fanfare by the hundreds of millions and of which only one or two, if any, reach their goal, potential drug molecules may be expected to swim mightily for a while, but that's it. That no one at Vertex had ever "had a drug" wasn't especially alarming—many people in the industry work their entire careers without one—though it worried some of the scientists, who wondered if they weren't being overly presumptuous.

Boger, of course, not only intended to prove that Vertex could make drugs but drugs that were in all ways superior to those of the world's pharmaceutical leaders. Besides working exquisitely, Vertex's molecules were to be prototypes for a more rational system of drug discovery and a death knell for screening. The competition with Schreiber was thus for him neither the big picture nor the big prize, however much it might rankle him. Boger had set himself not against an icon like Woodward but against the best-run, most admired scientific company in the world. When Boger had left Merck twenty-two months earlier, it had perhaps forty scientists working on FK-506. In time, that number would probably reach one hundred. That, far more than Schreiber, impelled him now.

A year after the labs had opened, Vertex's science by mid-fall 1990 was operating almost in full. There were groups making molecules and tearing molecules apart. There were researchers synthesizing genes and screening gene libraries, extracting proteins, growing them, crystallizing them, chasing their structures, simulating hypothetical structures on computers. There were people modeling molecular interactions and writing new software to try to design drug molecules from scratch. Each week the company tested dozens of new compounds to see how they bound to FKBP, whether they inhibited protein folding and whether they stopped, through any of a half-dozen mechanisms, T cells from proliferating in test tubes. There was an animal pharmacology lab where promising molecules were pumped into mice with skin grafts on their foot pads, a simple model for testing rejection.

Drug development is an iterative process, a backbreaking multiyear series of assays aimed at funneling molecules upward through an evolutionary gauntlet. At the lowest level, molecules are tested for chemical affinity to a target, usually a protein. Those that are successful are examined for biochemical activity. If they're active—if they change the functioning of a target—they go into cells. Those that somehow affect cells without simply killing them in a toxic onslaught are tested in mice, then rats, then rabbits, and so on, through dogs and primates and up to humans. Each compound's progress is adjudged, especially in the later stages, by its "therapeutic index"—a kind of cost–benefit review whereby medicinal rewards are weighed against toxicity. Eventually, Boger planned for Vertex to do all such testing itself. Now, however, the company was focused entirely on discovery. It was looking for new molecules that showed the potential—with perhaps a decade of testing, $200 million, endless tweaking, and the ever-present threat that it all might simply explode—to become a drug.

In August, Boger had promised Chugai that Vertex would have its own compounds that were as potent as cyclosporine in cell assays by the end of the year. It was an exhorbitant leap—a hundredfold increase in activity from Vertex's best molecules—and Boger's pledge made the chemists, whose job it was to make the compounds, shudder with misgiving. "We've got until Christmas to do what Sandoz hasn't been able to do in twelve years," Dave Armistead grumbled.

Armistead was the lead chemist on the project, a surprisingly thoughtful, gravel-voiced Virginian who had come to Vertex via Yale and Merck and whose soft-spokenness belied a macho sensibility toward science. Armistead likes to synthesize "ball-buster molecules." At age thirty-four, he power-lifts several nights a week after work, a discipline that has paid off with a bulging chest and arms like dock ropes. With piercing blue eyes, high ruddy cheekbones, and a spike of brown hair, he combines the feral, imperturbable air of a Native American brave with the rowdy menace of a Brian Bosworth, the football player. And yet though generally fearless about work and as much a natural chemist perhaps as Schreiber, he found Boger's arrogant predictions disturbing. Making new molecules is one thing; picking up two orders of magnitude of biological activity in four months is another. Armistead knew what Boger was asking, and he didn't share Boger's confidence that it could be done.

As de facto head of Vertex's chemistry effort, Armistead objected strongly to what he saw as two equally disturbing precedents. He thought that by setting expectations too high Boger was inviting Chugai to exercise an escape clause in its contract, a clause that tied its payments to Vertex's making "adequate" scientific progress. Why incite disappointment? Armistead wondered. He was also discouraged for philosophical reasons about the demands of corporate oversight. "Things are going to change now that we have someone to answer to," he said. "It's going to be a lot more like working for Merck. We're going to be making reports, and reports need results." Like Boger, Armistead believed that research was distorted when scientists were made to account for their efforts. "People at big companies are of the mind-set that science is just turning the crank," he says. "They want to be able to stand up at the end of the quarter and say, 'We've turned out these 200 compounds, and though they're all dead, we worked hard.' They're interested in generating data points. That's what's going to earn them a good rating."

But Armistead was far more troubled by the scientific issues. Despite Boger's statement to reporters on the day of the Chugai signing that Vertex "likes to go into projects where we are very sure of the biochemistry," the company was still, eighteen months into the immunophilins program, sure of disturbingly little. The scientists still didn't know how FK-506 worked and whether FKBP was its relevant target within the body. That no one else did either was scant consolation, as no one else claimed to need the information as badly or made such a point of having it. Immunosuppression was still a mystery with very few molecular clues. And though scientists at Vertex and elsewhere were trying to solve it, the chemists had nothing but their own hit-or-miss assay results to use in coming up with new compounds. They had no structural information to draw against: no lock, just other keys. Worse, they weren't sure what they were trying to make. They didn't know, for instance, whether blocking the protein folding action of FKPB had anything at all to do with suppressing the immune system or whether FK-506-like molecules weren't simply all toxic.

Such uncertainies were dark clouds fulminating on Vertex's horizon; indeed, one had already begun to break loose. There was now a gathering body of evidence that the protein folding hypothesis that Boger and everyone else had seized upon eighteen months earlier was incorrect, a red herring. Schreiber had raised the first public doubts a year earlier with his 506BD, which showed that, by itself, the binding portion of FK-506 didn't block the enzyme from working: In cells, 506BD was a dead compound. He postulated that the business-end of FK-506 must therefore be the portion that sticks out into space—the "effector domain," he called it. The implication was that those atoms mingled with something else, another "partner" protein, and that that was what accounted for the drug's activity. Boger disputed Schreiber's conclusion; he thought it simplistic. Indeed, Vertex's own assay results had raised a similar specter sometime earlier. By the time of the Chugai signing, the company had made several molecules that were substantially smaller than FK-506 and inhibited the enzyme activity of FKBP just as well but that were feeble immunosuppressants. Either the molecules weren't getting to their targets, or they were getting there but were irrelevant. In any case, they were far from being drugs. "We have more inherent potency than we have potency in cells," Boger warned the scientists. "We're at the point now where I would encourage chemistry to start being very careful of the questions it asks."

Addressing those questions without benefit of knowing which atoms on the molecule bound to FKBP or even whether inhibiting the enzyme was a meaningful goal left Armistead and the other chemists to practice the very science they abhorred: raw medicinal chemistry. In other words, trial and error: Make a compound, assay it, adjust it subtly to make another molecule, test that second molecule, compare how they behave, make a third molecule that maximizes the best features of the first two, test that, and so on. To the chemists, few results could have been more frustrating—or embarrassing. Not only did they not have the information that they needed to design drugs and that Boger had promised them, they were having to compete against Merck and the other large companies in exactly that area where the traditional drugmakers were most dominant: their ability to generate and test thousands of compounds. If medicinal chemistry was essentially a prerational, "monkeys with typewriters" approach, as Boger liked to say, the chances of producing a *Macbeth* strongly favored the firm with the most, and most experienced, monkeys. Vertex had five chemists on the project, with a couple of decades of pharmaceutical experience and not a drug among them.

Armistead and the chemists wanted intensely to be careful about the kinds of questions they asked, but in truth they had no choice. Until Navia and Yamashita solved the structure of FKBP and could show them how their molecules bound to it, and until the biologists, mainly Harding and his group, identified the true molecular target, they were flying blind. Contrary to Boger's drumbeating, they weren't doing structure-based design. They were doing its opposite, perilously outflanked and outnumbered.

This, Armistead believed, was the hoariest feature of Boger's pledge to Chugai. Armistead admired Boger greatly. He hadn't lost faith in him. But he was a realist, and Boger notwithstanding, Vertex's chemistry effort was up against a sheer, potentially ruinous, and deeply humiliating wall. "Nobody's going to buy an inhibitor of FKBP," he said. "We've got to figure out what else you have to do that will throw the final switch and light up the cell assays. That's what FK-506 and cyclosporine do.

"If we don't have that, we're just jerking off."

What they had, what they'd always had, was the structure of FK-506.

Born of a soil fungus, isolated from a beery black broth, purified and deposited in a culture bank in Japan where it was sealed away from competitors, synthesized in a pair of internecine tong wars, deconstructed by X-ray crystallographers, then suddenly soaring to become best supporting actor in *Science*'s 1989 Molecule of the Year issue, the compound was self-championing. Whatever quarrels there were about the importance of FKBP, there was none about FK-506. It was a drug, and it was a drug because of its architecture, its shape.

From the moment Armistead saw the structure of FK-506 three years before, he knew he wanted to synthesize it. It was exactly the kind of "big, sexy, macho molecule" that quickens the blood of synthetic organic chemists and that he had just had his first major success with as a Yale postdoc. That was October 1987, near the time of the crash on Wall Street, and Armistead had just been hired by Boger to join his nascent structure-based design group at Merck. Boger had little interest in the race to be first to make the molecule, but to redesign it he needed to know how it was assembled. Armistead leapt at the opportunity: "You've got some guy that flashes you this very provocative molecule and you know this isn't something you're going to put together in two to three weeks. This is going to be a major challenge. You knew you were going to get press for it."

There are various approaches for reproducing natural compounds, but the one that appealed most to Boger and Armistead, for different reasons, was a "convergent" synthesis. Says Armistead: "Boger thought—and I still believe this—that the inherent problems with toxicity and bioavailability [how efficiently a molecule performs within the body] with FK-506 were part of the structure of the molecule. Those problems aren't going to be solved, I don't think, by running a chemical reaction on FK-506 and taking that product and making a drug out of it. What he wanted to do—this was prior to the identification of FKBP, so there was no receptor to study—was a synthesis made of four major building blocks, so that you could take these four pieces and learn how to glue them together. Once you did that, you could take one of these pieces, make drastic changes to it, glue it all back together again, and see what came out of the process. The idea was to have a synthesis that literally converged on the total synthesis: We would build the components, glue them together, and have FK-506."

To Armistead, who would be trying to make the molecule, a late-stage convergence was preferable for another reason. "From an efficiency standpoint, if you've got sixty steps, and you try to do them in a row, you're never going to get there. A linear synthesis wouldn't work. You needed a convergent synthesis. Coming together as late as possible was the key."

From the outset, Boger's plans ran counter to both Merck's culture and Merck's needs. His charter, supported by the company's top executives, was to develop a new way to discover drugs, but Merck's success was based on generating large numbers of new compounds. A lengthy, uncertain, total synthesis meant that Merck would be left to sit idly in a prime area where the stakes were likely to run into billions of dollars and where its competitors were all charging ahead. "Middle management hated it. Hated it!" Armistead says. "It was not a speedy way to make analogs. As soon as Boger left, they chopped the nuts off of this and started doing analog work."

As Boger found out, there was another issue: His group wasn't the only one at Merck trying to make FK-506. Within the drug industry, there are two rival groups of chemists: medicinal chemists, who try to invent new drugs, and process chemists, who work on devising new, cheaper ways of producing them. Max Tishler, patron saint of process chemists, elevated the art tremendously, but at Merck and elsewhere, process chemists still struggle for respect. Their work is thought to be dull, unimaginative, unglamorous. Merck's process chemistry group, seeing in FK-506 the same opportunities that Armistead did, also had launched a major effort to synthesize it. A similar duel was meanwhile developing at Yale, where Schreiber and Sam Danishevsky, his department chairman (in whose lab Armistead did his postdoc), were also competing to make the molecule. Typically, if not surprisingly, the intramural competition at Merck was just as secretive and ruthless as Merck's competition with the Yale groups and others outside the firm. "They wouldn't tell us anything," Armistead recalls. "I'm over here banging my head against the wall, trying to solve a problem that's been solved in the building next door, and they're getting the same paycheck that I am. It was an insanely competitive situation."

Merck's process group narrowly won the race to synthesize FK-506, beating Armistead and his collaborators, who immediately dropped the project. Completed in the fall of 1988, the group's structure attracted international attention and dealt Schreiber, who eventually came in second, a small but significant blow. Then, in December, Boger announced he was leaving the company. Armistead, "very, very disappointed," was also shocked. He'd never heard of anyone quitting a major drug company to start his own firm, much less a fast-tracker like Boger. Historically, biomedical start-ups were launched by molecular biologists interested in producing proteins and other large biomolecules. They competed with each other in markets that were small and untested. But Boger was a chemist. He was starting a drug company, not a biotech company, which meant he would be competing not with other small companies but behemoths like Merck and Glaxo, with their stables of billion-dollar drugs. Armistead echoed the skepticism that prevailed at Merck at the time. "You're going to go out and start a *drug* company," he said. "I thought it was ridiculous."

But Armistead was intrigued enough to want to hear more—an attitude, he recalls, not shared by others in Boger's group. "In a big bureaucratic situation, there's a lot of rewards just for being there, for tenure," Armistead says. "Boger rewarded for performance. A lot of people thought his leaving was a big opportunity for them. He gathered us all together to tell us he was leaving and I remember one person saying, 'There is a God after all,' and walking out."

Now, at Vertex, a different sort of internal competition drove the FK-506 chemistry effort. When Armistead was weighing whether to leave Merck for Vertex, he enlisted his closest friend from graduate school, Jeff Saunders, then a chemist at Squibb. "I figured if I was going to jump off this diving board, I was going to take somebody with me," he says. Saunders, the son and grandson of chemists, was born to the trade. His careful thinking and commitment to doing his own benchwork impressed Boger, who, taking him to dinner, instantly offered him a job. (Says Saunders: "I wasn't unhappy until Boger told me I was.") To Armistead and Saunders it seemed ideal: Ever since they were students, they'd discussed starting a business, a custom chemical company or, reflecting another shared passion, a wine store. This was better: It had all the upside—the opportunity to work together and get rich—with little of the risk. Saunders, keener, wirier, with a diamond stud earring, was quieter than Armistead, less blustery and intuitive, more modest, but with a similar air of freelance bravado. Like brothers, the two were inveterate rivals.

Armistead and Saunders's friendly competition set the tone for Vertex's chemistry program. Working across from each other at the bench and at adjoining hoods, they attacked neighboring sections of the molecule—Saunders, an unusual pairing of oxygen atoms on one side; Armistead, a sugarlike ring adjacent to it. They weren't trying to design drugs, but to identify those portions of the molecule most involved in binding and inhibiting enzyme activity. They wanted to see how small a molecule they could engineer that would retain FK-506's potency in those areas, while changing the structure enough to make it patentable. As with Boger's original notion of a "scaffold redesign," each hoped not just to make a better FK-506, but a new one, one that nature itself might have made if it had intended the molecule to block the action of FKBP inside the human body instead of performing some unexplained role in the life of a fungus.

Chemists "run reactions." Because molecules are collections of atoms arranged according to physical principles that are well understood, knowing which reactions to run and how to run them define the chemist's craft. Even as recently as Woodward's time, a synthetic organic chemist often had to discern the optimal conditions and make the reagents himself before he (the overwhelming majority of organic chemists are men) attempted a new reaction. Now, however, after 150 years in which millions of facts have been amassed about hundreds of thousands of individual compounds, chemists tend to believe they can make a molecule into any shape they want from standardized parts. They're more like builders than architects, drawing their ideas from catalogs the size of big city phone books.

As Vertex began making molecules, almost all its early successes belonged to Armistead, who quickly found a substitute for the sugarlike ring. Marveled Saunders: "It was probably a half hour in Aldrich [a well-known chemical supply book]: 'What's available? What can I buy? What will it do for me?' " Saunders, meanwhile, was becalmed. Nothing he did seemed to work. He was running as many reactions as Armistead but couldn't make the molecules he wanted. In almost a year of twelve-hour days—of coming in on weekends when only he, Thomson, Yamashita, and Murcko were routinely in the labs—he submitted only four new compounds. None of them was successful.

"It bothered me a lot," he recalled, "not only because of the image it was presenting, but I wasn't producing. . . . It was annoying, and it got to be frustrating and embarrassing. It threw me off the pace. After six or seven months, I got so caught up with it that I was ready to drop it, but Boger said, 'No, don't drop this. This is worth doing.' "

Armistead also defended Saunders. They shared ideas, hung out together after work, drank, got into noisy arguments. In the lab, they razzed each other like mechanics at adjacent bays. But Armistead was no solace; he was cruising. In the company's formative days, he was establishing himself as one of its leaders. To the extent that Vertex had a fast track or any track at all within the swirling atmosphere of Boger's social experiment, Armistead was on it, traveling with Boger to Japan, representing chemistry at key meetings, speaking for the project. Saunders reacted painfully. "Dave's one of the best chemists I've known," he said. "Not the smartest. Not the best read. Not the easiest to get along with. But as far as being productive, there's no question. Unfortunately, I'm compared to him."

Saunders's confidence was shaken and he had sharp pangs of envy over Armistead's success, but the two never had a falling out as Saunders and Thomson had. They were rivals over status and position, but tied by an unspoken *lingua familiaris.* Armistead, bigger and stronger, seemed the competent and protective older sibling; Saunders, more sensitive, struggled erratically, quietly, trying to find himself in Armistead's wake. Indeed, it may have been as much for Saunders as for himself that Armistead was now angered by the impositions of the Chugai deal: Vertex was about to become a less forgiving place, particularly in chemistry. He feared what impact the new production demands would have on his unsettled friend.

Saunders viewed Boger's promise to have cyclosporinelike molecules in cells by the end of the year with an added dread. Not only did he, like Armistead, think it would be extremely hard to do, he felt he had the burden of a lost year to account for. He had barely made any contribution and now he was being expected to emerge suddenly from his slump and help make an almost mystical gain in activity.

And standing next to him, shoulder to shoulder, overshadowing him, was the ubiquitous, conspicuously more muscular, and occasionally goading specter of his closest friend.

"We're a lot alike," Saunders said of himself and Armistead one night after work. "I'll never deny that he's my nemesis, though.

"Dave's . . . Dave's always there."

Jon Moore sat—his beetle brow furrowed, square Lebanese-American face flushed, and football player's shoulders heaved forward—like a New Age keyboardist at a pale-green and beige seven-foot console, alternately adjusting dials and typing instructions into an unseen computer. An auditorium full of such instruments and it would look like Mission Control in Houston, but Moore sat alone, in half-light, in a narrow room dominated at the far end by an enormous, gleaming stainless-steel cylinder lit by a single overhead spotlight. The cylinder housed an electromagnet so powerful that it tugged at the electrons on Yamashita's computer screen across the hall, blurring line after line of electronic type. Gangs of thick gray cables snaked across the carpet, connecting the device to Moore's panel and crossing a swath of yellow and black tape that marked where it was safe to walk without wiping out one's magnetically encoded credit cards or destroying one's watch. Moore habitually removed his wallet and timepiece and tossed them on the console before going to work.

Like Yamashita and Navia, Moore is a biophysicist—"kind of a weird undergraduate major because you don't really know what it is until you're a senior," he says. Like them, he was hired by Boger, at age thirty-three, to solve protein structures by a method neither as evolved nor as fashionable as X-ray crystallography. Moore is a proton nuclear magnetic resonance (NMR) spectroscoper—to crystallographers like Navia and Yamashita what a radiologist using magnet resonance imaging (MRI) was, perhaps five years ago, to those favoring more established computerized axial tomographic (CAT) scans. NMR spectroscopers and crystallographers both produce pictures of an unseen world by analyzing subatomic activity, and the competition between them is fierce, with NMR threatening to overtake crystallography, if not entirely, enough to make the rivalry uncomfortable. Like CAT scanners and MRI jocks, NMR experts and crystallographers outdo each other in pointing out the flaws in one another's work.

NMR is based on the theory of resonance. Like electrons, the nuclei of certain atoms, particularly hydrogen atoms, are considered to spin like tops. Put them in an electromagnetic field and they align themselves like so many bar magnets, resonating, or spinning, all at about the same frequency. Thus, for instance, there are about 1600 hydrogen atoms in FKBP, each with a positively charged nucleus called a proton, which, when placed in a glass sample tube and dropped into the maw of the electromagnet in Vertex's NMR lab, spin at approximately 500 megahertz, or 500 million cycles per second—the same frequency as the machine.

If all the protons resonated at exactly the same rate, it would be impossible to distinguish them. However, their spins vary ever so slightly, by about one five-hundredth of 1 percent. A hydrogen atom may, for instance, share an electron cloud with a larger, more muscular carbon atom, which exerts a drag on its spin. Or it may be near, in the roller-coaster structure of a protein, other protons that speed it up. By plotting the subtle shifts between protons, biophysicists like Moore try to figure out the location of each hydrogen atom in a molecule—a thousand points of light, so to speak, or a thousand distant stars. Knowing the chemical structure of the molecule, they then can map out its overall architecture by using the points as landmarks. (MRI, NMR's med-tech cousin, operates much the same way. Patients are immobilized and slid, like hot dogs on buns, into giant electromagnets that set every proton in their bodies whirring. Because water molecules, which have two protons each, resonate differently in different cellular environments, the technology makes it possible to identify, say, incipient cancer cells deep within tissue.)

Moore had gotten his first Thomson Unit of pure FKBP in August, several months after Navia and Yamashita. It was a month after he had arrived at Vertex, and the delay was unsurprising: Crystallography, having solved far more protein structures than NMR, is traditionally accorded such priority. Indeed, using NMR to solve the structures of large, hydrogen-rich molecules like proteins was still a new art, and Boger's decision to pursue Moore, one of its rising stars, something of a long shot. FKBP, with more than one hundred amino acids, was just beyond the point at which proteins were thought to be too large and complex to be solved by NMR. On the other hand, Boger knew the limitations of crystallography—it was useless without hard-to-grow crystals—and Moore had once solved the structure of a protein with more than ninety amino acids, the largest protein ever to be solved by NMR up to that time. Intense, brash, competitive, with the determination of a nose tackle, the squat, powerfully built Moore represented to Boger not only the opportunity to exploit a wide open and extraordinarily promising new field, but a backstop should Navia and Yamashita fail.

For Moore, the decision to work for a company was equally venturesome and uncertain. After getting his Ph.D. from the University of Pennsylvania, where he also went as an undergraduate, and doing two postdocs, he had expected to start out as an assistant professor, launching an independent research career. He'd recently received two prestigious offers—one from Florida State University, a leading center of magnetic research, the other from the Brookhaven National Laboratory on Long Island. But looking at the landscape of academic research, Moore recoiled. The more he thought about being an assistant professor and what it entailed—uncertain funding, chronic job insecurity, the need to sacrifice research in order to teach, dependence on better-known collaborators—the more appealing Vertex looked. He was particularly attracted by the chance, practically impossible for someone at his level outside of industry, to plunge directly into a hot area.

"In academia, you can't tread on the big guys' toes," he said. "It would be difficult for me to take on someone like Stuart Schreiber without the resources of a Vertex. I mean, think how long it would have taken to isolate protein on my own or with a graduate student, or even in collaboration with someone who might not have the expertise of a Thomson. You hear this scenario all the time: The guy you're collaborating with has one graduate student making protein, but she isn't showing up much in the lab. Stuff like that frustrates academic people to no end. You end up trying to find some little project where there's no competition and you hope you can squeak by without getting crushed."

If Navia and Yamashita could only guess at their competition,

Moore had known even before he accepted Boger's job offer that his main rivals would be Schreiber's group. NMR was one of the developing technologies that Schreiber had also brought early on into his lab, and as early as mid-March, a graduate student named Mark Rosen had begun a first gamut of experiments with a batch of Schreiber's enzyme. Rosen had never solved a protein structure before, and Boger had made much of his inexperience in recruiting Moore to Vertex and convincing him that Schreiber could be beaten. Still, with Schreiber's luck as a talisman, Rosen and a colleague had advanced quickly. By September, just as Moore was beginning his work in earnest, they had already located about 700 of FKBP's 1600 protons—more than enough to deduce a structure. In August, by assembling snatches of structural information, they had seen on their computer screens the first rough images of what researchers call the *gross folding topology* of the protein—its outer shape. Like the reconstruction of an ancient piece of pottery from smashed and buried shards, NMR structures are assembled bit by bit, and Schreiber's group was still months away at the least from having a refined structure. But they were several months ahead of Moore, who, having no idea of their progress, now set out alone to catch up with them.

"If you've never done a structure before, getting from assignments to a structure is a ton of work. If you have done a structure before and have the right software, you can cut a lot of corners and do it much faster," he said.

"Joshua did a very good job of making me think that 'Ah, Schreiber's never done a structure before. They'll never get anywhere on it.' Whether he knew that or not, I don't know, but his job was to sell me. I took it as a challenge. I motivated myself thinking about Schreiber."

Navia and Yamashita had no idea that Schreiber, along with Clardy, was also now competing with them directly, though even if they had it wouldn't have propelled them any faster. They already had their demon: Merck.

Scientists, unlike, say, athletes, conjure their own competition. They seldom know precisely whom they're competing with or where they are in the race. They hear things—rumors, reports—but the information comes at a distance and with a price: It may be exaggerated, deliberately misleading, or simply false. The atmosphere in a lab in the throes of a heated project is insular, secretive, xenophobic, superheated, and paranoid. Like the crew of a submarine in enemy waters, its scientists grimly hunker down at their stations, sweating anxiously, listening, awaiting the next depth charge. Even Boger, normally cavalier, now picked up each week's *Science* and *Nature* with trepidation, expecting to read something that would suddenly and irremediably derail him.

With crystallography, he had reason for concern. Merck had been rumored to have crystals as much as four or five months before Vertex. More, the scientist who grew them, a young former associate of Navia's named Brian McKeever, was well known to be one of the field's true wizards. It was McKeever who'd produced the crystals of HIV protease that Navia had used to solve its structure, the one scientist Boger had failed to recruit in his tong war with Rahway. Now, throughout the fall, as Navia and Yamashita tried to grow better crystals of FKBP than the ones that had first diffracted marginally in July, and as Yamashita sought to derive from an enormous mass of data the first glimmerings of a structure, McKeever's specter was constantly with them.

Once scientists crystallize a protein they can eventually solve its structure, but getting from one to the other requires several translations in data. Yamashita by now had collected millions of pieces of data about FKBP, skeins of numbers that gave him the map coordinates for each atom on each molecule of protein as it was frozen in space. However, by itself such information was useless. None of the atoms was prominent enough to stand out against the others. He couldn't tell where one molecule ended and another began. It was all one endless, undifferentiated mass. To interpret his numbers, Yamashita needed some identifying mark, a solid, stationary point from which to proceed.

Commonly, scientists overcome this hurdle by growing crystals that incorporate bigger, bulkier atoms—so called heavy-atom derivatives. Imagine, as author Horace Freeland Judson has proposed, the crystal lattice as "vulgar patterned wallpaper." Heavy atoms are distinguishing marks—the "tip of a rosebud, eye of a bird"—that show up at repeated intervals. Two or three such marks, overlaid, make it possible to discern the simple repeating pattern within the overall matrix, in other words, the outer dimension—the global shape—of a single molecule.

Navia had predicted that he and Yamashita would have heavy-atom derivatives of FKBP within two months after they first crystallized the protein; structure, a month later. However, there had been mounting delays. The first heavy atom they used—platinum—acted like a bowling ball dropped into a fissure of ice: The stress around the crack forced the crystals to break apart. With other atoms, it had been impossible to get them to bind to the molecules, or if they bound, not to change the shape of the protein. In an extraordinary and provocative piece of science, Thomson had managed to make FKBP unfold—so that it was just a linear chain of amino acids with all its atoms, including those normally buried deep inside, laid bare—then snap back into its original conformation, without losing activity. It was like a child's party blower: Thomson uncoiled the protein, laid in a heavy atom, then rolled it back up. Discouragingly, it still didn't provide Yamashita the help he needed.

Each time a heavy atom didn't work out, Yamashita had to try to grow new crystals under new conditions. He then had to collect an entire data set, which took about a week, before he could generate computer maps showing whether a derivative would work. The entire turnaround took about three weeks, and each failure plunged Yamashita into a wider desperation. He was working all the time, staying at Vertex most nights, napping for forty-five minutes or an hour on the floor by the X-ray generator before rousing himself with a cigarette and a Pepsi and hurling himself back into the numbing cycle of growing new crystals, collecting new data, crunching new numbers, making new maps, encountering new failures. As it had for Thomson before him, the world now receded, disappearing completely behind his obsession with his work. Like Thomson, he believed Vertex was depending on him utterly and that he couldn't fail. Feeling isolated, he also now began to see the world as increasingly conspiratorial.

Everywhere Yamashita saw rivals, and everywhere he fought them privately by subsuming himself in plots, morality plays, agonies. He was sure his principle opponent, Brian McKeever, would know how to overcome the problems with heavy atoms and that Merck was therefore unbeatable. This made him distrust Boger, who had told him Merck couldn't win, and Navia, whom he blamed for much of his misery. As Yamashita saw it, Navia was gravely ambitious. Certainly, he was impatient. Pushing hard for results, he often was too agressive with crystals, destroying them in the process of his experiments and leaving Yamashita to repeat the work. He used a lot of protein and, when they got in his way, could be careless with people, especially John Thomson. Yamashita believed Navia was determined to take credit for all crystal structures solved at Vertex, regardless of his role, and thus he—Yamashita—would never be able to advance within the company or his career no matter how slavishly he worked or brilliantly he performed. The months of failure with heavy atoms—months during which Navia was often away, attending NIH site visits and shuttle launches and business meetings, or was focused on other projects—left Yamashita in charge of a struggling program, but with no ultimate authority or hope of reward. Meanwhile, he was eyeing Jon Moore, whose progress he also feared. Losing to Moore, though not as ruinous as being beaten by McKeever or oppressed by Navia, would still deprive Yamashita of his great opportunity to be the first to solve the structure of FKBP. Swept up in this competitive torrent, Yamashita lashed himself more furiously.

For Yamashita, in this state, emotions became another kind of enemy, and he fought them just as hard. Ever since his earlier disappointments with growing the first protein crystals, he had affected a steely resolve. "The highest level of feeling I'm coping with right now is hunger," he'd said during that period. Now, throughout the fall, whenever someone asked how he was doing, he answered in a low, dulcet voice: "I'm stable." Thomson, who saw perhaps more of him than anyone else, questioned Yamashita's notion of stability and assessed him rather as *metastable*, the scientific term for that last moment of tenuous equilibrium before a system spins out of control and crashes.

When he wasn't acting dysfunctional, Yamashita was affecting and warm, laughing easily at his predicament and going out of his way to bolster others. He was particularly solicitous of Navia, who, despite his resenting and fearing, he also, in the Japanese tradition, honored and revered. Once, having to adjust the X-ray beam, he said: "I wouldn't want Manuel to do it. He's at an age where it would be unwise for him to spend too much more time at the machine." (Noting that experimental crystallographers risk leukemia as a result of heavy X-ray exposure, he later remarked: "You have to die sometime.")

If Yamashita dealt with his frustration by repressing it, Navia was sharp edged, volatile. More than his work with either FKBP or HIV, his first love now was a method he had devised for using enzyme crystals to speed up and improve chemical reactions. Early trials had been so promising that Boger had begun talking about spinning off a new company just to market the technique, which had tantalizing applications for both the chemical and drug industries. Having already solved a significant number of crystal structures, Navia was looking, like Schreiber and Boger, to widen his mark on science, and he saw himself potentially fathering a new technology with his observations about using enzyme crystals as biological agents. Others were less persuaded. Two consultants flown by Boger from England to review the work found the work commercially unfeasible. After one particularly brutal session, one of them concluded, "I wouldn't say it's been blown out of the water at this stage, but it's been in for a pretty serious dismemberment." Navia was distraught. Publicly, he welcomed the criticism, though others, especially Yamashita, began maintaining a respectful distance, knowing his tendency to show his disappointment in sudden and unpredictable ways.

Driven yet inseparable, Navia and Yamashita were drawn together more and more with comic intensity. Wary of each other's feelings and perhaps their own, they were gushingly polite toward each other, even as privately they seethed. "After you, Doctor," Yamashita would say, passing through a doorway, Alphonse to Navia's Gaston. "No, no, Doctor, after you," Navia insisted. Beyond the conflicting clichés of their birth—Navia, the tempestuous, hotblooded, Cuban-American; Yamashita, the inscrutable, stone-faced young Japanese-American—they were proving to be deeply incompatible. Despite their often genuine pleasure at being together and sincere mutual respect, something like hatred also now ran between them. Thomson, who liked them both but was becoming increasing annoyed with their relentless calls for more protein, now began referring to them in private as "Abbott and Costello."

In September, a rumor raced through Vertex that sent shock waves through the labs. It was that Merck, in combination with researchers at Yale, had produced the ultimate experimental result for determining FKBP's biological relevance, a so-called transgenic deletion, or gene knockout. Boger and the immunologists had long thought that the only way to determine absolutely whether blocking FKBP with FK-506 caused immunosuppression was to give the drug to an animal that didn't have the protein. If the animal's immunities were suppressed, it would prove conclusively that another receptor than FKBP was the drug's true target. Now, the scientists began hearing from several sources that the Merck/Yale team had engineered such an animal. By deleting the gene for FKBP from mice embryos, the researchers reportedly had been able to "invent" an FKBP-less species. The shock was that the animals were said to be just as susceptible to FK-506 as normal mice.

Though the rumor was unconfirmed—no such experiment was ever reported in the literature—the implications were thunderous. If FKBP wasn't the right target, then all Vertex's work was suddenly voided. Everything. What good were better inhibitors of a protein that was biologically irrelevant? Of what value was Thomson's Herculean extraction of the protein or Yamashita's dogged search for heavy atoms? Yamashita heard that Merck was shutting down its program in FK-506 and implicitly believed it. Boger, forced to quell Yamashita's distress along with everyone else's, correctly pointed out that without published evidence the rumor had to be treated as false although he, too, was concerned. "You wouldn't just hear that there was a paper out," he told a small group consisting of Navia, Yamashita, and Murcko. "You'd hear the sky tearing." Echoed Murcko: "The psychic cry of fifty anguished Ph.D.'s over the ether. It's very unlikely that anyone in the world has that kind of evidence and anything else is speculative."

Confronting the possibility that FKBP was irrelevant, beset by the problems with heavy atoms, assailed by the consultants over his immobilized enzymes, needing to go gingerly around Yamashita lest he snap and be gone, Navia was combustible. Witty and gracious one minute, he was explosive the next. On a day in early October, he slammed down the phone in the lunchroom after an escalating disagreement with a software vendor and launched into a tirade that even some of the others who'd grown accustomed to such outbursts found eye-opening.

"I'm stuck here until Friday with a pike up my ass while this marketing asshole is going to take until the fucking end of the week to tell me he can't give me what I need," he yelled, his Jesuit training squelched. "If it's a marketing problem, that's his fucking problem. . . . A bunch of goddamned used-car salesmen." He stamped the floor, kicked two chairs, and jerked a thirty-five-pound plastic jug into the water cooler. Then, just as suddenly, he caught himself, apologized, and, straightening his tie, headed for the men's room. "I'm going to stick my head under cold water so I don't have a complete fit," he announced contritely.

Yamashita, by comparison, seemed calm during these periods. He maintained the demeanor of the stringently controlled rational man. It was only after work, going out with Thomson and Laura Engle to drink, that his desperation poured out. He was working as hard as anyone at Vertex, as hard as Thomson had worked, and was making no headway. Returning to Vertex to sit through the night at his computer or at the hookup on his kitchen table, he still had no key for unlocking the numbers that would reveal the structure of the protein. He was bereft, exhausted. He decided to go to Hawaii to visit his parents for two weeks at Christmas, but as the date of his departure drew close, he still wasn't sure he could leave. On the day before, he worked through the night, then, coming home and showering, fell asleep for five minutes on his couch.

He recalled: "In my dream, I thought, 'I can't go away. Brian will beat me. I've got to cancel my reservations. I can't go. Brian will beat me.' "

CHAPTER THIRTEEN

Scientifically and financially, Vertex has three modes: project, protoproject, and what Boger calls "preprotoproject." Immunophilin research had been by necessity in full project mode from the minute Boger conceived it because it was all that he had. Its larger shortcomings—notably the question of whether Vertex had the right target—were blindly forgiven as Boger needed to identify quickly a field in which he could claim to have a lead. Businesswise, the strategy worked. Boger had been able to sell the project at a premium. Now, as he weighed the imperative of a second project, he knew it wouldn't be nearly so easy.

"This is the most important decision we'll make in the next twelve months," he told the scientists. "If we blow the second program, it could be fatal."

Prone to hyperbole, this time Boger wasn't exaggerating. Industrywide, fewer than one in ten programs yields a drug: No company, no matter how smart, can survive on a single project. Worse for start-ups, they are often marginalized to the most speculative, high-risk areas of business and research or sandwiched into niches. Small, emerging drug companies are like first-time horse players with gravid ambitions and dire means: They can't bet enough on favorites to win big so they try to parley intriguing long shots. Boger had wagered almost everything he had on immunophilins.

With the race still on, the windows were now closing again. He needed to place a bet.

Vertex's one protoproject—AIDS—had failed to move him. Navia's "hallucination," which started it, ended dismally, a bad trip. Two chemists, Roger Tung and Dave Deininger, spent four months making compounds based on his computer models only to find that they were dead—inactive—against HIV protease, the enzyme Navia predicted they would block. Meanwhile, a compound first made by the chemists to inhibit FKPB showed some inadvertent promise but not enough. Unimpressed, Boger had begun to think more about what HIV protease could do for Vertex than what Vertex could do to the enzyme. "We're not going to do a smoke and mirrors project again; it won't fly," he told the scientists at a meeting in September. "But there's the possibility that we could use HIV protease to field a larger aspartyl proteinase program. That makes it interesting."

That Vertex might use AIDS as a stalking horse to raise money to bootstrap other projects struck several of the scientists as cynical and disingenuous. Boger had no such qualms. Science is rarely, if ever, linear, and information isn't necessarily most useful in answering those questions that elicit it in the first place. Indeed, the brightest facet of the drug industry's on-again, off-again optimism about making AIDS drugs was its recent successes with related systems. Renin, the protein-cleaving enzyme on which Boger had done his career-making work at Merck, was a perfect example. No renin blocker had ever become a drug despite a decade of intensive research efforts at Merck and elsewhere. But trying to inhibit renin had prepared the industry for HIV protease, which is strikingly similar.

Boger: "One of the things Merck did by publishing and patenting early [with renin] was to greatly stimulate the search for variations of this model if only from a patent-busting point of view. Consequently, there's an enormous literature on renin. Yet the ultimate payoff for all of this might turn out to be HIV protease. There are groups all over the world, in all kinds of laboratories, all set up with a huge amount of methodology and experience, ready for HIV protease. If renin inhibitors never get developed, it'll be a perfect argument for how applied science is really basic science and how there's no distinction between them."

Boger's interest in developing HIV into a project had another impetus: Nissin. The giant Japanese noodlemaker with the tomb-like labs that Aldrich had likened to Dr. No's fortress was keen to "make a play" in AIDS. The company had set up a lab to screen compounds across the river in the biomedical Manhattan surrounding Harvard Medical School and had invited Vertex to submit promising molecules. In the conventional wisdom, it was extremely late in the business development cycle of the disease. AIDS was already crowded with players. Those looking for strategic alliances had long ago found them, and Vertex was perhaps a year from having done enough science to approach a bona fide drugmaker. But if Nissin, which sold $1 billion a year in ramen noodles, a company whose TV pitchman in Japan is Arnold Schwarzenegger, was interested in HIV protease, Boger was more than willing to craft a compelling story.

"What would nail it up and make that a chipshot," he told the scientists, was if Vertex could stop the proliferation of the virus in cells. "Of course, if we can do that, internally we'd all believe we'd have a renin inhibitor, too."

Around this time, Boger flew to San Francisco for a week-long conference on aspartyl proteases, the class of protein-splitting molecular scissors that includes renin and HIV protease and which he calls by the more formal proteinase. He seldom went to scientific meetings anymore unless he was asked to speak, but Vertex's interest in HIV promised to make the trip fruitful. "It's everybody," he told Aldrich. "I'll get a real snapshot of where everybody is."

"You going to do any business development schmoozing while you're there?" Aldrich asked.

"I don't have much to bring to the party."

"That never stopped us before."

"It just means I'll have to be more mysterious. I'll bring my veil."

In fact, the meeting was surprisingly productive. The overwhelming obstacle to making drugs that were protease blockers remained the orthodoxy that the best inhibitors were, far and away, all peptides, those bulky chains of amino acids easily dismembered by the gut. However, two large drug companies, Roche and Abbott, now challenged that dogma. They had made compounds that blocked HIV protease in the lab but that were less than strictly peptidal—"funky," Tung called them. Boger was intrigued. Calling Tung from the conference, he authorized him to begin making the molecules, to explore them as possible leads.

Even more tantalizing were reports of another protease that Boger saw as tailor-made for Vertex, cathepsin E. Researchers had just identified the enzyme as cleaving a key protein in the chemical pathway that causes high blood pressure. As Boger knew from long experience, hypertension was the world's largest, richest, and most competitive market for drugs. That market included some of the best drugs ever made, and yet because of its incomparable demographics (older, insured, mostly male, chronic, at risk, and concentrated in industrialized countries) had a seemingly insatiable appetite for new ones.

Here, from a business standpoint, if not scientifically, was the precise FK-506-like scenario that Boger had told the scientists not to expect again. If Vertex could prove, with a small, self-sustained effort, that cathepsin E was a relevant target for drugs and if it could make nonpeptidal protease inhibitors based on Abbott's and Roche's compounds, it might suddenly have a lead in an important new area of research that matched its position in immunophilins a year earlier.

Upon such leads, real and imputed, Boger knew big money was to be raised. He began thinking that with the right sort of deal, he not only could support cathepsin E, but bootstrap a protoproject in renin as well. He might even be able to float HIV, which was beginning to cost $3,000 per day, until that, too, got on its feet. Boger had vowed, as he had with AIDS, to cede aspartyl protease research to the big drug companies, but now the logic of bundling a major effort in the area became all but irresistible. In his mind, as he flew back to Cambridge, he quietly upgraded cathepsin E to a preprotoproject. Quickly, he began formulating how to sell it to the scientists.

"That's a religious question," Boger said, hoisting his size 13 brown oxfords onto a chair and tugging at his beard in a pose of mock introspection. "Is it possible to go faster than the speed of light? Yes. Are we going to do it in the next few years with the money we have? No. It's a question of resources. If you're asking me whether I think we can design a pill that will inhibit an aspartyl proteinase in a limited amount of time with a limited amount of money, the answer is, yes, I think so."

Murcko, the questioner, shrugged. He was exasperated. For months he had suspected Boger of dragging his feet on HIV. If Vertex was going to be working on it—and he was sure it would be—why wasn't the company attacking, now, before it fell even further behind? From his one-hundred-hour weeks of modeling inhibitors at West Point, Murcko knew the kind of resources Merck and the other big companies were bringing to the problem. It was obvious to him what Vertex needed to do. But Boger had refused to commit more than five people full time to the project. Murcko thought Boger's caution went deeper than mere prudence. He was trying to pin Boger down, excise from his calm demeanor some nodule of doubt that would explain his reluctance, get him to say that he thought the goal of the program was impossible and that he was looking for a nonautocratic way to end it. Boger, ever indeflectable, wasn't giving.

Boger came to this meeting, several weeks after his trip to California, with an agenda that, if not hidden, was less transparent than Murcko and the rest of the scientists in the room would have liked. A regular Wednesday morning meeting of the Aspartyl Protease Project Council, Vertex's working group on HIV, the session was intended to review the group's work and discuss its now three potential targets. Like all Vertex project council meetings, it was closed-door, with the casually dressed scientists sprawled around a table in the conference room. The room, barely larger than the table and a dozen or so chairs, is decorated functionally with two blond credenzas, a small whiteboard, and Chugai's Mount Fuji pictures, its telling gift. Boger, preferring "flat" management, created the councils to speed the flow of information while eliminating the need for middle managers, although Murcko, especially, suspected him of using them to handle the scientists.

Murcko thought Boger was trying to engineer the overthrow of HIV protease by cathepsin E. Boger would never admit it, he thought, but the AIDS project threatened him. Unlike FK-506, which Boger had initiated himself, HIV had been championed by the scientists over and above his resistance. Despite Boger's constant invitations for dissent, Murcko thought he was incapable of ceding control on such a crucial matter as choosing a project.

Boger's concerns, in fact, were less psychological. If he favored cathepsin E, it was only because Vertex's progress in AIDS had been too inconclusive for him to try to sell to a prospective partner. Once Vertex announced a project, as opposed to a protoproject, the company was committed long term: perhaps $5 to $10 million a year for five years; more, possibly much more, after that if the effort generated a drug candidate. Boger desperately needed to launch a second project soon but had yet to see anything in HIV protease to justify that kind of investment. As always, the icy short-term pressure of raising money subsumed the vagaries of morality and social good. Vertex wouldn't pick the project that would benefit the most people or target the most pressing need; no company would. It would pick the project with the best overall chance of success, which first meant having a good story to tell investors. In that arena, by late 1990, AIDS was a tough sell. Perhaps the toughest.

Murcko was not alone in trying to dislodge Boger. Roger Tung, another former Merck chemist, also now saw in HIV protease, if not a superb drug target for Vertex, a notable opportunity for himself. Tung, thirty-one, like Armistead, is unabashedly ambitious. Like him, he had jumped from a promising career track at Merck and was trying to regain, along the alien footing of Boger's social experiment, a similar one at Vertex. A "sort of first and a half generation, mixed background, American-Japanese and -Chinese," Tung also was motivated by a redemptive streak to equal Yamashita's. His mother's father, a Japanese trade liaison before World War II, lost everything and moved with his family to the United States; they were interned in Montana, where Tung's mother, now a writer of technical books, spent part of her childhood. His father's father was chairman of the board of the Bank of Hong Kong; his father the lead engineer on one of the most popular mainframe computers ever designed at IBM. As a boy growing up in upstate New York, Tung had watched his father falter in his own career because he wasn't assertive and because he didn't have a

Ph.D. "He hit the proverbial glass ceiling," Tung says. He was determined not to be so constrained himself.

Early on at Vertex Tung had come up against Armistead's predominance in immunophilins and it had frustrated him. "I'm a very headstrong person," he said. "I don't like the idea of being subordinate." Thus HIV became an escape hatch. For months he worked slavishly to make the Janssen molecules that had inspired Navia, not because he thought they'd work—he didn't—but as a "way of getting myself into a different area, where I wouldn't be jostling with Dave and where Dave was predestined to come out on top." But Tung's independence took its toll. Demanding and intense, he alienated most of the other chemists with his high-handedness. His coal black hair became filigreed with gray seemingly overnight as he labored over the bruising synthesis of the Abbott molecules. Though he considered fighting HIV "the most important problem right now," he had to defend a position he had deep reservations about, namely, that Vertex could actually design an AIDS drug. Tung is a skeptic—"I live to be proven wrong," he once said. Yet he had put himself in the contradictory position of saying he thought Vertex could do structure-based design in AIDS if only to advance his own ambitions. "I want to find out whether the science we espouse so freely actually works," he would say. "I want to know if we tell the truth."

Now, on the council, Tung joined Murcko in appealing to Boger's vainglorious pride. He pointed out that the Abbott compounds, which Boger favored, suggested that nonpeptidal HIV inhibitors might work as oral drugs. In the company lore, that meant that Vertex ought to be able to assemble ones that were even better. Boger—disingenuously perhaps, for he derided big companies as much as anyone in the room—responded that Abbott was a large, successful drugmaker and that having a molecule of such potency gave it a compelling lead, a lead that Vertex might be wise to respect.

"But they're not doing anything with it," Tung said.

"You don't know what they're doing with it," Boger snapped.

"But they've got a history, and it's not good," said Navia.

Cathepsin E, Boger pointed out alternatively, was virgin territory: Vertex would be starting out equal to the rest of the industry.

Boger obviously favored such a competitive position and said so. But he was swiftly rebuffed by several of the scientists, who noted correctly that the reason there was no competition in cathepsin E was that no one had proven that it was a drug target. Whatever HIV protease's difficulties, the prospect of becoming totally, embarrassingly, ruinously wrong wasn't one of them.

"I don't need to know the biological use of the enzyme to be excited about it as a project," Boger said. "I just have to know we're not chasing the rainbow.

"Listen, if I told you I had a harmless pill that, if you took it every day and had a heart attack, would reduce your chances of permanent heart damage by 40 percent, wouldn't you take it?"

"The FDA's never approved anything like that," Navia said.

"Yes, it has," Boger said. "Aspirin."

Of course, aspirin had been in use for more than a century and no one yet knew that a cathepsin E blocker would be effective, much less safe. But that didn't impede Boger's argument. He liked these exchanges, was good at them, saw them as healthy. He also liked riling the scientists, who, he thought, had become complacent about the inevitability of working in AIDS. Stirring the pot further, he now mentioned that besides being implicated in blood pressure diseases, cathepsin E was found to have a role in regulating the immune system: In other words, an inhibitor might not only help prevent heart damage, but help stop organ rejection and cure autoimmune diseases.

"An immunosuppressive hypertension agent?" Navia said, chortling absurdly.

"Why not?" Boger said. "One-stop shopping."

At this, Navia's incredulity shifted into annoyance. As Vertex's senior scientist and ranking voice within the pro-HIV group, he was the one person Boger deferred to and the one Boger often seemed the least willing to confront. Dismissing Boger's fictive cathepsin E inhibitor as "a drug looking for a disease" and questioning whether the enzyme's immune function was crucial or inadvertent, "a feature or a bug," he denounced the idea out of hand.

"The FDA hates multiindication drugs," he said. "One drug, one disease, that's how they like to do it."

Boger cut him off: "I don't think the FDA cares about anything unless an animal falls over and dies. The FDA doesn't sit with a biochemical flow chart, waiting to say, 'This could happen.' They're not that interested."

Not quite a stalemate, the meeting ended tersely: Vertex would bootstrap the two programs in parallel. Tung would continue working on the Abbott compounds, which they would send along with Vertex's own molecules to Nissin for testing, while the biologists tried to produce enough enzyme for Navia to crystallize. Meanwhile, Dave Livingston's enzymology group would begin the series of experiments to prove cathepsin E's legitimacy as a drug target.

Boger was pleased. He had managed to get cathepsin E onto the table without permanently alienating the HIV group. At the same time, he had sent them a signal that they had to move fast if they wanted the project to go forward. Having set in motion a nonauthoritarian approach to decision making, he now found himself more and more having to channel the insurrectionary impulses of the scientists. It was a difficult act to balance. He still thought AIDS could be disastrous for the company. On the other hand, he had allowed some of his best scientists to imagine themselves in charge of a major structure-based project in HIV, a project that Merck had launched amid exorbitant hopes and international acclaim, but on which it had been making little visible progress. As Aldrich put it, "We can tell a great story. We've got the Merck first team here."

"Six months from now," Boger announced, "we have to know which horse we're going to be riding."

"What's so magical about six months?" Navia asked, still piqued.

"Nothing's so magical. Eighteen months from now we're in trouble; six months we're not. But in October 1991, we run out of money unless we go back to the board. That's where the six months comes from. If we sit down in March with a prospective partner, there isn't a prayer that we'll have a deal closed even by November. But at least we'll know we have a front-burner contingency to raise some venture capital.

"June," he said ominously, "would leave us with an almost irrevocable decision either way."

It was mid-October, and Boger still thought it wasn't too late for

Vertex to enter either area. Two months later, with Tung still struggling with the Abbott compounds and Livingston mired in cathepsin E, he was less sure.

"I have this nightmare that I'm going to read in *Nature* the results of an experiment that we could have done that definitely proves cathepsin E's biological relevance," he said at a meeting of the project council. "At that point I'm just going to writhe on the ground and gurgle."

Late fall came and went with a depressing lack of progress, not only in chemistry and enzymology, but in all the labs. Failed experiments, equipment losses, recalcitrant vendors, unverifiable results, artifacts, squabbling, jealousy, frustration, impatience, rage—all took their toll. It was as if the company had a virus, a malaise, that worsened with each pass. Six months earlier Vertex had run a T-shirt design competition, won, amid suspicion that business had again hijacked science, by Aldrich. On the back, the shirts said: "We don't leave success to chance." Now, though some of the scientists still wore them to work, a darker sensibility prevailed. "I'd rather be lucky than smart," moaned Navia, only half-joking and still stuck in the search for heavy atoms. "What's our credo?" Thomson asked Moore, leading a familiar call and response. Grunted Moore, borrowing a response from Murcko: "Grief! Pain! Angst!"

They had all experienced slumps and dry spells before, but for a year Vertex had been so productive, so surpassingly successful, that a Bogerite exemption seemed to float down and lift them in its embrace. None of them was deluded into thinking they were doing rational drug design, not yet. But now the problems were harder. The scientists felt themselves moving farther into the unknown, which, as they knew, was unknown for a reason. More than a few feared it might be unknowable.

Boger, as always, led by example. He was upbeat, dauntless, equable, tireless, and expansive. Flipping between business and science, between existing work and planning ahead, he was in Zen field general mode, calmly and gregariously attending areas away from the main action that he was certain would be crucial later on.

Like Tishler, he kept "all the threads in his own hands" but seldom pulled on the ones that the scientists expected or wanted him to. For instance, although he had told Chugai that Vertex would have molecules that were immunosuppressive in cells by the end of the year, he put no pressure on the chemistry group. He stayed out of the labs completely, concentrating instead on making new slides or micromanaging the growth of the project councils. "I don't think there's a lot of understanding about what I'm trying to do organizationally," he said on a day when anxieties were racing about the general sluggishness in the labs. "I think everybody wants desperately for Mussolini to come in and run things. But I'm not going to do that. The big issue here is how I can evolve a radically different organizational structure into something that works. Believe me, if I turn out to be wrong, it'll take forty-five minutes to fix."

Part of this elusiveness was Boger's determination not to interfere with researchers he believed to be at the top of their fields. Part, too, was an appreciation for the rhythms of science, rhythms that in almost any other endeavor would be maddening, months and years of unrelieved failure punctuated, if one was fortunate, by sudden leaps ahead, and if one wasn't, by prospects of everlasting darkness. Patience and detachment were necessities. Boger had predicted clearly in September that Armistead's group would have cellular activity. He had even called the type of breakthrough, saying, "It'll go in a jump. It won't creep down 5 percent at a time." Having done so, he then stayed out of the way, keeping things, as he said, "indeterminate." The emotional rigors of maintaining such a calm distance amid soaring stakes and widening failures unnerved some of the scientists, but Boger loved them, especially in periods like this one when nothing seemed to go right. "It's got to be hard," he said in December as the inertia in the labs seemed to thicken. "It's got to be a little scary or I lose interest."

Individually, the scientists would have been happier with a few breaks and some help. If Boger thought that by setting them wheeling together in the right motion Vertex would accomplish great things, most of them were up against deadlines and competitive pressures or driven by company and career goals that made them think only of themselves.

They were strapped, oversubscribed. Harding, for instance, was now mainly responsible for determining FKBP's role in immunosuppression and whether or not it was the right target. It had long been assumed that FK-506, like cyclosporine and many other drugs, bound tightly to more than one protein within cells, that there were other FKBPs of varying sizes and shapes and that they, perhaps, not Harding's original protein, triggered immunosuppression. The search for new immunophilins had attracted a formidable field, and Harding, as codiscoverer of cyclophilin and FKBP, had a major stake in maintaining his position within it. Not only would discovering another FKBP, or, better yet, a family, further his priority, it would establish his independence from Schreiber, for whom the stakes were just as high or higher. There was also the patent income. Harding earned thousands of dollars in royalties every time Yale and Harvard licensed FKBP to a drug company. The money had become an important stopgap during the past two years when he and his wife, Robin, had moved from New Haven to Boston, buying a house in the suburbs at the height of the New England real estate boom. Robin had given up her job as a financial officer in a hospital to make the move, and the couple had a new baby.

Harding devised a series of experiments to try to pull related proteins out of Thomson's thymus extract with synthetic antibodies, but he found it nearly impossible to do them. He was busy screening compounds for cellular activity and helping Debra Peattie's group search for the gene that encodes the enzyme—everyone's work but his own. Typically, he was frustrated and angry with himself. Self-abrogating to a fault, he began behaving like a martyr, complaining, dithering at the bench. He was especially jealous of Schreiber, his erstwhile collaborator, who as an academic had no responsibilities other than to pursue his own interests, pander to his own goals.

"In biology, even in the best circumstances, if one day out of five is productive, you're doing well," he muttered. "I'm just not doing any science."

Structure.

Jon Moore glimpsed it initially in snatches, like the first faint shadings of a photograph plunged into a developing bath. A crook here. A hairpin there. Suddenly a straight run of three- or four-billionths of a meter. After weeks of sitting morosely in a darkened office next to Boger's or in an open area outside the modeling lab, wrestling with the positions of hydrogen nuclei, he began to develop a reasonable certainty about a prize sliver of spiral. Embedded in the blackness of his computer screen, it was, Moore thought, a stretch of helix near the beginning of FKBP's amino acid chain.

It was a few days before Christmas. Moore had the wan, subhuman cast of a cave dweller. He also was wildly encouraged and straining to check his optimism. The architecture of FKBP was beginning to tumble out before his eyes, and no one else had published it yet. Schreiber's minions still hadn't beaten him.

"Fantastic!" Navia said, entering from the lunchroom. Mired in the search for heavy atoms, he had found it irresistible to visit on occasion: Macy keeping an eye on Gimbel.

"Three strands and a 'bit o' helix,'" Moore reported.

"You may pull the fat out of the fire for our end," Navia said. "A low-resolution structure for us would be tremendously helpful. If you have a ribbon diagram [an overall map of a protein's fold], we can plug it in and short-circuit this heavy-atom crap."

Navia had begun to face the grim possibility that it might be years before he and Yamashita could solve the structure with the approach they were using. The prospect had made him consider other options. Recently, crystallographers at NIH and elsewhere had pioneered a new approach for orienting crystal data. Called *molecular replacement*, it involved using a piece of NMR structure of the same protein as a map, a guide. Navia, who would take any shortcut he could get at this point, saw in Moore's emerging images his best and perhaps last hope not to be shut out with FKBP. His self-imposed deadline of Thanksgiving had come and gone. He, like Thomson, had stuck his neck out, and heads had begun to wag.

"If we can pull this off, it'll be unique," he said, "It's never been done for real. It'll put an end to this X-ray versus NMR crap." Navia is as violently competitive as anyone at Vertex, though now—Moore wondered if it was for his benefit—he denounced the absurdity of scientific rivalry. "Two fucking methodologies that are absolutely complimentary, yet you hear people at meetings . . . they're infants."

Moore was noncommittal. He didn't have enough structure yet to be of use to the crystallographers, and even if he had, he wasn't sure whether what he was seeing on his screen was correct. He'd still located only about half the protons he needed for a complete diagram. He said nothing, and Navia quickly changed the subject and the tone.

He told Moore a story about being at a costume party and not being recognized.

"Gee," he said in character, exaggeratedly scratching his head. "Too bad Manuel missed the party."

Then, in another voice, he said, "Gee, Clark, Superman was here, but where were you? You missed him. How come whenever Superman's here you're not around?"

Navia loves to mock dumb wonderment. He was now hopping from foot to foot, flipping his tie and smiling widely. "Even as a kid, I thought, What's the matter with these people? They're so thick? Can't they see everytime Superman appears, Clark vanishes?"

Moore laughed, enjoying the diversion. Working so closely with crystallographers who were racing to solve the same structure as he was a new situation, one he was unlikely to have encountered at a university. It was awkward. He was pushing ahead with work that could make his career. Yet if he could help Navia and Yamashita, Vertex would benefit, even if he didn't. He decided to concentrate on his work and play it by ear.

Two days later, piecing together scraps of structure, Moore glimpsed for the first time a major piece of the molecule's backbone. It looked like the back of a baseball glove—a web of five parallel strands—and comprised about 40 percent of the enzyme. It was enough perhaps to try a molecular replacement experiment, but he nonetheless withheld it from Navia, and Navia didn't ask for it outright. Boger, assuming as always that the scientists would seek collaborations that were explicitly in their own interest, said nothing.

In her lab coat, worn most days over a dress with a Mickey Mouse pin on the lapel, Patsi Nelson looked like a mildly eccentric small-town pediatrician. At thirty-nine and a mother of three, she brought an effusive maternal selflessness to Vertex's craggy, egotistical emotional climate. She was warm in a place that's usually raging hot or ice cold.

Nelson is an immunologist and, by culture and training, a Californian. She came to Vertex after a postdoc at Stanford, a stint at the Scripps Institute in La Jolla, and several years at Gene Labs, a West Coast biotech company, where she worked in AIDS research. The job brought her in contact with infected blood and instilled in her a keen respect for the dangers of her work. "I always treat *everything* as if it's contaminated," she says.

Nelson's compassion was pronounced, and unlike Navia and the other men in the company, she didn't try to exorcise her feelings about her work. Many women, competing in a male-dominated world like science, adapt with an underground humor of their own. "He had two arms of research and I was one of them," Nobel laureate Gertrude Elion once said, speaking about co-prizewinner George Hitchens. Nelson took her science personally, felt its thrills and torments deeply, and assumed others did, too. "Oooh, I feel so bad for the chemists," she would say when their compounds turned up dead in her cell assays. Watching Harding struggle, she said sympathetically, "Poor Matt. I wish I could help him out."

Throughout the fall, the chemists delivered Nelson several new compounds a week, and, as Boger predicted, none was much better than the others. His promise to Chugai loomed ever larger, and Nelson worried for the chemists. She wanted desperately for their molecules to succeed and was clearly pained when they didn't. One group of molecules, submitted in late November, particularly distressed her: They inhibited FKBP as well as FK-506, but when she and Harding put them in with human T cells, they clumped and fell out of solution. It reminded Harding of one of those watery Christmas scenes that you shook up and then watched as the snow swirled gently to the bottom. Nelson was dismayed: The molecules weren't even getting inside the cells. Even if they were active, there was no way of knowing. Either way, they were useless.

In mid-December, Armistead submitted a compound he thought might overcome the gap between enzyme inhibition and cellular activity. By now he thought he had done all he could by making FK-506 smaller and was adding chains of atoms to those redesigned pieces of the core that were the company's best blockers. He was trying to occupy the same space as FK-506 by attaching different atomic subgroups, an effort akin to a sculptor's experimenting with different ligatures for the arms of a heroic statue. On one of the molecules—the 367th compound produced at Vertex and submitted for evaluation—he replaced the original crumpled horn of FK-506 with a symmetrical V shape. It was three carbon atoms long in each direction and had a ring of atoms at each tip. On paper, it looked like a crooked rabbit ears–type TV antenna or, as Boger elegantly suggested, a semaphore, one of those signals made, usually on the decks of ships, with outstretched flags. Armistead, eyeballing the distances, had gotten it out of the Aldrich catalog.

When Nelson got back the data, she was ecstatic. The molecule was one hundred times more potent than Vertex's previous best inhibitor, exactly the gain that Boger had predicted to Chugai and an extraordinary improvement. Tested several ways, it was just as active as cyclosporine. Yet such a huge leap in potency innately worried her. It could be an artifact, an error in her cell-counting equipment, anything. Normally she would wait and retest the compound the following week, but the magnitude of the improvement demanded immediate confirmation.

Each week Nelson got a delivery of T cells from a blood bank at a hospital that culled them from whole marrow, but it was closed until after the holidays. She couldn't wait. Showing Harding the data, he agreed that it was too important to sit on, and so he volunteered his own blood. "Poor Matt," she recalled. "I bled him. The first time I went through his vein, but he wouldn't let on how much it hurt. Finally, I got enough cells to test the compound again."

The results were identical. Nelson immediately went to her computer and composed a Christmas card for Boger. On the overleaf were two graphs showing V-367 and cyclosporine spiking, then declining in similar fashion.

"It's beautiful data," Boger exulted, "and the best news is that there are more compounds behind these that'll make 367 more active just by improving solubility. We've started a series that should enable the chemists to quickly bring up the potency by another factor of ten."

Boger was soaring. Everything was as he'd foreseen. In a little more than a year Vertex had generated novel druglike compounds. In three months, it had gone from having poor cellular inhibitors to having a molecule ready to test for immunosuppression in animals. Another jump like the last one and it would have molecules as potent as FK-506. And that was *before* having structural information about FKBP, which, with Moore's apparent success, would also now soon be available. It was all just as he had said it would be, all as he had sold it. Brandishing Nelson's Christmas card, with its simple line drawings of Vertex's compound and provocative data points, Boger said: "If this piece of paper fell into the wrong hands, I'd be worried for the next six months about somebody derivatizing our molecules and beating us. If we had structure data, I'd say, 'Go ahead, try.' We'd be so far ahead we'd be unstoppable."

The year thus ended as it had begun, as Boger had said it would, triumphantly, though now the triumph was less a matter of Boger's speculations and more of hard-won science and its attendant good luck. It had taken a "slugfest," as Boger called it, in chemistry to get the company to where it was. Yet now that it was there, the aura of invincibility that had seemed to disappear with the Chugai party suddenly returned. There were those who were still wounded: Navia and Yamashita, for instance, who smiled tautly when Armistead and Saunders began calling the crystallography lab "the chemist's smoking lounge." But even Thomson was resurgent. Recovering from his siege with FKBP and his pique at the depredations of the Chugai episode, he arrived the following night at Vertex's Christmas party in an outlandish double-breasted tuxedo and bearing a tray of elegantly prepared thymus fillets.

"Fuckin' A!" announced Armistead.

"Cyclosporineland!" Saunders said, brushing jealousy aside.

"We're cruising," said Aldrich, who knew the reward for scientific success was more, and more expensive, science, and thus the need for more money, more business development, more and bolder storytelling.

CHAPTER FOURTEEN

On the road together at night, Boger and Aldrich plotted strategy. Boger typically neglected to eat breakfast or lunch, and when they got to a restaurant finally at 9 P.M., after a day that started at dawn with their racing to catch a plane and often ended with a third repetition of his slide show, he was famished and voluble and frequently hoarse. Chewing zinc lozenges—Boger never takes cold remedies, only metals—he and Aldrich usually ordered fish. Then they stretched out, two pinstriped tyros in soft chairs, to organize and refine their mission. "Pulling money out of other people's pockets," Aldrich called it.

There was an inevitability to their discourse. Vertex needed to raise enough money to run perhaps two presidential campaigns. But neither of them wanted to go back to the VCs. The preferable route—the only route, Aldrich would say, as the company began to burn more and more cash—was to raise its value to the point where other, less privileged investors could be brought in.

This was the route he and Boger discussed. When the scientists got together they talked about science, competition, and women and what they would do when they got rich, but Boger and Aldrich talked voraciously, consummately about money: how to leverage it, what to trade for it, how to position Vertex to get more. There were deals, but deals were insufficient. The real money—the $200 million or more that Vertex would spend before it ever sold its first pill—had to come from the one place such sums were available: Wall Street.

Boger knew as much about the stock market as he did about everything else. He'd studied the development strategies of small publicly held companies, analyzed the fluctuations in their stock prices. What he hadn't garnered on his own, he cribbed from his brother Ken, who'd taken more than a dozen such companies public. Aldrich knew more about financial instruments, but on strategic matters they were equally adept and utterly in agreement.

"You always want to go public when the market thinks the time is right, not when you have to," Boger would say, Aldrich nodding in agreement.

Scarcely was that less of a consideration than now. In the absolutely best of circumstances, Vertex was still at least five years from marketing its first drug. In January, the cost of running the labs climbed to more than $30,000 a day, only about a third of which came from Chugai. That left $20,000 to be incinerated every twenty-four hours, 365 days a year, until 1996—a minimum bonfire of $35 to $40 million. Yet that figure was misleading. It assumed, first, that the company stopped growing: a suicidal assumption, since without launching several more projects soon, Vertex was staking its future entirely on a few molecules that might be immunosuppressive in mice and an AIDS project with less than a half-dozen disgruntled scientists. It also neglected the skewed economics of drug development, where 70 percent of the overall cost arises *after* a drug candidate is identified. As promising as it was, Vertex was still a deep, dark hole financially, strictly for investors who could stomach such emptiness. Who else would even be interested? Who, in a regulated market, could conscionably sanction a sale to anyone else? What underwriter would see a profit in one?

Poor as the timing for Vertex's raising large amounts of money seemed internally, it was worse elsewhere. Wall Street's monolithic indifference to small biomedical companies was nearly as suffocating now as it had been sixteen months earlier at the Vista. Nineteen ninety had been a miserable year, if not for biomedical stocks, for the markets in general. The Dow had its worst year in a decade. More, many big investors were in deep psychological retreat prior to the January 15, 1991, deadline for the Gulf War. Big drug firms had done well, as had some of the first generation of biotech companies, which now, after more than ten years and hundreds of millions of dollars, were finally bringing a few promising drugs to market. But such success was double-edged: The young companies were widely thought to be overvalued, and even their own executives had begun cashing out. In October, for instance, Benno Schmidt sold 10,000 shares of GI, the Harvard offshoot he'd help found and served as chairman. Then, in November, J. H. Whitney, Schmidt's venture capital company, unloaded another 41,000 shares of GI—22 percent of its stake. The biggest blow came in the first week of February, when Abbott Laboratories sold its entire 6.4 percent position in Amgen, the darling of the sector. If the smart money was leaving, who but suckers remained? Wall Street surveyed the new crop of junior biomedical companies—so long from paying out, so uncertain, so confusing—and saw only what Moore and the other scientists saw: pain, grief, angst. "Coming into the first part of the year," a veteran investment banker would recall, "we all thought we were going to slog through another sloppy recession year. The market just seemed pathetic." Said another: "We were wondering how we were going to fund this industry."

This was Vertex's unsparing dilemma: It was burning through capital at a startling rate, had no second project to sell, and its only option seemed to be to return to those investors who demanded the lowest valuations and exacted the highest price. Aldrich, unsurprisingly, was anxious, stern, jittery, and impatient, particularly with the scientists, who he thought were becoming lazy. It bothered him when he came in on weekends and the parking lot wasn't full, as it had been at Biogen at a similar stage. Worse, he felt, unlike a year ago, he had nothing new to promote and the scientists weren't giving him anything. Then, he'd said: "In this business I've sometimes had to sell technology that was difficult to believe in, sometimes involving weak patent positions or marginal science. Here that's not the case." But he never liked AIDS as a project, even though he'd pushed Boger hard to formalize it, and saw little hope, at least in the short run, for cathepsin E. He needed to rebait his hook.

Boger, as usual, failed to match Aldrich's distress. "It's too hard," he said equably, "to keep looking for things that might not be there." Science—data—would provide.

In early February, following a desultory month in the labs, the Immunophilins Project Council met to discuss publication of its work. Starzl and Fujisawa had recently announced the first international conference on FK-506 to be held in Pittsburgh at the end of August. Abstracts—brief descriptions of work to be presented—were due by the end of the month.

Clinically, Starzl still dominated FK-506's progress but not so totally as when only he and his surgeons had the drug. Nearly a dozen other medical centers in the United States and Europe had now begun randomized trials with discouraging results. Outside Starzl's hands, the drug was a tempest, a demon. Contrary to his reports, FK-506, like cyclosporine, seemed to go straight for the kidneys, causing acute renal failure in up to 25 percent of all patients. The FDA, inundated with crisis calls, had to install a hot line to assuage irate transplanters, many of whom felt that Starzl, Fujisawa, and the agency had misled them dangerously. Starzl, still maintaining that FK-506 produced only low-level nephrotoxicity, believed the error was in overdosing; using Fujisawa's protocols, the other centers were giving severalfold more of the drug than he and his group. The FDA, agreeing, was moving quietly to salvage the trials. Boger, expecting the conference might well become a brawl over safety, naturally welcomed the dispute. He'd always wanted FK-506 to succeed but never so soon or so well that it precluded the next generation of drugs.

For Vertex, the conference was a chance to, as Boger put it, "plant the flag." He knew that Merck and Schreiber, still the putative leaders in basic FK-506-related research, would use the conference to showcase new work and that it was crucial for Vertex to do the same. The company needed to show some muscle. "I'd like to see ten Vertex presentations," he said, purposely setting a higher number than any of the scientists thought possible.

Several of them blanched. Though critical to the company, publication was also fraught with professional risk. For instance, Thomson was preparing to disclose how he and his group extracted and purified FKBP—a prospect that horrified Navia. "Until we solve the structure," Navia snapped, "I don't want to tell people how we crystallized the protein. We'd be cutting our own throats."

Boger disagreed. To him, publication was a paradox. Scientists publish to announce new work and establish authorship. Within industry, they also publish to frustrate competitors, befog imitators, and titillate investors. Boger had made aggressive patenting, the publishing arena most critical to Vertex, a hallmark of its chemistry effort—"throwing barrels of nails off the back of the truck," he called it. But he thought the company risked little by promoting its other discoveries as early and prominently as possible. "As long as we don't put up compounds, we're not cutting our throats," he told Navia. "There's a way to do everything else.

"There's two situations where you want to be aggressive about publishing. One is when you're behind; the other is when you're ahead. When you're ahead, you demoralize the other guy. When you're behind, you have nothing to lose. It's when you don't know that you want to be most careful."

Boger knew that no other issue would be more volatile among the scientists. Socially, Vertex was still as fractious as a five-year-old's ego. As long as there had been no data to report or take credit for, it had been relatively easy to maintain a patina of unity. But now each scientist's desire to control the disclosure of his or her work—to keep proper score—was about to come in acid conflict with Boger's need to use Vertex's science for his own goals. That those goals were less than purely scientific and frequently so crass as simply to lure investors set Boger irremediably apart from them. In this one area he was as much a natural enemy as Aldrich, who coldly referred to the scientists' most exquisite insights, their noblest experiments, as "product." By rights, their science was in fact fodder. It belonged to the company, not themselves. But Boger knew he couldn't simply expropriate it as McDonald's does the exertions of its burger flippers. Disarming Navia, he was trying to quell an incipient mutiny before he needed to do so by stronger means.

Throughout January and February, Boger searched Wall Street's impenetrable geometry, looking for new angles and thinking that he recognized a powerful one in the sidelining of big institutional investors. Health portfolios, traditional high earners, were now as much as 40 percent liquid; this meant that fund managers, who make their commissions on quick growth and high turnovers, were miserably parking tens of billions of dollars in low-yield cash.

Those dollars, those fund managers, Boger thought, couldn't stay out of the market long: They'd be pent up, bored, itching to get back in. Like sunstruck surfers scouring the horizon on a calm day, they'd be susceptible to the least sign of a swell.

That needn't be Vertex itself. Boger knew that small biomedical companies rove through the financial world in virtual lockstep. When one bolted, they all did; when one tripped, the rest collided like automatons. Certainly it had been thus in 1987, when Genentech's disappointing sales of its much ballyhooed clot dissolver, TpA, dragged the entire biotech sector down 39 percent, plunging it into the near coma of the past three years. One or two successes among the industry's flagships, Boger knew, would have the opposite effect, jumpstarting a free-for-all that would rain money on everyone.

There were other interim routes. Wall Street's cold shoulder had led three small companies recently to escape the thrall of venture capitalists without going public. The companies, not much older than Vertex, had raised more than $70 million by selling stock to qualified investors—people deemed by the Securities and Exchange Commission (SEC) to have earned more than $200,000 for at least two years or to be worth more than $1 million. So-called private placements, these sales had been at higher valuations than either venture capitalists or institutional investors were willing to pay.

Wall Street's gloom was fickle, Boger knew. It could turn to rapture overnight. Spastic, mercurial, spectacularly irrational, its judgments were the very opposite of the analytic, information-based, sharply realized science he was trying to accomplish in Vertex's labs. And yet the investors were out there, reeling like the sun. Boger might disdain them, but he respected them implicitly.

On February 22, 1991, Amgen, for the past few years a bellwether, received approval for the first of a new class of genetically engineered drugs that boost the immune system's ability to fight infectious disease. The molecule was only the company's second drug in eleven years, yet it couldn't fail to arouse analysts and investors. It was being promoted for cancer patients, not to cure their cancers but to close a therapeutic loophole. Because chemotherapy and radiation kill white blood cells, the risk of secondary infections is one of the chief drawbacks to most conventional treatments. Used as an adjunct, Amgen's drug allowed doctors to push more aggressive therapies by raising the threshold of what patients could endure.

If ever a molecule was made for Wall Street, Amgen's GM-CSF (granulocyte macrophage colony-stimulating factor) was it. Piggybacked on the ever-growing cancer market, a short-circuit for costly hospital stays, and with a huge "off label" potential as an all-purpose immune booster—it could have sales, some analysts predicted, of up to $750 million a year. Wall Street had run Amgen's stock up more than 300 percent during the past twelve months in anticipation of the FDA action. Now it exploded.

Vertex was at the far end of the continuum from Amgen, but Boger could feel the tug of universally rising expectations as Amgen's stock price nudged $100. Wall Street bestowed great value on companies that finally made it to the finish line in the drug-approval process: Amgen's market valuation would soon rise to more than $4.5 billion. Even as its stock overheated the following week, Boger saw in the frenzy just the jolt that Wall Street and the industry had been waiting for.

He had little respect for Amgen's science—"Not bad for a few cloners," he said, echoing a remark he had made a year earlier when Genentech sold 60 percent of its stock to Swiss-owned Roche Holdings Ltd. for $2.1 billion. But the sudden lurch in momentum was unmistakable. Wall Street's iron facade was suddenly rent with new fissures, new angles of attack. Fresh money was beginning to pour in. Boger's calculations instantly began to reflect the change.

Schreiber, coveting not money but greatness, continued to publish intimidatingly. In January, he was sole author of a prestigious review article in *Science*—"Chemistry and Biology of the Immunophilins and their Immunosuppressive Ligands [high-affinity binders]," which, as its title suggests, sought to cover the world of cyclosporine and FK-506 under a single, encompassing imprint—his. Though it contained little new data, the article laid out what had become the central focus of Schreiber's research. He now believed that the drugs worked not by inhibiting protein folding but by interfering with a far more complex process called *signal transduction.*

Cells, biologists know, are composed of three regions: surface, nucleus, and cytoplasm, the aqueous space in between. Much has long been known about the nucleus, which contains DNA and most of the other machinery for controlling cellular reproduction. More recently, scientists have clarified much of what occurs on the cell's surface, where molecules "talk" with other molecules in their environment. For instance, with T cells, they know that when a foreign body presents itself, surface proteins identify its structure and relay the information to the nucleus. The nucleus, in return, designs antibodies that are assembled in the cytoplasm and dispatched to the surface to help repulse the intruder. This is how the body defends itself. What's not known—and what has been of surpassing interest to molecular immunologists—is how the messages about the exact shape of the interloper and, later, the blueprints for the antibodies are transmitted protein by protein in parallel and interlocking cascades through the cytoplasm. Thus *signalling* or signal transduction or, as Schreiber invariably now called it, "the black box of signal transduction."

Without clearly identified actors like DNA and surface receptors, immunologists looking inside the cytoplasm had become frustrated. They could make out bits and pieces of its biochemistry but were still far from understanding its overall design. Schreiber, however, believed he possessed several vital pieces of the puzzle. To the biomedical community at large, cyclosporine and FK-506 were significant primarily as drugs. But they were perhaps even more tantalizing as submicroscopic searchlights, as probes. By latching tightly to cyclophilin and FKBP, which are found not only in T cells, but in virtually every cell in the body and in every organism, they illuminate the regions around them. For more than a year, a postdoc in Schreiber's lab had been probing furiously with both complexes to locate their immediate partners. To Schreiber, the prize was no less than the answer to one of the most intriguing and competitive questions in biology: How do cells talk to themselves? How are they wired?

With the *Science* paper, Schreiber was planting his flag over a major expanse of cell biology and moving further and further from his moorings as a synthetic chemist. He was not alone in his theorizing. A relentless acquirer of information and contacts, he was picking up biology on the fly from a cadre of prominent collaborators, most ambiguously two prominent immunologists at Harvard Medical School, Steven Burakoff and Barbara Bierer, who also had connections to Vertex. Burakoff, through Schreiber, had joined Vertex's SAB before Boger and Schreiber's falling out and had remained on the board since; Bierer, Schreiber's former protégé, was a paid consultant.

Schreiber and Burakoff had begun collaborating on FK-506 in 1988, shortly after Schreiber's arrival at Harvard—a productive arrangement for both. Burakoff gained access to Schreiber's small molecules and understanding of chemistry; Schreiber, to Burakoff's medical school connections and insights about T cells. "Stuart was insatiable about biology," Burakoff would recall. Branching out—"dabbling," Burakoff called it—in AIDS, they soon developed a molecule that briefly looked so promising that it was reported in more than 300 newpapers, landed Burakoff on *Good Morning, America*, and led Harvard to spin off a new company before the compound proved to be useless.

But the big prize for both of them remained signal transduction. "We're only now learning the instruments that make up the orchestra," Burakoff would say. "What we haven't been able even to approach is how the orchestration occurs and how doing all that gets coordinated to produce a symphony." Schreiber, with his characteristic bravado and scant experience with biology, was more optimistic. "The prospects for fundamental insights appear promising," he concluded his paper in *Science.*

Others, of course, were less impressed. Responded Boger, caustically: "There are those who would argue that firing a bazooka into the main switching box of the phone company doesn't tell you how the phone company works."

In the beginning of February, Schreiber followed up his *Science* paper with a letter to *The Journal of the American Chemical Society (JACS).* His coauthors were Burakoff and Bierer. Together they suggested the existence of several other potential targets for FK-506—other FKBPs.

Harding was distraught. The search for other binding proteins had largely been his responsibility at Vertex, and now here was Schreiber, whom he'd taught to discover proteins, announcing in collaboration with Bierer and Burakoff, with whom Harding presumably ought to have been collaborating, evidence of perhaps four new immunophilins. Harding had feared that by working in industry he would forfeit his place at the forefront of research, and now it was becoming so. Worse, he felt powerless to stop it. He literally had given the company his blood and yet was no nearer the individual goals he had set for himself. He was especially pained by how it looked. "Steve and Barbara are supposed to be on our side, and we don't know what they're doing."

Boger was restrained by comparison. It was a source of intrinsic, unspoken tension between him and the scientists that while he wanted as desperately as any of them to have certain information, he cared far less, in most cases, who discovered it. Clearly, no other question was now so critical to the company as whether Harding's and Merck's original FKBP was the protein a Vertex drug must block in order to be immunosuppressive. However, who answered it, other than the obvious benefits to morale and for purposes of positioning the company with investors, was not nearly as important to Boger as the fact that an answer be gotten and that it quickly be made available to Vertex, preferably on favorable terms. He was annoyed that Burakoff and Bierer were still collaborating with Schreiber and that they hadn't told him in advance of their discovery, but he was uninterested in firing any more SAB members. He was much more concerned with how to use their data.

According to the *JACS* paper, there were apparently four more FK-506 binding proteins in T cells. Whereas Harding's and Merck's original FKBP weighed about 12,000 daltons (12,000 times as much as a hydrogen atom), Schreiber and his colleagues had found significant binding with proteins with molecular weights of about 13,000, 30,000, 60,000, and 80,000. Schreiber still believed FKBP-12, as he now called it, was the relevant target of both FK-506 and rapamycin, but the protein unquestionably had competition. A family's worth.

Boger suffered no public doubts, though he might have, given the implications. First, there was the obvious matter of which of the proteins was most relevant. It could be that blocking just one of them was necessary for immunosuppression, or more than one, or all, or that inhibiting one was necessary but insufficient. If there was one major protein, as Schreiber assumed, what was the role of the others? It could be, too, that the side effects observed by Starzl and the other transplanters resulted from FK-506 being soaked up not by the lead target, but by the secondary ones, which—it was not yet known—might be concentrated, say, in the drug-sensitive tissues of the brain and kidneys. How did you design a drug to hit one receptor but not the others? You'd need structures for all of them. You'd need to go back to the beginning: Thomson slogging through four more protein extractions; Yamashita and Navia finding the right crystallization and heavy-atom conditions for four more proteins. Boger had adopted FK-506 as Vertex's inaugural project, in part, because its goal appeared to be straightforward: to design a better inhibitor of a known enzyme. But this was orders of magnitude more difficult. Even if you had the five structures, it might take years to develop the biological understanding—What did each protein do? In which of the body's cells was it found? Did the others work through still other partners, as yet unknown?—before you could even attempt to design a new molecule. "A nightmare of terror," Murcko now called it.

There were business horrors as well. What if Schreiber discovered the true target first and applied for the patent? Vertex might eventually have to go back to Harvard to try to license it. What if the problem simply became too complex to warrant a structure-based approach? What good were Vertex's best molecules, its draw for investors, if there was no clear rationale for developing them? What about Chugai? Wouldn't the discovery of another target release the company from its contract?

These had been unspoken fears all along, and now they flailed to the surface like divers ahead of a shark. Boger moved quickly to allay the damage. As he had eighteen months earlier when the burst of sudden publicity about Starzl's success with FK-506 threatened to preempt Vertex's entry into immunophilins, he declared this latest development favorable to the company, a vindication of its overall approach. In fact, he told the scientists, a family of putative receptors was good news. It leveled the field. The scientific obstacles were the same now for everyone else—most notable, Merck's superior resources and head start meant even less than before. More, the essential structures of the new proteins all had to be similar to FKBP-12, otherwise FK-506 wouldn't bind to them. Thus Vertex's structural work wouldn't, couldn't, be wasted. As for business, he told them, the ultimate prize had not changed: Vertex didn't have to produce a drug that was better than FK-506, only different. Its leads remained solid.

Just as Boger believed in publishing early to demoralize and confound one's competitors, it may have been that Schreiber hoped to force Vertex and the rest of the field into a costly, time-consuming reexamination and retreat, one that would help him widen his lead. But Boger wasn't thrown off. He proposed only minor course corrections. He authorized Harding to focus his efforts on finding the actual proteins that Schreiber, Burakoff, and Bierer had postulated. Within weeks, Armistead would attach a small atomic tether to FK-506 so that Harding could set up a column similar to those he and Thomson had used earlier to fish for FKBP. Otherwise, the Schreiber "juggernaut," as Moore now began to call it, failed to intimidate or dismay Boger. "If I didn't have compounds that work, I'd be depressed," he said. "But I have compounds that work. I'm elated. It makes it harder for anyone else to come in behind us."

After the appearance of the *JACS* paper, Moore called Schreiber at his lab. The two had never spoken before, but Moore wanted more information about protein conditions that Schreiber had described in his manuscript—a routine request. Schreiber, ever effusive and bearing individual Vertex scientists no particular malice, was happy to oblige. He was often like this: willing to reveal more than was inquired about, so disarmingly frank when it seemed he ought to be secretive that one wondered why. What must he be concealing? Moore mentioned that he was trying to solve the structure of FKBP by NMR. Schreiber chattily revealed that his group had assigned more than 1000 of the protein's nearly 1600 hydrogen atoms—Moore had located fewer than 700—and left it at that.

Moore was puzzled. Schreiber was relentless about publishing, often rushing into print with data that others thought less than earthshaking. Yet Schreiber seemed to be telling him that despite having more than enough assignments to solve the structure, his group hadn't yet "closed the door." The two months since Moore had first seen the backbone of the protein had been grueling. He had buried himself in work, coming in at 9 A.M. and often leaving after 11 P.M., including weekends. He'd told his wife, Lonnie, not to expect anything from him, and she hadn't, staying up wearily to meet him with supper and a beer as he dragged in and crumpled in front of the TV. By the end of January, he had fifteen rough structures approximating the presumed actual structure, but he needed more data. Every day he expected to hear that Schreiber's group had won, and every day, when he hadn't, he had pushed himself on by telling himself he'd been granted another day's reprieve. He couldn't understand why Schreiber hadn't simply finished him off.

Sitting in the modeling room next to Boger's office, Moore attacked his data with a murderous zeal. He was going as fast as he could. Within weeks, he now estimated, he would have a single consensus structure. He had become like Thomson, not as nihilistic, but locked in the isolation chamber of his own unbridled compulsion. He was oblivious to the fact that Yamashita had grown intensely jealous of him, could barely stand to be around him because he now thought Moore had won. All he wanted, saw, tasted, was the endgame of his battle with the enzyme. For the moment, nothing and no one else existed.

"Everyone feels that they should be able to do it themselves," he would say. "I'm just as guilty as other people in that I wanted to do this structure with as little outside intervention as possible, not because I didn't trust other people, but because it was something I wanted to do and to prove I could do as a scientist. All during my training I always had to take my data and feed it off to someone else so that they could do the refinement or the structure determination. I said, 'Fuck that.' I hated that as a postdoc."

And then his computer would go down, or he'd have some trouble with software, and all his pumped-up underdog's resolve would come hopelessly unraveled.

"Sack of shit," he would croak, typing furiously, watching the computer devour his structures. "I have no life."

CHAPTER FIFTEEN

"Altruism," Boger spat, "has no other evolutionarily justified motive but self-interest. Selfless altruism is an oxymoron; it's impossible. For me, the best motivation for going into AIDS, where motives are just so suspect, is that we think we can make a difference. The science took us there. The opportunity took us there, which I think is more honest than getting in either because we wanted to make a lot of money or because we wanted to save the world. Martyrs are very selfish people. People who do things for grand motives elevate themselves perhaps inordinately high."

Boger sat game-faced at his desk, deep in fluorescence. It was Februrary 28, 1991, after 8 P.M., long beyond dark at the cusp of Cambridge's direst months. During the past several hours he had personally edited the scientists' Pittsburgh abstracts, chosen the fonts for them, photocopied them, tinkered with a balky copier, supervised their mailing—"micromanaged," as Murcko unreassuredly put it, the company's coming foray into gladitorial science. Now he was racing to get home to see Amy and the boys before leaving at five the next morning for ten days in Japan.

It had not, in the end, been a hard decision to make AIDS Vertex's second project. The scientists had produced encouraging data, the business climate had improved, cathepsin E faded; time had run out. But the choice had required from Boger a 180-degree conversion—an uncommon occurrence. For two years he had sworn, correctly, that Vertex had no place in HIV research. Now he was embarking with equal commitment to sell an AIDS program in Japan, his third "death march" there in twelve months. Packing his slides, rifling through the mounds of paper on his desk, he had the no-nonsense paraintensity of a skilled pilot heading into the teeth of a storm in a small, untested plane.

"We backed into this because we thought we could do it, so our motives are clear," he said. "This is not cold and calculating, ride-the-bandwagon stuff. This is science."

If Boger was defensive, it was because of the numerous personal and scientific demons he'd had to exorcise in reversing himself. As Susan Sontag and others have suggested, diseases and the responses to them mirror society. With AIDS, the scientific reaction has been at once strikingly productive and shamelessly self-serving. Boger didn't have to be an altruist to be alarmed by the "suspect" intentions of many of those he was now joining. AIDS research was rafted with them. Indeed, they had informed much of his earlier resistance.

At first, science had simply been uninterested in AIDS, the search for what was killing young gay men in California and New York suggesting a routine microbe hunt on the fringes of scientific respectability. Forty years after penicillin and streptomycin, new infectious diseases, somewhat justifiably, were considered washed up, passé. Cancer was then science's scourge of choice. Through the War on Cancer, brokered and then built into a federal powerhouse by Schmidt, researchers had discovered that they could develop and sustain large labs for pursuing their own interests as long as they could demonstrate some connection, however tenuous and remote, to treating the disease. Biomedicine was transfigured by the bonanza. Finding that they were rewarded chiefly for doing science that was visible, researchers concentrated on areas that invited attention, because attention brought money and money, more research. Competition—for money, reagents, credit, priority, and access—became not just the main thing, but everything. The drug industry, following suit, was similarly changed. Cancer research was a great story. Wall Street ate it up. No pharmaceutical portfolio was complete without a major anticancer effort, regardless of what there was to it.

The first rash of AIDS deaths in the United States was reported within weeks of Ronald Reagan's election as president. Fateful as that was for those with the syndrome, it exonerated even further the new era of scientific self-interest. Dr. Robert Gallo, the government's chief AIDS researcher, set the tone. Heavily promoted by the White House, which at first simply ignored the disease, he announced triumphantly in April 1984 that he had discovered the virus that caused it, then descended through a nine-year spiral of alleged scientific misconduct and personal disgrace that ended with his conceding that he had gotten the virus from French researchers who'd discovered it first. Loudly and publicly obsessed with winning a Nobel Prize, receiving $100,000 a year in patent royalties on a discovery he didn't make, Gallo ended up all but quitting science to defend himself from multiple investigations by Congress, NIH, the National Academy of Sciences, the media, and the General Accounting Office.

Willful and paranoid, Gallo became synonymous with the drive for individual credit in AIDS, but the atmosphere in drug research was by 1987, if anything, even more venal. In March, the FDA approved the first drug for treating AIDS—AZT—a twenty-three-year-old compound first designed as an anticancer agent and rescued jointly by researchers at the NIH and at Burroughs Wellcome, a British-owned drug firm best known for making Sudafed, a cold remedy. The molecule was primitive, toxic, and, having first been made under a federal grant and never patented, gathering dust in the public domain. Desperate for input from the drug industry, which still saw little opportunity in AIDS, NIH had pressured Burroughs to submit compounds it thought might slow the replication of the virus, and Burroughs had sent AZT along with several others it had identifed from the scientific literature.

Burroughs had studied AZT's toxicity, but most early testing of the drug was done at NIH. Still, in 1985, after government researchers determined that the molecule was active against the AIDS virus, the company asked for a patent. When the FDA licensed the compound two years later, AZT instantly became one of the most expensive long-term drugs ever sold, costing each patient about $8000 a year. By the time Burroughs's patent was issued in 1988, the company held a seventeen-year government-sponsored monopoly on the only approved drug for a disease that was fatal in every case, a drug that analysts were predicting would become a billion-dollar seller by 1992, a drug that the company hadn't discovered and hadn't paid for.

Science follows money. As long as there was little spending on AIDS research and the drug market looked relatively small and unpromising, most scientists and drug company executives deemed Gallo's and Burroughs's transgressions a sideshow, beneath comment. But the gravity of the contagion and the forced realization by the federal government and the pharmaceutical industry that AIDS was exploding worldwide and would not go away suddenly by 1987 reversed most researchers' indifference and unleashed a tide of scientific interest. NIH funding for AIDS research began to swell, multiplying more than 600 percent in three years. Researchers who formerly concluded their grant applications with "and its possible application to cancer" now learned to write "and its possible application to AIDS." The drug industry, witnessing the controversy over AZT with more jealousy than revulsion, but having little success historically with antivirals, tentatively began launching its own protoprojects in AIDS.

Whatever else, AIDS research had become a compelling story, and in the way of all phenomena taken up and amplified by the media, it now became fashionable. Venture capitalists, without a hot new commodity since microchips a few years earlier seized on the disease. Entrepreneurial scientists, dazzled by the sudden new opportunities for money and fame and indulging a taste for business, accommodated them. Together they quickly started new companies—AIDS companies—then just as quickly sold them to Wall Street. By 1987, a frenzy erupted. A British investment firm headed by Lord Rothschild invested in fourteen such companies. A small vaccine company promoting the work of Jonas Salk, who'd steadfastly refused to patent his polio vaccine, attracted so much money that its founders had to send back checks. Scientists became rich and famous without much more than a hopeful-sounding idea and a willingness to speculate on its chances. That few of them had any experience with making drugs or bringing them to market, that more than two-thirds of the companies were doomed to fail, that no one had any idea where the money would come from to keep them going, all were blithely overlooked in the frenzy to position the new firms.

AIDS at last had captured the interest of scientists, but scientific interest itself doesn't induce progress. The search for the cause of AIDS had been relatively narrow and straightforward—tracking down the right microbe. Curing it, however, would require an extraordinary new understanding not only of the virus and how it works but of the immune system and its complex reaction to it. How was science to proceed? Technically not a disease but a syndrome, AIDS is hydraheaded by nature. It's riddled with nightmarish features: a highly mutable virus; a long, deceptive latency; cofactors—other agents—that accelerate its timing and spread; multiple diseases rampaging simultaneously through different organs; a devastated, ineffectual immune system. Singular solutions—"magic bullets" like vaccines or one-shot therapies like antibiotics—were unlikely. The manifold challenges before the newly engaged research community were more like those associated with putting out a forest fire than capturing a sociopath. They required a coordinated effort on several fronts and were visualized perhaps better from the air than on the ground. Now that science had been enlisted in AIDS research, how to manage it remained elusive and unclear.

"Science," says Nobel laureate David Baltimore, "is much better at solving problems of its own devising than those it is asked to solve." At the time of AIDS's ascendancy in research, Baltimore was director of the prestigious Whitehead Institute at MIT, a leading scientific policy chief and statesman, and already—though the world would not know about it until later—embroiled in the publication controversy that would eventually force him, after a celebrated congressional fraud investigation that absolved him of misconduct but not misjudgment and arrogance, to resign as president of Rockefeller University. Baltimore, working to some degree in AIDS himself, tried, not uncharacteristically, also to become its tribune. By the time of the run-up in AIDS funding, he'd taken to publicly beseeching other researchers to overcome their aversion to managed research, put aside their obsession with credit, and sublimate their own urgent career choices and business imperatives to "a sense of responding to a national need." In short, he urged a crash program for AIDS—a Manhattan Project, he called it. The A-bomb project's biomedical cousin, the wartime pencillin program, perhaps was a more apt precedent, but scientists understood what Baltimore was proposing. Under federal leadership, the nation's best scientists would put aside competition and join hands in a common cause. Hardly subversive, Baltimore's notion stemmed from an established wartime practice: the temporary suspension of business as usual.

Scientists show their disapproval in an odd way. Something considered promising or important will be challenged vigorously, even harshly. Something deemed uninteresting, thus unworthy of pursuit, is met with a dismissive silence. Crows ostracize each other by cackling raucously; scientists, by turning to whisper something or staring at their shoes. Such was the reaction to Baltimore's plan. No one supported it. The implicit message was that money alone sufficed to marshal research and that the helter-skelter of the patent fight and the dominance of strong labs over weaker ones was preferable to any coordinated system for determining what science does and how it does it, even in AIDS.

This was the world Boger was now entering: a world of self-selection, secrecy, competition, and greed—none of which Boger abhorred, but which he believed also invited great posing and mendacity. There had, of course, been acts of extraordinary science and great humanity within AIDS research, but charity, Boger thought, was for those who could afford it. "Money-grubbing corporations need to look selfless," he said starkly in his office, fixing on a slide with his jeweler's eye. "I think Roy Vagelos and Ed Scolnick think it's one of the responsibilities of their research wealth that they have to act altruistically."

Merck's entry into AIDS research in late 1986 indeed accomplished much of what Baltimore with his appeals to common sacrifice could not. It was the arrival of the cavalry, not the soapbox rantings of a single voice of conscience. By then Merck was both the Arnold Schwarzenegger and Mother Teresa of American businesses. Its stock was in the midst of one of the great sustained surges in history; ultimately it would rise more than 500 percent in five years, twice as fast as the Dow during the greatest market run-up of all time. It would soon donate enough avermectin, a veterinary drug, to wipe out African river blindness. Within a year, Merck would be voted America's Most Admired Corporation by *Fortune*'s annual poll of executives, displacing perennial overachiever IBM. (It would eventually win the honor seven years running, hanging banners in Rahway saying America's Most Admired Corporation and, in later years, using the designation prominently in its want ads.) Brilliantly innovative, spectacularly profitable, yet governed seemingly by a deep humanity and sense of responsibility, Merck had managed to retain much of Tishler's and George Merck's heroic altruism while growing at 20 percent a year. Merck and Company remembered World War II if no one else did. "The Miracle Company," as *Business Week* described it in a flattering cover story, seemed the perfect antidote to the squalid carnival in AIDS research.

Swaggering, Merck helped redeem the search for AIDS drugs by pushing it to a higher level. CEO Roy Vagelos, a highly competitive M.D. with a penchant for brilliant science and aggressive play—he was always photographed for magazine profiles in a kayak or in tennis garb—announced that he was "damn optimistic" that the company would succeed. Shunning the normal secrecy around research, he encouraged Merck's scientists to talk openly about their progress and share vital data before publication. AIDS was still considered a small market, and Merck risked losing what little edge it might have by being open, but Vagelos gamely insisted it didn't matter. Merck was the industry leader. Its rightful place was at the forefront of important science, regardless of the risks.

What Vagelos knew, of course—and what soon drew other big drugmakers into the field—is that the drug industry, and Merck especially, now had cause for confidence, a promising new target. AZT worked by blocking reverse transcriptase, an enzyme common in cancer viruses and a difficult target for drugs. However, in early 1987, a young Harvard-educated molecular biologist at Merck's West Point labs, Irving Sigal, and his colleagues had been dissecting the virus for alternate drug receptors when they discovered something extremely hopeful, the salient role of an aspartyl protease in HIV's reproductive cycle. Here—HIV protease—was the key to Vagelos's optimism. Vagelos, as director of research and then CEO, had made enzyme inhibition Merck's main scientific area. Largely because of Boger, it had led the field in renin, an exceptionally close relative of the viral protease. Indeed, the company's next big drug, a prospective billion-dollar seller for treating enlarged prostates, worked by blocking a proteaselike enzyme. Vagelos's bravado in AIDS was more than generic: The company was strongly in its element.

Sigal championed the project. Two years younger than Boger, he was, if anything, broader than Boger, having been trained as a chemist before making a string of exceptional discoveries in biology. His father, Max, had been director of research at Eli Lilly, and Sigal seemed destined to rise at least as far at Merck. Brilliant, single-minded, passionate, brusque, abrasive, he'd begun acquiring by age thirty-three an extraordinary degree of influence within the company. Indeed, now, as Sigal began assembling his program in AIDS and Boger, his structure-based design effort, the two unavoidably became rivals, heirs apparent to the helm. Not surprising, there was little love lost between them.

Sigal was formidable, a powerhouse. "Irving was the kind of guy who, if he said, 'I'm going to deliver protein on such and such a day, in such and such a quantity and such and such a purity,' you could bet on it," says Navia, whom Sigal had recruited to solve the enzyme's structure (Boger, who had just invited Navia into his group, "cleared the decks" for the move). "He delivered. He had the right focus. He wasn't doing this because it was going to be a tour de force for his lab, but because he wanted to go for this disease."

By the summer of 1988, with Sigal driving the project ahead and Boger, unbeknownst to anyone at Merck, edging closer to leaving, Navia and Brian McKeever began the experiments that would lead three months later to their solving the crystal structure of the protease. They were now in competition with a second group at NIH. The period was intense, strenuous, and marked by heated disagreements over strategy. Despite Vagelos's edict to share information with Merck's competitors, Navia worried (as he would later at Vertex) about publishing Merck's crystallization conditions before the structure itself was finished. He and Sigal quarreled bitterly. After one such argument, both of them left, exhausted, for the December holidays.

It was the last time Navia would see Sigal. Returning from

Heathrow Airport in London four days before Christmas, he was killed when Pan Am Flight 103 exploded in a fireball over Lockerbie, Scotland, killing all 258 passengers. He was thirty-five.

"I was devastated," Navia would recall three years later. "I knew that this program was really going to have a problem. The guy who was the soul of it was now gone. These terrorists view these airplanes as being filled with nameless, faceless things. But I know that there was one guy on that airplane who probably more than anybody else could have, by now, significantly affected a disease that has the potential of exterminating the human race."

Whether Sigal's death or Boger's defection two days later had a greater impact on Merck, both came to symbolize the company's sudden vulnerability. As a martyr of AIDS research, Sigal had not "elevated [himself] . . . inordinately high," but others had, and now, like Navia, they were bereft. Merck had been able to impose a new discipline and public-mindedness on the field in part by pointing it in a vivid and promising direction. That direction had been promulgated especially by Sigal and Boger, who believed equally that one could stop the virus by tailoring molecules, atom by atom, to block a key piece of it. Now, both of them were gone, and though Vagelos and Scolnick professed no public doubts about the company's prospects, others did. Eight years into the epidemic, Merck's AIDS efforts seemed to outsiders to bog down in familiar ground—an inability to make protease blockers that were smaller and less easily metabolized than peptides, which were both bulky and weak. Not quite back where it started, the search for AIDS drugs again seemed daunting, implacable, stalled, demoralized.

Boger's initial reluctance to go into AIDS at Vertex had reflected this new uncertainty, even though, as Aldrich continually noted, the company had inherited Merck's "first team"—Boger, Navia, Murcko, and Tung. That, of course, was an exaggeration, but without Sigal, it was also true that perhaps no other group of scientists so embodied the reasons for Vagelos's original optimism. Now, after reconsidering and after proving that the Abbott and Roche compounds indeed blocked the replication of the virus as effectively as AZT, Boger was becoming as confident as Vagelos had been. When the chemists began synthesizing two classes of novel inhibitors that were equally active, all Boger's litmus tests were suddenly and irremediably met.

"Now that we're up and going and can see what's possible," he said, "we have a responsibility to make this work. This is possible *now.* We don't want to cede this opportunity until the time when big drug companies get around to it."

But had Vertex, by delaying its decision, been too late? Belated timing was as oxymoronic to Boger as selfless altruism. With sufficient data, there was no being wrong in his world. "When I'm in decision mode, I let events take control. You shouldn't decide things when you don't have enough information. Because of my worry about this, I set a very high standard about what had to happen before we could make a firm decision. I waited."

In fact, Boger's timing now looked canny. Small companies like Vertex that were chasing AIDS while trying to survive had in the past been forced to sell to corporate partners at the earliest possible moment. The costly, uncertain development gauntlet for experimental drugs demanded it. But AIDS patients, with nothing to lose and no future, had demanded and won a vastly truncated timetable for making new drugs available. In December, Merck announced that it had begun testing a new reverse transcriptase inhibitor in Europe that was found in a natural products screen scarcely six months earlier. The time-value-of-money implications of such a compression for a small, cash-starved company were fathomless, beyond reckoning. Rather than being hopelessly behind, as Vertex had appeared to be as recently as three months ago, the company (Boger could now state confidently), whose compounds had yet to be tested in animals, trailed the industry leaders by less than a financial quarter or two. Vertex had become competitive practically overnight.

Novel inhibitors, Merck's first team, a greased track for development, record turnaround time—these were the elements of Boger's story for the Japanese, the elements he had to have. He knew he could sell them. What he didn't know was whether anyone besides Nissin, the family-run noodlemaker and perennial shopper, was buying. "He's about to present this cold to more sophisticated companies, shall we say," enzymologist Dave Livingston, a biotech veteran, observed as he watched Boger prepare. "We haven't done a reality check on HIV, and now we're about to do that in a big way." And unlike with Chugai, there would be no Benno Schmidt to broker the introductions.

In Japan, Boger and Aldrich called on nearly a dozen companies, scrabbling by cab and train between their headquarters in Tokyo and Osaka. Their reception was thoughtful but ambiguous. A courtesy call early in the week to Chugai, which hosted a lavish banquet for them, convinced him that the company was still pleased with its investment. Meeting with Nissin the next day, Boger assessed the prospects for an AIDS deal at fifty–fifty. Altogether encouraged, he and Aldrich returned to Boston with the self-satified glow of travelers who had courted hostility by bartering for antiquities in a strange country, only to return with a bounty of new leads.

But by then something far more reaching had happened to reshape their world, something that would have implications beyond a productive sales mission. On March 7, 1991, as Boger and Aldrich were midway through their trip, a Washington appeals court ruled sweepingly in favor of Amgen in its five-year patent battle with GI over EPO, the long-disputed antianemia drug. Overturning a lower-court ruling, the decision gave Amgen a complete patent monopoly in the United States over the sale and production of EPO. Amgen's stock, already widely thought to be overvalued, soared on the news, rising $12 to $113. GI's shares sank $21.75, to $40.25.

Wall Street might not know a clone from a clown, but a rout it understood. With the victory, Amgen's stock was now on its way to rising a stupendous 900 percent in two years. Where else could one make that kind of money? Where else could a few smart people (aided by sharp patent lawyers) make something that could produce $1 billion a year, as EPO was expected to, and be protected by the government—the *government*—from competition?

For a decade Wall Street had stood by like an abused suitor waiting for the market in small biomedical companies to make up its mind. Would they pay out or wouldn't they? Now, at least for the victors, the answer was thunderous.

Boger could feel the temblors all the way in Japan. He had left a week earlier with Wall Street groggily shaking off years of indifference, and now he could feel its heat rising. By the time he returned to Cambridge two days after the ruling, the landscape was reverberating wildly. Several companies not much older than Vertex and, Boger believed, nowhere nearer to making money, announced big private placements. Others filed to go public. *Public.* For years the acid test for such companies had been having a drug in clinic, a moneymaker near at hand. These companies weren't even close. Two months earlier the institutional investors who buy up most initial stock sales wouldn't have thought it worth the cab fare to meet with them. Now, Boger heard, they were tripping over each other to get in on the action.

The billions of dollars in investment capital that Boger had watched roiling on the sidelines was now beginning to burst the dikes of caution and pour into biotechnology, all of it looking for the next Amgen. Like everyone else in his position, he shifted his sights instantly toward the tsunami. He discreetly forgot about HIV and Japan and began assessing the larger sweep of events. "I have no problem saying the company is worth $35 to $40 million right now," he said, reviewing the suddenly goiterous evaluations being placed on the other companies in Vertex's class. "Is it worth $120 million? If somebody else thinks it is, I'd be a fool not to agree."

The EPO ruling had immediate and volatile consequences. In addition to GI, the other big loser was Chugai, which had licensed EPO in the United States and was expecting to become a global player on the strength of the drug's American sales. The setback would undoubtedly slow Chugai's growth and make it at once more dependent upon—and more anxious about—Vertex. "I'm glad we had dinner with them early in the week," Boger joked. "If we'd seen them Friday, we'd have been lucky to have gone out for noodles." When some of the scientists openly wondered whether Chugai might be forced to pull out of immunophilins, Boger assured them that things weren't that bad. A few were not encouraged.

But Boger was way past them now. He could see the changing business geometry as clearly as he could visualize the whirling of molecules in space. Dozens of small companies like Vertex would soon be circulating through Wall Street, looking for capital. They would have their hands out like guests at a party or, in another sphere, like small molecules in the cytoplasm of a cell. Each would shake a lot of hands, just as molecules do in a sense, their atoms, their fingers, touching, probing, interpenetrating. A few would clasp harder, more enduringly. These would be the winners. The markets would welcome them in, and through their connection, they would succeed, success meaning above all else survival. Their binding, as with molecules, would be strictly competitive—that is, those with the highest affinity, that fit the best, won. As for the others, they would be stripped, discarded, as Boger believed GI would now be. "They're about 300 people too big," he said. "They're going to crash and burn and the smart money will pick up the pieces."

Boger had always said the time to go to Wall Street was when Wall Street was ready. Now was that time. Whether the company itself was equally prepared was something neither he nor anyone else had the luxury to consider.

John Moore gazed dully at his computer screen. Five theoretical structures of FKBP were overlaid before him like traceries. Where they overlapped, the lines erupted in a luminous purple, pieces of spine. Where they didn't, the filaments of individual atoms burst into an angry matrix, like threads springing from a rotting twill. Overall, the protein's shape was now clear, though similes for it revealed as much about the beholder as about the molecule. Al Vaz, Vertex's facilities manager, thought it looked like "a crushed beer can"; Boger, "a hermit crab shell."

In their sameness the individual structures told a compelling story, but it was their differences that now most troubled Moore. Unlike crystallography, NMR measures molecules in their native state, in water, where they flop about like jellyfish. To get a high-resolution picture, Moore needed another ten high-quality structures at least—perhaps another week or two of painstaking work. Only then could he average them mathematically, cleaning up the discrepancies on the screen. Knowing the approximate location of all of the atoms was not enough. A publishable structure required exactness, certainty. The field, still dominated by crystallographers, didn't tolerate unresolved noise. One didn't design a drug to fill the core of a beer can.

Moore was now consumed with finishing the structure and submitting it to a good journal. "It would be very simple for somebody with twenty people in his lab to make the kill at this point," he said on March 11, two days after Boger returned from Japan. He had worked all weekend and was pushing ahead as he had for the past six months, alone. Increasingly exhausted, he'd now begun to doubt that strategy. "The way people conceive the field, it's that Schreiber's done everything. We'd like to change that," he said. "But I've put myself in a vulnerable position. I may end up very disappointed."

Although he never thought he would win in the first place, Moore's expectations remained firmly grounded, but that wasn't true of Yamashita. His behavior became increasingly erratic. In the two months since Moore had first identified the protein's backbone, Yamashita alternately was aloof and obsessive, dispirited and rash. One Monday in January, after working until 4 A.M., he said, "It's good that we lost. This is my work, not my life." A minute later he changed his mind: "I guess if I'm willing to work on this bullshit all night, I'm taking it seriously." He worked slavishly, leaving Vertex only on Sunday nights to volunteer in the emergency room at Brigham and Women's Hospital, a grim change of scene but one that Yamashita enjoyed. He began talking about becoming a doctor, fascinatedly reporting about the first person he saw die—a Salvadoran pizza deliverer who was shot in the head during a holdup. Yamashita had counted the money in the man's pocket—$78—and filled out his death certificate. He thought perhaps in medicine he could do something both important and satisfying.

His hands took on a different cast. His left hand turned yellowish; his right, raw and pink. He insisted it had nothing to do with his work, but it recalled a time in graduate school when he had to turn off a robotic X-ray beam arm that had gone askew and was scattering radiation throughout the lab. Then, his white cell count spiked for several days and he had to be hospitalized although the doctors assured him that he'd only absorbed a localized dose. "No scars," he would later say, "at least not any that anyone can see." Temperamentally, he was now subject to long periods of strained composure broken up by jagged fits of rage. "It really helps," he said, "to be intensely violent for about five minutes." To settle himself down, he'd begun going out drinking regularly with Thomson and Laura Engle, who listened sympathetically but didn't know what else to do for him.

Yamashita was also working alone. Propelled as much as Moore to prove himself, he had rejected any help from Navia, who, recognizing his need for independence, had vowed to let him make his own mistakes. It was far from easy. With each successive failure with heavy atoms, Navia had to restrain from simply imposing himself. Boger supported Navia's forbearance, it being in keeping with his social experiment. But there was no question it was delaying vital information. "We don't go after structures because of some implied need," Boger said. "We go after them because they help us know things we can't know any other way." Without that data, Vertex's chemistry effort was stalled. Not only were the heavy atoms not working, but Vertex's crystals were behaving badly. For two weeks at the end of February—two weeks during which Yamashita's behavior became steadily more obtuse—he barely could grow them at all despite using the same conditions as before. He was distraught. He told Murcko that he thought the situation was now hopeless, that he would never get the structure, and that he was quitting crystallography. "It's too mind-wrenching," he said. "I need something more predictable with a higher success rate."

Desperate, he finally took Navia's suggestion to change his mother liquor, a move that at once stabilized the protein and gave Yamashita sudden confidence that he not only would solve the structure, but soon, perhaps in the next two months. "We now have a basic crystallographic problem like other crystallographers have," he declared in early March, relieved finally to have something go right, "not one unique unto ourselves."

It thus was in a rare moment of dual optimism—Moore's and Yamashita's—that Boger first heard the rumor that Schreiber and his collaborators had defeated them both. According to Boger's information, Schreiber and Martin Karplus, the other former SAB member let go by Vertex, had submitted an NMR structure to *Science* at the end of January, then, a week later, he and Clardy had delivered an X-ray structure. Back-to-back, the two papers purported to show not only what FKBP looked like, but how FK-506 affixed to the protein, how the two molecules bound.

It was an astonishing coup, so strategically bold and technically brilliant that even Boger had to concede that Schreiber now "owned the field." As Moore had guessed after his phone call to Schreiber at the end of January, his group had finished the NMR structure months earlier. However, they had waited for Clardy to catch up and confirm their findings.

"The final refinement went with lightning speed," Schreiber would recount. "We were working around the clock. Clardy and [graduate student] Greg Van Duyne came up here with all their data on February 6, 7, and 8 or so, right around my birthday. I put them up at the Charles Inn. We'd come in in the morning, work on the paper, analyze our data, go out to Bartley's Burger, work, have dinner working on the paper, come back, work until the wee hours of the morning."

It was the existence of the two papers, still under review, that Boger first heard rumored on March 12, 1991, more than a month later. Two days after that, on March 14—"Black Thursday," the scientists would call it—he confirmed the story and broke the news to Yamashita and Moore.

"How are you, Mason?" he asked, poking his head into the modeling room where Yamashita was working. The visit was so unusual that Yamashita suspected at once that Boger hadn't come to swap salutations.

"I'm fine, Joshua. How are you?"

"I'm fine."

"Oh," Yamashita said, "did Merck beat us?"

"No."

"Who beat us?"

That it was Schreiber and not Merck so disjoined Yamashita that he was oddly subdued, dispassionate. "I was worried because Mason had taken it all quite well," Boger would recall. Moore characteristically cursed a few times and went back to work.

Boger's mood was stout. He has a relentless talent, as Saunders puts it, for "finding the pony"—an admiring reference to the eternal pragmatist, who, led into a roomful of manure, reaches for a shovel and begins to dig. Inheriting from his mother a rigid intolerance, as his brother Ken observes, for "anomie," he never broods or pities himself. Now, he saw at once the great opportunity in Schreiber's triumph. Schreiber and his collaborators had found the structure first but that didn't mean they had won, not officially. Publication, not discovery, was the ultimate test of winning, and Schreiber's structures were not yet in press. Thus, as Boger saw it, there was still time for a tie. Schreiber might have won, but that didn't mean that Vertex had lost. The company had until Schreiber's publications were accepted and in print to submit its own, something Boger was convinced that no one else, not even Merck, was now as close to doing. Within minutes, he managed to convert Schreiber's rout into a scenario where a photo finish between them was still well within reach and quite possibly Vertex's alone to lose.

Boger instantly began engineering Vertex's comeback. Schreiber's strategy may have been impeccable, but holding the NMR paper had left him exposed. "Schreiber knew we were working on the NMR structure. I told him," Moore said. "But he wanted to have the X-ray paper, too, because it told a more complete story. That's what left the door open for us. That saved our asses." Boger's strategy was to have Moore rush to finish his structure and submit it to a competing journal within four weeks. Meanwhile, Moore was immediately to give as much of the structure as he was sure of to Yamashita, who would try to use it to sidestep the need for heavy atoms through so-called molecular replacement, the experimental shortcut first proposed by Navia in December and avoided by all parties since. With Moore's paper planting the flag, Yamashita and Navia would try to complete the X-ray structure by the time Schreiber and Clardy's paper appeared in print, thus legitimizing Vertex's claim to a tie.

"It's first and goal," said Boger, who didn't care about football but had picked up on the national rage for its metaphors during the Gulf War. Happily in Schwarzkopf mode, he assured the scientists they'd have every edge, every resource they needed, "every computer in the place"—he tapped the turbomouse on his desk—"including this one."

• • •

It was a giddy, bullish moment in American history. A week before, the members of Congress had jumped to their feet to lionize George Bush, chanting his name and war-whooping on the floor of the Capitol. The Democrats had worn American flag lapel pins, but they were bested by the Republicans, who paraded out small American flags that showed up better on TV. The Democrats had had to beg for a share of the flags so that they, too, as the *Times* reported, "could wave to the folks back home." A corresponding delirium on Wall Street prompted similar depths of patriotism. During the seven weeks of the Gulf War, the Dow had surged more than 500 points. Frenzied investors, suddenly making money again, spread their largesse among all sectors of the economy, big and small, profitable and nonprofitable. That the nation remained deep in recession and that the war had failed in its central objective of extirpating Saddam Hussein barely touched the good-time fever radiating from the nation's power centers. America was a winner again. The country was drunk on unexpected success. Improbable delusions were not only licensed, but nurtured, as if it didn't matter that you could see the artifice behind them, so powerful were the pleasurable images they conveyed.

For the emerging biomedical industry, it was showtime. The suddenness of Wall Street's and the nation's insobriety was matched by a near instantaneous parade of strapping young companies that had rehearsed for just such a moment, such a splurge. Each had a sizzling story about miracle cures, new technologies, riches beyond words, and each, like the Republicans and Democrats, was determined to capitalize on the rapture of the moment by trumping its competitors and upping the stakes. Encouraged by Wall Street, which had begun to enjoy the return of rich commissions and fast-growing portfolios, the companies rushed to announce bigger and bigger deals, earlier and earlier in their development, with less and less science to back them up. Companies no older than Vertex—and no closer to profitability—were now raising upward of $40 million in public offerings. They were being valued higher than manufacturing companies ten times their size with millions a year in profits.

On Black Thursday, the day Boger confirmed the news of

Schreiber's victory, an article in the *Wall Street Journal* announced the latest "mega-start-up," as Boger called it: a company rushed into existence to exploit the new climate on Wall Street and that had been able to raise in its first round of financing several times the $10 million that Vertex had been launched with just two years earlier. Details were scant; there was little to tell. According to the *Journal*, the company was yet unnamed, would be located on the East Coast, planned to develop "pharmaceuticals that piggyback on the cell's own mechanism to fight disease," and had raised earlier that week $30 million. The source of the information was—startlingly, if not surprisingly—Kevin Kinsella, whose Avalon Ventures was said to be launching the company with New York investor David Blech, a thirty-four-year-old former stockbroker and part-time musician who had already started and held large stakes in nearly twenty biotech companies. Blech's system of practically giving away stock to brokers and prominent business leaders who then have a stake in promoting his deals had earned him more than $300 million while placing him at the center of a network of influential investors, including Gerald Ford, Bill Gates, and former Citicorp Chairman Walter Wriston, all of whom sat on the boards of Blech's firms. According to Kinsella, the new company, which had no employees or labs and was still months away from opening, already had a market value of $45 million.

"If that's the standard for $45 million," Boger snorted, "then we're up around Amgen."

The *Journal* failed to account for how Kinsella calculated the extra $15 million, though Boger knew. Kinsella, who was still on Vertex's board and a major stockholder, had told him. The "piggybacking" was actually a reference to signal transduction, the new hot area in immunology. And the company had assembled an extraordinary SAB—"SAB of the Gods," Boger called it—that included several Nobel laureates and, more to the point, Schreiber, whose high-visibility work gave the field much of its cachet. Indeed, the company had been a direct result of Schreiber's firing. Angered by the loss to Vertex, Kinsella had hurriedly flown east in October to hear Schreiber's side of the story. Still miffed, and after inviting Schreiber to discuss his bitterness with the full board (Schreiber declined), he asked Schreiber what else was new.

Schreiber burst into his stock speech about signal transduction. Sensing a huge new opportunity that instantly overshadowed his regret at losing Schreiber at Vertex, he and his partners began at once assembling the new start-up that he now was flogging in the press and on Wall Street.

Boger had been stunned by Schreiber's involvement. He didn't view the company as a threat—"They're not going to be a factor for a long time," he declared—but he despised "the cuckolding aspect" and was astonished that Kinsella would set up another company so close in form to Vertex that they would likely end up cannibalizing each other.

"It's a grotesque situation," he rankled. "Kevin thought it was a terrible mistake to boot Stuart off the SAB. He wanted to keep him out of the hands of the competition. So now they're the competition together."

Echoed Aldrich, "I don't know how the guy sleeps at night."

It had been enough for Vertex's scientists to be beaten again by Schreiber scientifically, but that he now might also be competing with them in drug design all but unnerved many of them.

"Regardless of what you think of the guy, he isolated FKBP, he cloned FKBP, he expressed FKBP," Armistead said. "He's got an NMR structure and a crystal structure, and he's a *chemist.* Now he can do what he's best at. Hell, I think he can compete with us. I think he can compete with anybody."

"We've always said that the ones that are going to win in this area are the ones with the most information," said Livingston. "Stuart now has a lot of very good information that, in the right hands, could be very dangerous to us. I hope he sells it to Wyeth Ayerst, but there's nothing in principle to stop him from selling it to Merck."

Or, it went hauntingly unsaid, to Kinsella's unnamed company. It hadn't been lost on Boger that Schreiber might press Harvard to license his FKBP-12 structures exclusively to his and Kinsella's new firm as he'd once been willing to do with the protein to Vertex. The race for the structure was therefore now even more of a showdown between him and Schreiber. Not only were they combatants over enshrining chemistry and the primacy of immunophilins in that cause, but they were conceivably now rivals in the arena of drug design—the one area Schreiber had always sworn never interested him, the one area where Boger's ambitions were greatest. Before, he and Schreiber were failed collaborators. Now they were at war. As with all scientific conflicts, Boger recognized that the ultimate outcome might be settled not in the laboratory, but in court. He asked Ken to begin reviewing all possible legal claims against Schreiber and Kinsella.

Suddenly, Vertex was a much more dangerous place emotionally—dark, coiled, apocalyptic.

"Joshua thinks he's Christ and Stuart is the anti-Christ," explained Thomson, who himself took a more mordant view. He came to work the next day wearing a T-shirt that said Shit Happens. "To cheer everybody up," he said.

CHAPTER SIXTEEN

When Yamashita was eleven or twelve, his father, on leave from Vietnam, gave him a copy of Thomas More's *Utopia.* He was living on an army base in Germany with his mother and older brother during the endgame of the Vietnam War—a slight, rootless, Japanese-American sixth grader struggling ambivalently to fit in. His father, a medical technician, was a stern, remote figure whom Yamashita revered but seldom saw, and the book, about an ideal pagan city-state where everything is happily governed by reason, was a common antidote to their travails.

A third-generation itinerant whose mother still worshiped long-dead forebears at a bureau-top *butsudan*, a household shrine, Mason knew his father up to that point less as an intimate provider than as an icon, someone whose family history informed his and whose experience seemed an objectification of his own confusion. His father, Hisao, grew up one of seven children on a small strawberry farm in Gilroy, California, Garlic Capital of the World. Midway through World War II, when he was eleven or twelve, the farm was confiscated and the family interned.

"They were in Tule Lake," Yamashita recalls dutifully. "Grandfather had a heart condition, and during the final days of the internment, he got pneumonia and died in the camp just before he was supposed to be sent back to Japan. During the time he was sick, he'd been asked by the American government to pledge allegiance to America or Japan, and he chose Japan. Thus the whole family got shipped off to Tokyo even after he died.

"Tokyo," he continues, "had been destroyed. But Dad could speak English, so he got a job as a janitor in one of the mess halls. His whole family was starving. Since he was an American citizen, he went to Hawaii, got a job on a pineapple farm, and sent back all his money to Japan to keep his family alive."

Conditions on the pineapple plantations were brutal, and the elder Yamashita, who'd lied about his age and had had little regular schooling, eventually escaped by enlisting in the army. Shipped back to Japan, he met Mason's mother. "She'd had a horrible life," Mason says. "Her father was an alcoholic and left when she was young. Her mother had been going to work one day on her bike and it was very snowy and she was killed by a train. Mom was adopted. The family went to Manchuria. The Russians detained them there after the war. Then she finally made it back to Japan, met Dad, and moved to the United States."

Yamashita's own childhood was equally bearingless and disjointed if less harsh. Raised on army bases while his father was in Korea and Vietnam, he lived mostly in Hawaii until age ten, when the family moved to Germany. "I was always in the shadow of my brother," he recalls. "He was an incredible star: straight-A student, though also a little peculiar. He loved to perform these weird experiments on me. I remember once he went to the library and took out this Time-Life book on child psychology. He made me terrified of tornadoes by putting these images into my mind—a straw driven through a tree by the incredible force of the wind, frogs swept up from a pond by a waterspout, then raining down dead several miles away.

"He's phenomenal. In many ways he was more my father than my dad was. The first bad thing that ever happened to me was my brother leaving to go to Caltech when we were in Germany."

Utopia (literally meaning nowhere) launched Yamashita's moral education. Written between 1515 and 1517 during the flowering of global exploration, it was an attack by More against the greed and injustice of a Christian Europe recently grown genocidal in its thirst for power and wealth.

"I was more like Rich [Aldrich] then—a Republican—and Dad

would try to moderate me, try to make me more compassionate. I thought it was stupid."

He pauses: "It was strange. I personally think I would have killed myself if I had to go through what he went through. I would have thought that given his history, he'd have turned bitter, sour, but he didn't. As I get older, I'm getting to feel much more like him."

As an adolescent, Yamashita moved to San Francisco, then back to Hawaii, where he attended public school. "It was so boring. A lot of the time I would just stare out into nothingness, do nothing, for long periods of time. I probably was strange." His mind, however, was far from quiescent. He agonized over the right way to live and belong. He now was reading Camus's *The Plague*, in which the ultimate measure of a good life is altruism: "the passionate indignation we feel," Camus wrote, "when confronted by the anguish all men share." Meanwhile, he fell in with a grade school friend, a Pentecostal, and the two of them spent their Saturdays as amateur soul savers at a run-down shopping mall.

Torn, Yamashita ultimately renounced the church. "In *The Plague* there's a priest who admonishes his parish that the plague is a judgment placed upon them by God," he explains. "He eventually gets sick and dies, but mysteriously, he has no symptoms. Tarrou [Camus's main witness], on the other hand, has the most painful death. He has both bubonic and pneumonic plague. I think Camus is signifying that the religious man is shielded in his shell: He will live and die not really having lived, whereas Tarrou, having stripped away this shell, lives a very hard life, but a much more fulfilled one.

"It was a very difficult time for me, because the church was trying to drill me with rules. They were trying to build that shell around me. But I wanted to live life and suffer pain."

He enrolled at the University of Hawaii, burning with earnestness, intensity, and, soon, dismay. Emulating Camus's hero, he hoped to go to medical school, but got a C in sophomore biology. "I decided," he says, "to become a chemist instead."

Here was the tentative resolution to all Yamashita's wandering: "It intrigued me to think that there are billions of these little things that are actually doing reactions the way you want them to. I didn't believe it at first, because I had been trained so heavily only to believe that which I could see. But people were telling me to believe it, on the basis of spectroscopy [that branch of physics that includes both crystallography and NMR]. I have to say I had my doubts. To this day I look at crystals and say, 'They're nice, clear, very pretty, but are they really ordered? I mean, what am I really looking at?' "

In Yamashita's moral universe, science, truth, doing right, his ambivalence toward his brother, and atoning for his parents' suffering thus now merged. He became a scientist, it seems, out of an adolescent moral imperative. Science was stringent. It was built on order. That order resided seminally, ultimately in the smallest of objects and minutest of forces—the realm of the subatom. And yet if one was entirely careful, absolutely precise, one could see it, know it, control it. "We have such clumsy hands by comparison," he says, "but these hands can actually direct billions of molecules to all do the same thing—to pick up a hydrogen atom or drop one. It's amazing." Yamashita began to see in the interconnectedness of subatomic forces and matter a perfect paradigm of human behavior, for his own transitive existence. People are attracted and pulled apart. They're governed by simple, irremedial bonds: "the only certitudes they had in common—love, exile, and suffering," wrote Camus near the end of *The Plague*. In precision lay truth, even redemption.

There was only one great proviso: You had to be absolutely right. Otherwise, everything fell apart, collapsing hellishly into entropy and abomination.

Putting his trust in spectroscopy, Yamashita took his leap of faith.

With the backbone of the protein provided by Moore, Yamashita attacked the final stage of solving the structure with astounding resilience. Moore, as planned, finished first, making his last assignments in mid-March and rushing to write up his results. But Yamashita dogged him incessantly, narrowing the gap. Commandeering a space in the darkened modeling room next to Boger's office, he built his first crude "map" of FKBP-12 in less than two weeks, half of what would be considered a fast time.

It was lonely, painstaking, grueling work. Because X rays breeze through atomic nuclei, what diffracts are clouds, or "density," of electrons, like the chalk outline at a murder scene, not the atom's corpus but its shell, its aura. On a computer screen, this density is represented three dimensionally as contoured electronic "nets," holes in space. Packed together, touching in spots, they crimp and blob, twist and billow, like the oily "amoebas" of a sixties light show.

Yamashita, wearing battery-powered 3-D glasses and his ever present headphones, plunged himself into this subuniverse for twelve- and fourteen-hour shifts, seven days a week. He started in late afternoon and finished, dazed, as the first arrivals came in before 8 A.M. Knowing how the protein folded made his job feasible but no less trying. The screen was a fathomless black, superimposed with a dense matrix of billowing mesh and fragments of a twisting, coiling skeleton of 1600 atoms. The idea was to lay the structure end to end within the mesh, like a ship in a bottle, without breaking it or contorting its shape. Yet the matrix wasn't continuous. It was erratic, broken into blobs. The effect was like piecing together the skeleton of an extinct, unknown beast within the windswept fragments of its ghost.

It got worse. Yamashita's "chain trace"—the first rough map of the protein—was only a hypothesis. It had to be proven, proof being subjective at best, given all the biases and limits of trying to simulate unseen events with incomplete knowledge, imperfect understanding, and a group of technologies that, though improving, remain far from ideal. Not all density, for instance, is protein. With FKBP-12, thousands of individual water molecules floated within the recesses of the enzyme, each with its own electron cloud. Rotating a section of protein into a pocket of density, one could easily be outside the parameters of the enzyme without knowing it. Crystallographers inevitably find themselves having to explain away much unaccounted for density and defending maps that crisscross empty space. Computers check their assertions: Do they meet the criteria for how such atoms are known to bond? Are the bond lengths and angles "legal"—plausible? Yet, as Navia says, crystallography is a "garbage in/garbage out system. You need to do a tremendous amount of refinement to purge yourself of your original sin."

Working on the edge of this abyss, Yamashita chose to evade the moral terror of looking for truth through the bent eye of a needle, but there was a physical terror he could not avoid. Molecules, of course, exist in three dimensions; cathode ray tube images, two. The experience of depth comes, in part, from the ability to zoom in and out, up and down. Once inside a molecule, it's frightfully easy to lose one's orientation, slip into an unfamiliar plane, slide past the molecule's rim, get lost, hit a void. Zooming for hours through a mass of undifferentiated density can induce a harrowing vertigo. It perhaps didn't help that Yamashita accompanied himself with high-decibel rock and roll. One of his favorite anthems now was the Cars' "Just What I Needed," with its telling line: "Doesn't matter where you been, as long as it was deep." By the time he stood up to leave most mornings, he had a look of shell-shocked delirium. He was indeed in another place.

In fact, though outwardly detached, inwardly he was happy, at peace almost. The despair of the heavy-atom search was now behind him. After months of noisily resenting Navia, Boger, Moore, and his erstwhile rival, Merck's McKeever, he was at the center of attention. Any new protein structure was still an event—only about 300 had been solved. An important one like FKBP-12, he knew, guaranteed major press. And it was all his. In graduate school he'd have had to hand off his structure by now to someone else for refinement, but Navia, considering map building a "one-man operation," had let him alone. He even sensed from Navia a hint of envy, teasing in reply: "If this is what you want, you're going to have to take a $70,000 pay cut and start working eighteen-hour days." (Rejoined Navia, affording to be gracious, "Mason did an amazing job, but 99.99 percent of the credit is going to come to me. That's just the way it is.") Yamashita had become a crystallographer because crystallographers were kings. Imperiously, he now felt himself taking on his proper mantle, an assumption that wasn't always kind. "People will read my papers and Jonathan Moore's papers," he said, "but they'll give John Thomson's papers to their technicians and say, 'Here, get me some protein.' It's sad. John's busting his gut in the wrong field."

By the end of March, Yamashita was all but done, having only to complete his refinements. Rushing to have something on paper, he spent the weekend of March 30 and 31 writing a draft. Exhaustedly, gleefully, he brought it to the Immunophilins Project Council, Vertex's working group on FK-506, on the morning of April 1, a painfully sunny Monday at the beginning of what Nantucketers call—for its deceptive innocence, its sudden thaw in emotions, its unexpected venom—Hate Month.

The council was irritable and cross from the week before. The progress toward solving the structure had taken its toll primarily in the punishing demand for more protein. For months a crisis had brewed. Thomson, beleaguered, finally brought his frustrations to the council. "They want 200 to 300 milligrams tomorrow," he said, reviving an old complaint about the crystallographers. "We in biophysics have wanted that kind of quantity from the beginning and could literally use a gram tomorrow and haven't got a sniff of it. I'm a biophysical chemist and I've done very little biophysical chemistry for two years." Navia, furious, tried to remain politic and in control. Here he was asking for protein so that Vertex could crystallize FKBP-12 and FK-506 together, not for himself but to see how the two molecules fit, design a drug, catch Schreiber, beat Merck, validate Vertex's position before the world—everything that was most important now—and Thomson was hectoring him about his own experiments. Navia deftly proposed that Vertex hire at least two more people to prepare protein, but Thomson, suspecting correctly that Navia wanted the new hires to be in Thomson's group, thus consigning it to being even more of a protein "service" than before, suggested that Navia's group make its own protein. Navia exploded. He refused to let the issue die. Finally, Boger had to reprimand them both. "There's only one goal for all of the science that's supported by this company," he said, "and that's to produce a drug. These," he said, referring to Navia's and Thomson's concerns, "are foreign concepts to me." Stewing, Navia left the meeting to kick chairs and pound on tables; Thomson, to seethe and grouse. They hadn't spoken in a week.

Yamashita imagined his paper producing triumph and succor both. After the initial success with 367, the chemists were again stalled for lack of data. The company had pulled itself back into striking range of Schreiber. An X-ray structure was money in the bank. Handing out copies of his manuscript, he felt that he had single-handedly, heroically fixed several long-standing problems all at once, made things whole again. Children of concentration camp survivors often describe a deep longing to mend their parents' broken lives, rescue them from their shattered past, heal their pain, heave themselves on it like a grenade. Yamashita seemed to be acting on such an impulse. He seemed to think he might even reconcile Thomson, his friend and confidant, and Navia, his boss and father figure, much as a child of a broken home imagines patching up his parents' distressed marriage with a glowing report card. Generously, innocently, he listed the two of them as coauthors as well as Boger, Moore, Murcko, and several others, including the whole crystallography group, which primarily had provided moral support.

"Twenty-five guys, twenty-five cabs," Jim Rice, ex-Boston Red Sox slugger and malcontent, once said digustedly of his famously uncollegial team. Such was the lack of unanimity, of common cause, in the council's response.

Because no one but Thomson had read Yamashita's draft, the discussion focused not on the paper's content, but its existence: Now that Vertex appeared to have solved FKBP-12 both by NMR and crystallography, what publication strategy ought it pursue? Like Schreiber, Boger and Navia optimally favored back-to-back submissions, presumably to *Nature*, which would counter Schreiber's acceptance into *Science*. In tandem, the two papers together would tell a more complete story, they argued, and rebuff Schreiber point for point. Moore, however, objected. His paper was nearly finished. To hold it up for perhaps several weeks while Yamashita continued to refine his data, he said, would be "suicidal."

"My exact words were, 'I've thought about that [dual submission] and I reject that alternative outright,'" he said. "'I reject it on the grounds that if those papers are submitted together to *Nature*, the odds are very good that they will throw out the NMR paper or tell us to condense them. Basically, we were doing the same structure of the same protein, just by two different methods, and we were using the NMR structure to get the X-ray structure, so why not incorporate the NMR information into the X-ray paper?' Joshua did not like that response."

Boger, in fact, had other concerns. On Wednesday he and Aldrich would fly to New York to meet with Goldman Sachs to discuss raising tens of millions of dollars through a private stock offering. The frenzy on Wall Street for new biomedical issues had become wild, voracious, unpredictable, and Vertex needed desperately to get in on it or be left behind. The value of a well-timed, high-profile publication like Yamashita's crystallography paper, with its tantalizing suggestion that Vertex was now but a stone's throw from the Promised Land of structure-based drug design, could not be overestimated or oversold.

"It may be the difference between our ability to raise $10 and $20 million," Boger said.

There was a cavernous silence. Though he hadn't said so directly, Boger seemed to be implying he thought Vertex should withhold Moore's paper not for any scientific reason, but because Yamashita's would bring a better price. "That opened a lot of eyes," Moore said. "People were wondering, 'Are we doing science, or what are we doing here?'" Boger had raised the point, as he often did, only to be provocative. But to some, particularly Thomson, to whom science was an honor, an ethical imperative, the baldness of the equation—science and money—was devastating, beyond the pale.

Thomson bristled. For weeks his body language at council meetings had indicated a ripening contempt: slouching in a chair at the far end of the table, arms crossed, Ray-Bans cocked, he had sunk into stony opposition. Now, frustrated in his own work and his ongoing quarrel with Navia; bitter about the apparent auctioning of Vertex's scientific integrity; both critical and resentful of the crystallographers, whom he saw as having wasted protein with impunity, and yet who he thought were being coddled and elevated by Boger despite the fact that Yamashita's paper was nowhere near as evolved as Moore's; insulted by Yamashita's inclusion as coauthors people whose contribution to the structure he considered irrelevant and at being "used as a technician" despite his own backbreaking efforts with the protein, Thomson leapt peevishly to Moore's defense.

"I don't understand why NMR is being treated like the ugly sister," he scowled. "I myself am the only one here to have read both manuscripts, and Mason's still needs a great deal of work. Read it. You'll agree. It's not ready yet."

Boger cut him off. Whatever he might have implied, it was now clear that what he intended was just what Thomson was urging. He wanted everyone to read both papers, including Jeremy Knowles and Don Wiley, a crystallographer at Harvard. As the two most respected SAB members and those most familiar with enzyme structures and with *Nature*'s publication policies, he sought to blunt any further animosity by enlisting their counsel. Wisely, he thought the scientists needed a cooling-off period and saw no further advantage in discussion. Reluctantly, Moore and Thomson agreed to hand-deliver both papers to Harvard that afternoon.

Yamashita, however, refused to let the matter die. Though he and Thomson were friends, he felt unfairly criticized. Waiting for the meeting to end, he approached Thomson and demanded that Thomson explain more fully his objections to the paper.

"What do you really mean, John?"

"Nothing," Thomson said, walking away, trying to avoid a fight. "It's just not polished."

"Look, guys," Navia said, inserting himself between them, an unlikely peacemaker. "Let's settle down."

"No," Yamashita insisted. "I want it settled now. I want to know what John's thinking."

Thomson wavered. The conference room was emptying. He preferred to talk to Yamashita alone in private. He was rushing to beat the time difference to place a call home to Melbourne. But Yamashita persisted.

"You're obviously not going to let me go until you have something," he said impatiently. "All right. I want to know what half the authors on that paper did."

Yamashita jolted: "Ah. *That's* it! That's what this is all about."

Who would receive credit was not Thomson's sole complaint or even his main one, but it was the one upon which Yamashita seized. Having pressured Thomson into confessing it, he now started attacking him on it. Thomson recoiled, shaken. Of all Vertex's scientists, he had been perhaps the least self-serving, working ruinously at the most unglamorous tasks, chaining himself to the bench, a staunch company loyalist. Yet he now he felt himself being painted into a corner as a credit monger. Blindly, he snapped. "I don't want to be the person around here who's remembered for taking a *Nature* paper away from four people," he told Yamashita. "Do what you want. Just leave me out of it. Take me off the paper. I refuse to be listed as an author."

Raving, Thomson found Moore and the two of them collected Moore's manuscript and stormed out of the building toward Harvard. Whether they failed to bring Yamashita's paper deliberately or simply out of neglect, the point was the same. The sulfurous emotions that had once been reserved for Schreiber now flashed over inside the company. People were no longer talking, not because they hadn't succeeded, but because they appeared to have succeeded too well. "I pride myself on fairness," Thomson muttered in the car ride over, shaking his head. "And suddenly I'm the asshole."

Rubbing his face in his hands, Yamashita meanwhile sat alone in the darkened modeling room next to Boger's office. He was desolate. Everything had gone spectacularly wrong. In his manuscript, he had been careful to credit Moore for the piece of NMR structure that had enabled him to solve the X-ray structure, yet he had ended up looking as if he were trying to usurp him. He'd named a large group of coauthors in what he felt was a spirit of communalism that others would support and admire, yet had only managed to offend Thomson without whose protein there would be no structure and without whose ministrations he'd have known only grief over the past six months. He had tried to be moral, and everyone was furious at him.

The slender cord of scientific truth that was Yamashita's moral lifeline was starting to fray heavily against the jagged edges of scientific practice. Being right was one thing, but one also had to win, and Yamashita's moral code abhorred competition. He was in over his head and felt himself drowning. Unable to work, hating himself, he left Vertex and went home, resolving to quit the company, to quit research, as soon as his stock vested and he could afford medical school. At least there, he told himself, he would know the rules.

"I don't think this new age of scientific collaboration is going to work," he lamented.

• • •

Racing between meetings, Boger swept into Aldrich's office late the next morning to announce Wall Street's latest folly. "Regeneron," he snorted derisively, "$99 million."

Regeneron, a three-year-old biomedical start-up, which by its own admission was perhaps a decade away from making any money, had gone public—gone "out"—that morning at $22 a share. It sold 4.5 million shares, raising nearly twice as much money as it planned. Wall Street, already drunk, had chugged the offering like a college student inhaling beer through a funnel on spring break.

Aldrich glowered. He didn't have any more faith in the stock market than in Harvard. Regeneron had hit the market at the right time with the right story: It was working on treatments for the latest vogue disease, Alzheimer's, and had a $50+ million research partnership with Amgen. In the current climate, that was like having a deal with God to make immortality pills.

Yet to him and Boger, the company's "internals"—its proprietary technology, competitive base, and, most particularly, progress toward developing a drug and making a profit—were unimpressive. It had unclear patent positions on two nerve growth factors: natural proteins that *might* help reverse the deteriorating brain cells in people with Alzheimer's and Parkinson's diseases. But Alzheimer's especially was going to be a nightmare for any company trying to develop a treatment. No one had a clue how the disease worked. Did Regeneron have any evidence that it could make an effective, safe, deliverable drug to stop the rotting inside people's brains? Proteins worked well in test tubes but were notoriously hard to deliver within the body, requiring shots directly into the diseased area. Was Regeneron going to shoot its drugs into patients' frontal lobes? How many people would buy such a drug? How would it be tested? The only way to calculate whether a drug against Alzheimer's, where the clinical picture is clouded by spontaneous remission, actually works was to examine brain cells for the hard plaque associated with the disease. That meant waiting for hundreds of test subjects to die before one knew whether one had anything. It could be decades before one had enough data for the FDA. Meanwhile, the whole area of neuroactive drugs was awash with litigation: depressives and insomniacs suing for tens of millions of dollars because, they claimed, top-selling sleeping pills and antidepressants had made them psychotic; families of homicide victims coattailing, piling on; personal injury lawyers, seeing in the drug industry a huge, elephantine target, advertising for cases; sympathetic juries.

If one studied the fine print in Regeneron's prospectus, all this was implied, but few, it seemed, bothered. At that moment, the company's value was a bloated, absurd $341 million, which worried Aldrich far more than its shabby internals. To justify such a valuation, Regeneron would have to be a Fortune 500 company years before it ever turned a profit. As that was extremely unlikely, Regeneron's stock was all but inevitably headed down, perhaps right away. Now that the money on Wall Street had been lured back into risky biomedical plays, Aldrich dreaded it being scared off again by a sudden free-fall.

"If Regeneron tanks," he said, looking up desultorily from his computer, "it may burn the market."

Boger nodded. By his own admonition, the key to raising money on Wall Street was getting in and out at the right moment. Now, the duration of that moment, which he, like everyone else in the industry, had failed to anticipate, looked as if it might be tauntingly brief.

"It's going to turn very fast," he predicted. "A couple of billion is going to get soaked up and that's going to be the end of it."

Speed was everything now, speed and size, for Regeneron had also dramatically raised the ante. Tens of millions of dollars were suddenly insufficient when some small companies had several times that amount. The new, richer, more muscular start-ups could buy better scientists, add more projects, retain the value on their discoveries, all while withstanding the preproduct financial drought that yawned before the whole sector like the jaws of death.

"You can say what you want about Regeneron," Boger said, "but they're not going to die anytime soon. They can make mistakes for ten years. They may have nothing now, but they've got a lot of time to improve on that position."

Vertex had no choice but try to scrabble up to this new financial tier. But how? Boger thought it still best to try to raise the money privately. Compared with Regeneron, he thought, Vertex's internals were sterling, blue chip: It might be no closer to having a drug (though Boger doubted it), but it was going after proven targets with the kind of small molecules of which all previous best-selling drugs were made. The company was pursuing clearly established markets without the sort of capricious patent situation hanging over it that had decimated GI and could clothesline a protein company like Regeneron.

And yet Boger didn't delude himself about where the company stood or where it was headed. It was years away from profitability. Neither of its two programs could yield a drug. Its burn rate, though not irresponsible, was exorbitant and was destined to become astronomic. When it came to money, Boger had grown up in a tight, conservative world. His father's mother, from thick pioneering German stock, had an iron-fisted rule—"Don't spend capital"—so penurious that she had refused to bail out the family even when his father's squandering had caused it periodically to become strapped. With the Regeneron offering, Boger recognized instantly the sudden rise in stakes but was not about to exchange his veil, as he might compare it, for pasties and a G-string. He would make the case for a \$40 to \$50 million private offering to Goldman Sachs, enough to get through the next few years, and put off going public until Vertex's own internals were stronger. He would move quickly but not be stampeded. He would let events take over.

Boger favored keeping Vertex private for another reason: control. Publicly held companies were subject to intense scrutiny: from shareholders, analysts, regulators, the media, and ultimately the unwashed public itself. A parade of outsiders suddenly looking over one's shoulder. How's the company doing? What's it discovered? When will it have a drug, make a profit? Boger considered such questions anathema to science. More, like many scientists, he disdained the right of nonscientists to ask them. He was an affirmed elitist, saying, "The only problem with autocracy is that there aren't enough autocrats." Boger intended Vertex to become a major pharmaceutical company, a giant. He had left the best public company in America because he thought Wall Street's myopic insistence that it grow 20 percent a year had forced it to become cautious, constricted, myopic itself. As he had at Merck, he knew what he had to do to change the world, but it had to be done right or it wasn't worth doing, and public inspection, public *expectation*, could only interfere. Good as he was at selling, Boger couldn't glad-hand, and going public, whatever riches it might induce, would force him to placate forces he disrespected, exhort people he scorned. "I couldn't do anything where I had to deal with the idiot public all day long and smile about it," he said.

To Boger, the larger prize was not wealth or fame but success, winning. His every decision was calculated to optimize for that fact. But Vertex was still far from optimal even for the inevitable, transitive step of becoming a public company. It was barely two years old. It had fifty scientists working on two early-stage projects in temporary labs. Its managerial infrastructure was unfinished. "I wish this was all happening a year to eighteen months from now," Boger said after the Regeneron offering. "I would feel more comfortable with a time-certain outcome. Going public now would accelerate our growing up. It might force me to hire more senior management, maybe a CEO—something I would deny vigorously up until the day we go public but that is true. At the same time, it's infinitely harder to attract people to a public company. You lose your legal ability to issue penny stock. The company's morale rises and falls with the stock price, the circus atmosphere increases . . ." He paused. "Then again, you shouldn't underestimate the value of having $50 million in the bank. You lose some people but attract others. No question Manuel would have been here faster if we had that much at the beginning."

Aldrich saw in the Regeneron offering the seeds of "another long nuclear winter in biotech" and fretted Vertex's lateness in getting to the market. "There's all kinds of product out there, cranking, soaking up money," he said. But Boger was strangely uplifted. If Wall Street thought Regeneron was worth $350 million, what must Vertex be worth? As it had at the Vista eighteen months earlier, Wall Street's breathtaking gall, its utter shamelessness, amused and intrigued him like some exotic peep show. "It's really wild," he said. "There's no rudder in the whole process."

Boger left Aldrich's office as he'd arrived, laughing. Laughter was his all-purpose antidote, his innoculation against an ambiguous moral universe and the ironic human failure to apprehend it: He who laughs outsmarts the world. A lot of people thought Boger's laugh arrogant, self-congratulatory, and didn't like it. And yet it was also a measure of his fearlessness: He who laughs expunges doubt, pain. In a male-dominated world, especially, he who laughs leads.

Leading now, Boger entered his next meeting with his laugh solemnly squelched. He usually laughed hardest with the scientists, Robin Hood merrying his men, but he had not thought funny yesterday's eruption of competitiveness. Scrimmaging was one thing, but this was game time, and Boger was furious. He thought the scientists were being infantile, selfish, and he meant to correct the situation fast before it spread. "Strategic thinking about a project is everybody's business," he'd said earlier in the day. "There's no honor or distinction in having the last *Nature* paper before Vertex goes under because it can't raise any money."

Boger had singled out Navia, Yamashita, Murcko, Moore, and Thomson—the five scientists most involved with solving the structure and, with the exception of Murcko, the most aggrieved and divided. Thomson, nursing his rage, had yet to come to work, though it was now after lunch. The others had spent the morning fuming and threatening to quit. Now, they were like schoolboys caught fighting in the yard, hangdog but unrepentant. Boger, waiving his customary bonhomie and the rational, we're-all-adults-here presumption of his social experiment, rapped their knuckles hard.

"No one here but me is not expendable," he snarled at them. "Rewrite, cooperate, and smile, or you're fired." He then told them that having read both papers, he didn't think either one was publishable in its present form. "We were all devastated," Moore recalled. "He said some things in that meeting to whip us into shape that were not true. He said Jeremy [Knowles] had read both drafts and said neither was in any shape to be submitted to any journal. When I asked him later what Jeremy had said, he said 'Oh nothing, your paper was fine.'

"Oh," Moore said, switching voices, "Thanks for kicking me in the groin and then saying, 'Ooops, sorry.' "

Boger was more severe than any of the scientists had ever seen him. They had all at one time or another, usually when they weren't getting their way or losing to Schreiber, asked for a Mussolini, someone to direct them, tell them what to do. He had steadfastly refused, putting matters back on them. Now, however, he was steely, cold. Navia asked to say something, and he cut him off abruptly.

"I don't want to hear a word from any of you until tomorrow," he snapped, turning on his heel and bounding out.

By then, however, the next move was clear. Moore's paper would go on to *Nature* alone, and all efforts would be made to ensure its timely and orderly departure.

"After leaving UCLA, I had an incredible pride in crystallography," Yamashita said. "I had extreme trust in it. I also realized that if you optimized for speed, like I unfortunately do, you could make mistakes and you could cover your tracks."

Perhaps because a certain amount of unknowability was forgiven and because Yamashita trusted his mentors, he had been able to abide this contradiction. He had been faithful. But now, as he raced to refine his structure, his faith began to erode, displaced, not surprisingly, by an angry despair. He was working constantly, eighteen hours a day or more, and pushing himself and his methods to where he could convince himself of the absolute factuality of nearly any assertion.

Observed Murcko, "In all the things we do with proteins, what we're really doing is calculating all the forces between atoms: the stretches and the bends, the nonbonded interactions and the tugs between charged particles. But not all the equations we use to describe those interactions are accurate. Some of them are fudge factors. Some of them are thought to be correct even though the experimental data they're based on are wrong, only nobody knows that because nobody's gone back and double-checked the experiments. Some are pure guesses. There are assumptions, biases. There's user error. There's imprecision in the hardware and software. Sometimes they're actually determined from lots of valid experimental data, but that's unfortunately rare."

Yamashita became deeply, inextricably depressed. Despite the precarious methodology, bona fide protein structures had indeed been solved by crystallographers but often only after years of painstaking refinement and with elaborate teamwork. Yet he was alone, rushing now to finish within weeks. Achingly, he pressed ahead, lashed by a fury if not to succeed, at least not to fail. On a bulletin board above his computer he placed a drawing from a children's book of dogs feverishly digging bones in a yard. "Dogs at work," it read, "Work, dogs, work." Another picture was of a little boy dressed in ancient warriors' clothes. "All Japanese raise their children to be as militant as possible," he explained. The modeling room—dark as a cave and lined with the halos of Coke cans, styrofoam food boxes, compact discs, software manuals, and twin cyclopean computer screens—smelled as earthy as a gym.

He quit smoking because after every cigarette he felt tired for ten minutes—it was slowing him down—but he continued to drink, smuggling a bottle of vodka into the lab and sipping from it late at night when no one was around. The accumulation of map building and vodka made him surly. On a night not long after the manuscript fiasco, he hurled a glass at the half wall separating Boger's office from the lunchroom, leaving a golf-ball-sized hole. On another night he lit a small fire in the lunchroom sink, quickly extinguishing it and laughing about it afterward. "It's been very painful," he said on April 11, 1991, with only 30 percent of the structure still refined. "I think I'd be willing to sacrifice a finger not to have to go through this."

It had been a hectic, disparate day, typical of the period and of science. In the morning, Harding delivered to Jeff Saunders, one of the project's senior chemists, the latest animal results on compound 367 and two others that were structurally similar, 398 and 426. "It's available," he said wanly of 367. Saunders beamed. Despite Harding's effort at downplaying, both of them knew the news was critical. The molecule had already proven as active as cyclosporine in cell assays. Now, fed orally to mice, it survived the gut, gently penetrated T cells, and was still detectable in their blood eight hours later. The animals hadn't died and were showing no ill effects, scratchily scampering in their cages two weeks afterward. On the evolutionary ladder of drug development, oral availability was a hugely important rung, the difference between a promising molecule and a salable pill or, perhaps more to the point, between major profit and interminable loss.

Boger was ecstatic. He told the chemists to scale up production of the three compounds for Chugai, which would test them in larger animals. "I told them that if I saw a requisition for four plastic canoe paddles from Herman's, I wouldn't bat an eyelash," he joked about the sudden need to go from mixing micrograms to mixing grams, a millionfold increase. By the end of the year, perhaps, the molecule, or more likely a more potent descendant, would be tested in beagles, even primates. It might not be a drug, but it was a candidate, and Boger wouldn't be lying to potential investors if he told them the company had a promising new immunosuppressant now substantially along in development. It was a lot more, he mused, than Regeneron could say. Yamashita, meanwhile, spent the morning soaking 367 into crystals of FKBP-12. He had already done the same thing with FK-506, and those crystals were now on the X-ray beam, relinquishing data. Once he finished the structure of the native enzyme, and if the soaking experiments worked, Yamashita would attempt these so-called complexed structures in order to show how the molecules interacted, just as Schreiber and Clardy apparently had done. Snapshots, they would give Murcko and the chemists their first hard look at the relevant topography for making a better drug.

Around midday, Boger got a fax from *Nature* in reponse to Moore's paper. It was a rejection. The editors had decided not even to send it out for review. Boger was appalled, incredulous. *Nature*, like most important journals, relies on outside experts to determine a manuscript's scientific merit, yet the British in-house editors—generalists, by and large—had ruled Moore's solution of the structure of FKBP-12 unworthy of such peer review.

To Boger, such a perfunctory rejection was not only stupid, but unacceptable. He resolved to change it. He wrote back to *Nature*. His letter was strident, barbed. "I took all the 'you twits' out before I printed it," he said. He wrote that Moore's structure was the first by NMR of the hottest, if not most important, biologically relevant molecule in the world and should be peer-reviewed on that basis alone. He also made two points that the *Nature* editors, for their own reasons, could only find irresistible. He said that Vertex understood that *Science*, the magazine's premier rival, had received two papers purporting to have solved the same structure by NMR and crystallography and that Vertex was within weeks of also having the crystal structure by molecular replacement. The implication, none too subtle, was that *Nature* was about to be beaten to press in a crucial, high-visibility area and that Vertex could save it with the X-ray paper provided it first changed its position and sent Moore's paper out to be juried.

Editors of all journals, not just scientific ones, generally and regally adhere to a policy toward unsolicited manuscripts that is a variant of divine right: Once they've ruled, especially in the current cutthroat climate where publication is everything, that's it. *Nature*, ineffably British, is known especially to frown on overtly pushy displays of scientific gamesmanship. Boger's ploy, however, worked. While Moore was still smarting from the unexpected reversal of fortune, Boger called to tell him that *Nature* had reconsidered. It would send his paper out for review after all. Conducted by fax, favored by a global information network that, like a shark, never sleeps, the entire refusal, negotiation, and reconsideration had taken just less than twenty-four hours.

Boger was steaming on, crosscutting now whatever stood in his way. He and Aldrich met that afternoon with Merrill Lynch, which had taken Regeneron public and was Wall Street's largest underwriter of initial public offerings (IPOs). Unimpressed by Goldman's figures for a private placement, they'd decided to conduct a "beauty pageant"—bring in a parade of investment bankers, let themselves be swayed. Regeneron had already "tanked," as Aldrich predicted. "Merrill pigged out," Boger said. "They bumped the price, sold all their subscriptions at the IPO, and left no aftermarket." Regeneron was now trading at about $15, down a window-jumping 30 percent in just ten days. Boger unsurprisingly thought Merrill's suggestion that Vertex also consider going public a bit arch and self-serving.

"I'm looking for the answer to a very simple question," Boger said scientifically after the Merrill visit. "I want to raise $30 million: What's the best way to do it? The best way to answer that question is to get people favoring one approach over another—to the exclusion of the other—so excited about it that we can get some real information."

The following Thursday, with Goldman on hold and Merrill "ready to go," Boger and Aldrich met with an investment banker from Kidder Peabody. Historically one of three or four Wall Street firms to specialize in biomedical stocks, Kidder is much smaller than either Goldman or Merrill. It also was recoiling from a well-publicized series of disasters. In 1986, at the height of the big brokerage frenzy, it had been bought by General Electric for three times what anyone else thought it was worth. Some months later, its star merger strategist, Martin Siegel, was implicated in an insider trading scandal. Another broker was hauled from Kidder's headquarters in handcuffs; the company paid the SEC $25 million, presumably to get the government off its back; losing money, it cut bonuses, inciting an exodus of top managers that *Business Week* likened to a "meltdown." Kidder had a good track record for taking small companies public, had scaled back, taken account of itself. But it was still listing badly. It was conspicuously overshadowed by Merrill and the other large companies in the recent surge of biotech offerings.

Kidder's investment banker was a mild Harvard Business School graduate with a brisk, officious air named Al Holman. In his mid-thirties, Holman broadcasts earnestness. A slender, well-dressed, boyish-looking blond who has the smooth manners of ambition yet who likes to roll up his sleeves, he came to Kidder right out of graduate school in 1980 at that moment when both biotech and the stock market began bristling with fabulous expectations, and had survived the depredations of the past few years well enough to become one of its few remaining stars. A vice president and partner, he ran its Japanese investment banking group, raised money for a variety of firms large and small, and maintained his own list of clients. He had taken eight companies public in the past ten years and had evaluated hundreds more, visiting most of them in person. Flying up to Boston from New York with Kidder's biotech analyst, Bob Kupor, he was instantly taken by Boger.

"If I had seen in the last year fifty companies," Holman would say, he knew within fifteen minutes, "Vertex was at the top of the heap."

Holman was "stunned" especially by Boger's management ethos. "A lot of small companies are started, they get forty or fifty employees, and all of a sudden the president has a corner office, and he has an assistant, and everyone reports to the assistant. I remember walking through the labs and Manuel saying, 'Here's my office,' and pointing to a drawer in a filing cabinet. I knew that would play well with my constituencies."

Returning to New York, Holman and Kupor swept everything else aside in an all out effort to win Vertex's business. "It was the most intriguing story I'd heard in five years, so the issue became one of figuring out whether or not it was real and if Josh was real or not," Holman says. "We came back to the office that night, put together four or five people, and worked through the night and all the next day. We spent the next twenty hours on it nonstop."

To Holman, the key was weaning Boger from the idea that the company was still better off staying private. He had seen the entire cycle of the biotech industry and knew that the market, even with Regeneron and a couple of other companies "falling out of bed on their IPOs," would not soon again so heavily favor going out. Regeneron "shook everybody up," he says. "People said, 'My God! Rather than raise money privately at $40 million valuation, I can go out and raise money at $100, $150, $200 million valuation.' That disparity had not occurred since the mid-1980s. Regeneron broke through the barrier. Here's a company that's doing a $200 million valuation *even after it's cratering*. It presented an umbrella for all sorts of companies to come out."

In Wall Street's perverse calculus, Regeneron had become not an embarrassment, but an asset, a selling point. Nor was Vertex's lateness any longer an issue. "It was a totally virginal story," Holman says. "They hadn't been contemplating an IPO for seven months and talked to eighty people. That gave us the opportunity to say, 'This is how, if we were you, we would position it.' We were all sort of stunned by the market opening up, and none of us believed it was going to stay open a long time. So our recommendation was, If you're really interested in raising money, do it now, and do it as quickly as you can."

Friday was Boger's fortieth birthday. The following Tuesday morning, five days after their initial visit, Holman, Kupor, and the rest of the Kidder team flew to Cambridge with several copies of a fifty-page bound booklet outlining the case for an IPO. There were comparisons with other companies, market analyses, a week-to-week time and responsibility schedule showing all the regulatory and sales deadlines for a fast-track public offering. There was a grueling road show schedule set for early July, taking Boger and Aldrich around the world in two and a half weeks to talk to investors, that had them sprinting at the end between two U.S. cities a day. "The ultimate death march," Aldrich whistled respectfully.

Far from boilerplate, each of the charts and schedules was what Boger called "real stuff," the precise information Boger had presumed to get through the "beauty pageant." One graph in particular impressed him. It showed three ascending lines: the stock market, which had gone up spectacularly since the first of the year; the pharmaceutical industry, which was even hotter than the market as a whole; and arcing high above the other two, a clarion, new biotech issues. It was a rare alignment: the most money ever to flood into the industry, at a time when capital could make a hefty return in far more reliable areas but was deliriously pursuing companies like Vertex instead.

"This configuration will never again be the same," Boger said wistfully. "I don't want to be trying to go out when notebook computers are leading the market."

The labs crackled simultaneously that morning with their own news. *Nature*, in a near record turnaround, had accepted Moore's paper, sending terse congratulations in a one-page fax. It was twelve days since its initial rejection. As Moore would discover, a majority of the reviewers deemed the structure of FKBP-12 of such widespread scientific interest that they had urged immediate publication, regardless of whether or not the crystallography paper was forthcoming. Moore was dumbstruck, floating. "I'm just going to wander around and let people congratulate me," he said.

Of all the myriad implications of Moore's acceptance, the most compelling was Vertex's sudden and immediate validation. Most small companies and many large ones go for years without publishing in *Science* or in *Nature*, which, in particular, has been both arbiter and house organ to the great scientific revolutions of the past one hundred years. When Watson and Crick discovered the double helix of DNA, they announced it in a 700-word letter to *Nature*. It was thus exceptional that Vertex, in its first publication, should stake its flag at such a tier, especially after getting the normally impervious *Nature* editors to bend one of their cardinal rules. It was fully the kind of aggressive, gate-crashing role that Boger had forecast for Vertex from the start, and now, having been achieved, it was like a rush of hormone to the system. Boger, deep in discussions with Kidder, gloated appropriately when he heard.

Each of the paper's four authors—Moore; Debra Peattie, who had worked on the molecular biology; Matt Fitzgibbon, Thomson's assistant; and Thomson—enjoyed a fresh surge in status. Thomson especially gleamed at the news, though it distracted from neither his work nor his moral struggles. A year after first isolating the protein, all but recovered physically yet looking older and more haggard, he was again working around the clock, processing twenty-five pounds of thymus, trying to isolate a half gram now of protein. Shy of laurels, he received the congratulations of the other scientists with a practiced shrug and a smile. He was still angry at the crystallographers, and, one sensed, nursing other grudges that even the powerful vindication of the most prestigious paper of his career couldn't quite unseat.

Throughout the morning, the structure papers—Moore's and Yamashita's—were the hot topic in the labs while Boger and Aldrich entertained Kidder's pitch for going public behind the closed door of the conference room. Starting out separately, the discussions quickly, inevitably, merged. Now that Vertex was going to have one, and most likely two, such papers in print and was within sight of the information it said it needed to design a new drug, its internals suddenly looked much more sound, sound enough to support, if not a Regeneron-sized deal, something close. Boger had always said the money was the chief reagent that Vertex ran on; now, it appeared, the company could raise $50 million or more if it was willing to capitalize quickly on its scientific achievements, however preliminary and incomplete.

Kidder, meanwhile, wanted to know just how meaningful the structure of FKBP-12 was—not just its short-term public-relations value, which was considerable—to the claim of actually being able to design drugs. The line between business and science, always tenuous, now vanished. Boger summoned Murcko, Navia, and Harding away from a meeting where Murcko was busily showing some new potential inhibitor designs based on Moore's structure. For the next two hours, they uncomfortably answered Kidder's questions about what they did and the likelihood that they would succeed. Boger had long ago made his peace with selling speculation, with putting a triumphant face on uncertainty, with prophesying in order to leverage reality, but Murcko and Harding in particular now swallowed hard. As scientists they were trained to trust data and data alone, and Vertex's structural work, though promising, was still much too inconclusive for either of them to feel confident enough to support Boger's boldest claims. They said so. A prophet who turns out right is a seer; one who's wrong is a charlatan. Neither of them had Boger's desire or incentive to cast himself in that role.

Yamashita, 70 percent done with his refinement and pressing ahead, was too junior and too busy to be called in on the discussion, but the repercussions reached him just the same. It was ironic: At the very moment he was doubting severely whether one could determine with absolute certainty the atom-by-atom structure of a protein, especially within the pressure cooker of industrial research, Boger was needing more than ever to make the case for the unimpeachable veracity of the structures he and Moore had generated. In fact, the irony was darker than that. Now that his was nearly complete, Yamashita could see that his and Moore's structures were significantly at odds. "The scientific community will see no particular problem that we're this different," he said gamely, attributing the distinctions to different methodologies. But he was also haunted by the disparities. Navia was beginning to check over his structure, but he was being dragged away more and more by business meetings. Meanwhile, Yamashita was factoring in the water molecules, one of the principle techniques crystallographers use to explain unaccounted for density—and cover up any mistakes they may have made. "Manuel's reviewing my map," he said, "but I'm rushing ahead with the waters now so I don't have any incentive to listen to him."

Beaten by Moore, still at odds with Thomson, doubting both his discipline and his structure and despairing alternately about contradicting and disappointing Boger, Yamashita continued all afternoon staring desperately at his computer. He was exhausted, beyond consolation. He'd been staring bleary-eyed at density for twenty hours, since the night before, and had continued throughout the Kidder visit. Finally, shortly after the Kidder team left, he crumpled on one of the pink settees in the lunchroom, curled up, and began muttering to himself. "Life," he said, "is a series of intractable problems." Laura Engle, sitting nearby, tried to humor him: "You'll look back on this problem a year from now and say, 'Hah, I thought that one was bad, look at this one.' " Yamashita wasn't assuaged.

"I'm having a *nervous breakdown,*" he said, much louder this time, pleading. "I just want this to work. I just want to go home. I've been having bizarre dreams about killing everybody in this fucking place. I want to leave this fucking place. It's all I see every day of my life. I'm tired. I'm very tired."

He moaned plaintively and then stopped.

"God," he groaned, "hates crystallographers."

Navia, witnessing this, came across the lunchroom. "Move over," he said in a fatherly manner to which Yamashita responded by quieting himself. As long his seniors needed him, it seemed, he would serve them.

"We must get the structure right," he said later that night, protectively. "Manuel will be destroyed if it isn't."

Yamashita's moral plank, his idealism, was breaking beneath him, foundering on imperfectability, which he took to be a violation of truth. He felt there was no longer any possibility of victory for him, for even if he finished the structure and tied Schreiber, he could only do so by cutting corners and compromising principles, by lying. It was one thing to fail oneself, another to be failed by one's ideals, one's God, in a sense. Yamashita's disillusion was absolute, and he saw all those who did not agree with him as hopelessly and evilly compromised.

Only his protectiveness toward Navia kept him going. Whatever his personal feelings, his disgrace, he felt a deep responsibility to shield Navia from both the pain of losing and from embarrassment. Ever the dutiful Japanese son whose father has suffered and whose lifelong job it is to atone, he now gave up on saving himself but not on saving Navia, who indeed had once gotten a structure wrong and had been stung by it severely.

That was with HIV protease. In his urgency to finish, to get the structure first and get it out to the world, Navia had misinterpreted a small portion of the density, about 15 percent. The region was away from the active site of the enzyme and had no apparent bearing on its biological activity; for purposes of drug design, it was both correct and sufficient. But biophysics is an exacting field. "Unless a physicist invented and built the machine that the experiment was done on," Boger once quipped, "he refuses to believe the result." Purists attacked Navia for making "serious errors" that along with other recent flaws had undermined the biological community's faith in X-ray structures as "gospels of truth." In some academic circles, Navia was considered an apostate, a fallen man. "It's a cautionary tale that we have to be careful in interpreting these things," Alexander Wlodawer, who ultimately solved the structure correctly using Navia's and McKeever's crystallization conditions, told *Science*. "We're not infallible, unfortunately."

The episode, coming shortly after Irving Sigal's death and the unraveling of the HIV protease project at Merck, affected Navia bitterly, though not as Yamashita believed. It didn't cause Navia to doubt himself, as Yamashita did. On the contrary, it reinforced—case-hardened—Navia's precepts for doing commercial science and for working in the drug industry. It made him more determined than ever to put a premium on speed and practicality, his ultimate reason, he says, for coming to Vertex.

"The world has changed," he explains. "The reason why people do structures now is not for the purpose of doing structures. People who say, 'Well, structures are sloppier now,' have to remember that the structures that people first worked on were practically minerals, and they worked on them for thirty years and spent a tremendous amount of time worrying whether or not the experimental data for each reflection had the correct shape.

"Now the process is driven by biology. One of the things that was assumed on the first day I arrived at Merck [in 1980] was that 'you're not going to work on a structure because the crystal is available, but you're going to work on this *problem*, for which we don't even have protein.' Well, when you do that, what you end up with are structures that don't diffract very well—and HIV protease is not a good diffractor. But what are you gonna do? In the old days somebody would have looked at that crystal and said, 'We don't want to work on it. This crystal's a piece of shit, it doesn't diffract well, I'm not going to be able to deal with it. I'm not going to get good data.' "

"Well," he snorts, "we're talking about AIDS, a planetary disease. The human race is going to go extinct. So what am I going to do, solve the perfect structure? Of course not."

Far from feeling that he had abrogated morality, Navia believed he was practicing a higher ethic. "My personal mission is to use this methodology to make drugs," he says. "I'm not an academic. When you work in a place like this, and you're cranking out a structure in four months or nine months, you're cutting a lot of corners. You're *supposed* to cut corners. My job here is not to provide perfect structural information. It's to provide adequate structural information—adequate for the purpose of having the Mark Murckos of the world do something with it and give something to the guys in chemistry. It's not to provide them with perfect crystal structures two and a half years from now; it's to provide them with adequate crystal structures now.

"This," he says, "is the essence of what I was trying to communicate to Mason—that for us to have taken a year and a half or two years to solve the structure of HIV protease would have been *immoral.* You can't. There are people out there dropping like flies . . .

"So you cut the corner, you get the structure, you buy somebody a month."

On Thursday, the day after Yamashita's eruption in the lunchroom, Navia and Murcko began analyzing his structure, now all but refined. As Yamashita caught up on his sleep and then began rewriting his paper, they examined the position of each atom, both as it tracked through the electron density and as compared with the known geometry of protein formation. Mason's interpretation of the density seemed fine given the generally low resolution. But they soon discovered "holes" in the structure—ten- to twenty-angstrom gaps between sections of the molecule that, if real, would make it impossible for it to exist. The twists and turns of the protein chain were similar to Moore's, but the distances indicated that such a molecule couldn't hold together, much less function biologically. "At first we thought it was a programming glitch," recalled Murcko. They went back and reviewed Yamashita's calculations. By the end of the day, however, it was clear: The structure was wrong, not just in a single region, but in its essential formation. Whatever the image on the screen that had been generated, it was not FKBP-12.

Navia and Murcko waited until the next morning to confront Yamashita. Navia, despite his equanimity the night before, was incensed and wanted time to cool down. His claim to caring only about the therapeutic application of X-ray structures had been subsumed by more temporal, more fiery concerns. He had let Yamashita work alone on the most biologically intriguing structure in crystallography not only because he had faith in him, but because it was convenient. It had allowed him to do what he wanted, especially to advance and promote his idea for using enzyme crystals as supercatalysts. But now that plan had backfired. *Nature*, in which he had published his own career-making structures, including HIV protease, was expecting Vertex's manuscript. The company was endeavoring to make the crystal structure of FKBP-12 its main stand-in for a drug as it tried to raise tens of millions of dollars to survive. Though he would deny its importance to him personally, the scientific world was waiting hypercritically to see whether he would redeem himself with a flawless piece of work. And all he had, after nine months, was an unpresentable failure. Kicking filing cabinets, cursing, he had exploded. He blamed Yamashita for resisting him, for being persistently egocentric and immature. He knew he had to be calmer when he confronted him lest he add to his already precarious condition.

When they told Yamashita, he first became defensive, then hostile, then—Murcko observed—"suicidal." He ranted for nearly an hour. Navia, who had ended up yelling at him almost from the outset, also blew up repeatedly. The two of them stood screaming at each other in the modeling room, toe-to-toe, their snarling intonations echoing through the labs like dogs barking under a bridge. Boger came in. He tried to get them to focus on the science, but it was no use. Months of hostility and frustration, going back to the initial attempts to crystallize the enzyme, poured out of them. Finally, Yamashita, hysterical, lifted up an upholstered steel chair and smashed it so hard against the floor that it buckled.

"You will all be stricken down!" he screamed.

It took Boger to stem the panic. Objectively, what had been lost were just the two weeks during which Yamashita had been refining the structure. Now that the problem had been recognized, it could be solved. Boger instructed the two of them to begin revising the structure together, anticipating that they would have to stall *Nature* for a couple of weeks but confident that their bargain with the journal would still hold. Meanwhile, he did a quick analysis of the situation and determined to prevent it from ever recurring. "I hate rules," he said, "but this one I'm casting in stone. No structure at Vertex will ever again be solved by one person. It's too hard. All this could have been avoided if Manuel had looked at Mason's work two weeks ago." Added Murcko, whose own work had now been set back indefinitely, only half joking: "This is because of the insufferably large egos that all crystallographers have. It's a universal truth."

Like a catamaran righting itself after a spill in rough seas, Vertex quickly resumed its course and fleetness. Within days, Navia and Yamashita began closing in on a structure that, if not perfect, was, in Boger's careful phrase, "good enough"; good enough to satisfy *Nature* and the crystallographic community; good enough to draw attention away from Schreiber and tantalize Wall Street; most important, good enough to begin to design drugs. Boger never seemed to make any moral distinction between truth and utility: What was true was useful, what was useful, true. Life was for him not a series of intractable problems, but of tractable ones, and he was solving them now as fast as they arose. Those he couldn't vanquish, he simply let ripen until he had more data.

He was certain now that Vertex would go public, indeed that it would be perilous not to do so. The market was holding: other companies had gone out after Regeneron and were selling out their shares at prices above their preferred range. A company's ability to raise money by selling shares derived from a single feature—its valuation—and, as Holman had noted, the gap between what Wall Street thought companies like Vertex were worth and what private investors thought was simply too great to disregard. Secretly, on May 1, Boger flew to New York to meet with Benno Schmidt to secure his blessing as the company's financial rabbi and chairman before he told the scientists. Meeting in Schmidt's conference room, the two were as usual in instant, universal accord.

"If you're thinking of raising fifty million dollars, raise seventy," Schmidt twanged. "If fifty looks good, seventy is going to look even better." But Boger, smiling, auguring, was already with him.

CHAPTER SEVENTEEN

Propelling both business and science, Boger had always behaved with the easy, sweatless confidence of a grand master playing chess with a dozen or so opponents in a show match. Now, with Vertex pursuing Schreiber, an IPO, a drug, a new project, and publicity for its first important publication, he seemed challenged for the first time. It was as if the games suddenly interlocked in three dimensions. A move in one varied the play in all. The time for each move was halved. Real money was on the table.

On Tuesday, May 7, 1991, less than a week after deciding to take Vertex public, Boger arrived in Cambridge before 7 A.M. with two goals utmost in mind. The previous Saturday, Schreiber had mentioned in a talk at Yale that his FKBP-12 structures would be published in *Science* that coming Friday. For Vertex to be able to claim a tie on the X-ray structure, that gave Yamashita and Navia a deadline of no later than noon Thursday, when advance copies of *Science* would be released to the media. After that, their assertion of independent discovery would be irrevocably stained. At the same time, Boger fretted being shut out in the initial burst of publicity that no doubt would follow Schreiber's papers. Moore's structure was scheduled to appear in *Nature* the following week, but *Nature* had a prepublication embargo that prohibited researchers from announcing their work without its consent. Violating the embargo and jeopardizing Vertex's relationship with the journal were out of the question. Also, now that Vertex was going public, it was barred from seeking any publicity that might inflate its stock price. Not yet technically in the federally mandated "silence period," the company nonetheless had to be careful about what it disclosed lest it delay Kidder's tentative offering date of mid-July until August, when Wall Street goes on vacation. Boger knew that sooner or later, probably sooner, the market in new biotech offerings was going to crash. He shuddered to think about waiting until September to go out.

He spent most of the day cloistered with Aldrich, Ken, and the Kidder people, working on the red herring, the prospectus that Vertex would submit to the SEC and eventually investors. At 10 P.M., he was still at his desk. With less than forty hours left, Navia, Murcko, and Yamashita were racing furiously next door to reconstruct the protein.

"We're going to have to work like crazy to get FedExes out tomorrow and get *Nature* to break its embargo and get the lawyers to give us an opinion that we're not illegally hyping the company so that the SEC won't put back our offering," Boger groused convivially. "It's ugly, I hate it, but it's part of the business."

Murcko, overhearing his lament, muttered, "We're going to have to get that guy back to being a scientist." But of course the matter was moot and, in Boger's mind, beyond distinction.

"Are you going to fire me for this?" Yamashita asked Boger at around 8 the next morning. He had been working all night.

"I have to wait for the data," Boger chortled.

Typically, Boger was 70 percent joking and half serious. He in fact blamed Yamashita for very little, thought he had done an extraordinary job under crushing circumstances, and would support him even if he failed again to get the correct structure. On the other hand, having X-ray structure mattered immensely now and Boger wasn't paying Yamashita, who had bounced back from his moral crisis with characteristic resilience, to fail indefinitely.

Far more urgent was the exquisite irony of Vertex's public relations conundrum. For almost two and a half years, Boger had sold the company's story mightily, doing all that he could to make it sound solidly grounded in science. He'd "worn out three pairs of tap shoes," he said. Yet now, at precisely the moment when it had a major discovery to announce; when *Business Week* and *Fortune*, in their expanding coverage of the biotech boom, had discovered Boger and were planning articles about him and Vertex; when the publicity benefits of positive media coverage would be incalcuable as the company went public, he was forced to be guarded, silent. To the salesman and the scientist in him, the incongruity was painful—"smothering the baby," he called it.

"I blew off Gene Bylinsky from *Fortune* this morning," he told Aldrich, gulping exaggeratedly in disbelief.

"Goes against the grain, doesn't it?" Aldrich said.

"It's galling. Anybody who wants to get around the SEC, to sell junk stock, to bilk widows and orphans out of their life savings, is not going to be hurt by these rules. The only people who are hurt by this are the ones who try to play by the rules. It's not protecting anyone. It's enough to turn you into a raving conservative."

Boger's first priority was to get *Nature* to lift its embargo. He might not be able to tell *Fortune* that Vertex was hard after Merck in developing a drug, but there was nothing in the SEC regulations to prevent the company from issuing a press release about a vital discovery with clear importance to the public health. The issue there was simply, if critically, one of timing. If Vertex announced Moore's structure at the same time as Schreiber, Karplus, and Clardy announced theirs, the media would report a tie and Vertex could exploit it accordingly. If Vertex was forced, however, to withhold its announcement for even a week, there would be no coverage to exploit, just the crumbs of a few second-week stories in the science press. Moore began the morning by calling *Nature*'s Washington office. Reminded that the journal's embargo was meant to discourage exactly the kind of commercial exploitation Vertex was proposing—exploitation that *Nature* believed did nothing to advance scientific inquiry—he charged back. He explained that *Science* was about to publish back-to-back structures of the protein. An hour later, a *Nature* editor in Washington called back to say the journal had changed its mind.

Moore sped to tell Boger, who called Ken, who told him that he and Kidder's lawyers, who were paid to be skittish about such things, now agreed that the SEC had no case to delay Vertex's offering. Hanging up, Boger strode triumphantly to the lobby, where a stack of press releases was waiting to be faxed to the *Wall Street Journal*, the *New York Times*, the *Los Angeles Times*, the *Washington Post*, the *Boston Globe*, *Business Week*, and perhaps two dozen other trade, industry, and scientific journals. "The blockade is down," he announced. "The ships are coming through. I want to blitzkrieg right now. When the press accounts come out, I want the footnote to say, 'There are also some people at Harvard who claim to have done this.' "

Boger hadn't won, but he knew now he hadn't lost. Reflecting on catching Schreiber, his overall glow was only slightly less boistrous than if he'd beaten him outright. "It's not quite what I want," he explained, returning to the lunchroom. "I want to rub his nose in the dirt and step on his head. But I'll settle."

To the scientists, *Nature*'s second reversal in a month was illuminating, improbable, and not without its jealousies. "This is the value of working in hot areas," Moore told Harding. "*Science* and *Nature* have no idea whether any of these structures are right, but they'd obviously rather publish shitty structures of important proteins than exquisite structures of boring proteins." Swept up in the tide of anti-Schreiberism, he added, "We're going to spoil his party. We're going to be a very sharp dart in his ass."

Harding was sullen, dismal. While the spotlight had shifted to Moore and Yamashita, he had been working since January to identify other FKBPs. Characterizing one at last, he'd just learned that morning that SAB member Steve Burakoff had a paper in press identifying the same protein. "They want to do their work, that's fine," Harding grumbled, referring to Burakoff and Barbara Bierer, Burakoff's chief collaborator and a Vertex consultant, "but there's never any sharing of any information." He was now working furiously to file a patent application to establish a priority date before Burakoff's paper appeared in print. It was another instance of Vertex being beaten by one of its own advisors, and Aldrich, furious, now wanted to fire the entire SAB.

Harding, long-suffering and perplexed, shuffled back to his desk, a pale shadow of Moore's and Boger's fire-breathing personae. It was to be for him another long, lightless day.

• • •

Throughout the morning and into the afternoon, Mark Murcko, Vertex's chief molecular modeler, performed "dynamics runs" on the protein. Now that he had a refined crystal structure of FKBP-12, the key questions were, What does it do and how does it do it? A simple snapshot wouldn't do. Murcko had to simulate the molecule's natural activity, bring it alive, posit its existence in not only space, but time. He had to "feel" it. "Karplus's group did this a month ago," he said. "They had twenty postdocs and access to an order of magnitude [ten times] more computing power. It's that against basically a couple of us, but who am I to be different?"

Sitting at a workstation diagonal from where Yamashita was completing his refinements, Murcko programmed it to simulate the protein's natural wriggling. Atoms gyrate in nanoseconds—billionths of a second. To slow them down, Murcko used the computer like an ultrafast strobe, freeze-framing every 0.3 picosecond (0.3 trillionth of a second). He then ran the images back in slow motion. They resembled the "washboarding" of bees, a herringboning, a shimmy. The entire dance lasted just a few seconds, though in it Murcko thought he saw hints of how the molecule might work.

He was especially interested in the active site. Most of the molecule shuddered only slightly; it was highly constrained. But at the opening of the hollowed-out core that Navia and Yamashita speculated was its "business end," the place where its atoms interacted with other molecules, including FK-506, Murcko noticed a remarkable flexibility. On one side was a group of atoms that he compared to a flap. At one point during the run, about midway, the flap swung open perhaps thirty degrees, like a miniature gate, then snapped shut. The implications for drug design were profound. If true—and Murcko was quick to volunteer his usual reservations and disclaimers—this movement in the flap area suggested several possibilities about the enzyme: that in binding it could accommodate larger groups of atoms than the size of its opening would normally allow; that the flap might act like a trap, holding the bound molecule in place once it had infiltrated the active site; most tantalizing, that the shape of both the protein and the small molecular binder, such as FK-506, changed significantly as the flap curled around the inhibitor. Murcko began to speculate that unlike in Schreiber's hypothesis, which was that part of the drug bound to the enzyme while the rest reached out and interacted with another protein, there was a subtler geometry at work. He thought the combined entity of drug and protein might form a new shape and that the new conformation accounted for its action.

In the race with Schreiber, Murcko's simulations provided a potent new thrust. Vertex had hoped only to catch Schreiber, but now, as Navia put it, "We've got our own story." Navia, particularly, was elated. "This paper is going to be a killer," he told Murcko. "It's going to be a lot more interesting than I had dared believe." Privately, he began to look forward to a full-scale debate with Schreiber, a high-profile slugfest that would avenge the split between them and climax, most probably, in a head-to-head confrontation at the FK-506 conference in Pittsburgh. There, with the scientific world watching, Vertex for the first time might challenge Schreiber's priority in the field by offering a new theory on the central issue of how the protein worked. At the very least, it now had something to offer as scientific interest shifted to structural biology.

The deadline for submitting the paper—it was now Wednesday afternoon—was less than twenty-four hours away. Yamashita was scrambling to finish the structure, though in fact his contribution—and the pressure he had felt—had already begun to recede. All he had left were a few refinements around the active site and to account for the water molecules in his density. Otherwise the onus was on Navia, who had worked until 3 A.M. the past two nights drafting the paper for *Nature* and would shepherd it through to completion. It was strange. Yamashita had wanted to solve the structure of the protein more than anything in the world. He had pushed himself dangerously for more than a year to do it, been beaten, recoiled bitterly, recovered, blundered badly, collapsed, recovered again, and was now within a day of being lead author on what was likely to be a career-making paper—a paper that Navia, his "nemesis," was writing for him and that seemed all but certain, based on prior agreement, to be accepted into one of the world's premier scientific journals. He had gotten, it appeared, almost everything he wanted, yet the pain of getting it had been so great that he felt no pleasure, only a comforting distance, as though he were alone in a life raft drifting from the scene of an excruciating disaster—the wreck of his own life. His prayers nearly answered, he was wary, numb, cynical, and oddly composed.

Besides cleaning up his final structure, he had one major matter left—authorship. As lead author, it remained Yamashita's responsibility to name those who would share credit with him, a choice that had previously opened rifts that he now hoped to close. He and Thomson had shakily repaired their friendship after their fight a month earlier, but Yamashita had not brought up the issue, and Thomson had continued to insist that he didn't want his name on the paper. Meanwhile, Thomson and Navia had reached an icy standoff, with Thomson complaining about what he considered Navia's ineptitude with the protein and Navia insisting that Thomson was blackmailing crystallography by holding protein hostage. Working again around the clock, Thomson had developed a severe case of conjunctivitis and wouldn't be in until after four—an hour away.

Yamashita sought Boger's counsel. He wanted Moore and especially Thomson back on the paper, though he understood Boger's determination to get the scientists to think less about individual credit and more about drugmaking. Boger was snappish; he'd heard Navia's complaints, and though he didn't believe them, he wanted to make sure Yamashita wasn't including Thomson simply to pacify him. "I own you and I own John Thomson, and there'll be no quid quo pros here," he said. Yamashita responded that he also didn't think Thomson was witholding protein, but that he wanted to share authorship with him so that Thomson would continue to talk and work with him. "If you do that," Boger said, "John's name won't be on the paper and John will be fired. Tell me how that's different from going into your dry cleaners and saying, 'This is a dangerous neighborhood, and it's going to be even more dangerous if you don't give me $500.' It's extortion." Yamashita left more confused than ever.

In the modeling room Navia was asleep in his chair, sitting upright, his tie still straight, looking comfortably ursine with his broad chin sunk to his chest. He was snoring lightly. Nancy Stuart came in. Stuart, Aldrich's chief assistant, had become Vertex's all-purpose factotum. Monitoring the project councils, developing marketing strategies, she strove to keep the scientists on task. At thirty-two, she was the person on the business side the scientists most trusted, having worked for a time at the bench herself and having a natural empathy for people. Gently, she tried to shake Navia awake. He didn't budge. She tried again, Rolfing him this time.

"I need to talk to you about our *Wall Street Journal* strategy," she said. "Everything has to go through [enterprise reporter David] Stipp. But you've got to call Waldholz." Veteran medical writer Michael Waldholz had covered the HIV protease story for the *Journal* and was one of the few reporters any of the scientists, including Boger, knew personally, much less respected. Sluggishly, Navia removed himself from his seat, from science, and placed the call.

Yamashita stared at a new field of density. He could identify the active site in his sleep, but he was looking for something else, evidence of Vertex's lead molecule, 367. By itself, the structure of FKBP-12, much as it had nearly destroyed him, was insufficient for drug design or figuring out how the molecule worked. The ultimate test was to solve the structure of the enzyme bound by other molecules, so-called complexes—in other words, to show the lock together with a variety of keys so that researchers could see and compare exactly which points made contact. Schreiber and Clardy had apparently already solved the complex of FKBP-12 and FK-506. Having complexes with Vertex's own compounds would show what changes the chemists might make to improve the activity of their molecules.

There were two ways to grow complexes, equally devilish and uncertain: soak the drug into existing crystals or, more tryingly, coax them into crystallizing as a unit. Yamashita had tried first to soak in the drug, but now, as he looked at the results on the screen, there was no additional pocket of density in the active site, no unaccounted-for netting. The soaking experiment had failed. Discouraged, Yamashita reported the result to Murcko, then left immediately for the cold room to begin trying to crystallize FK-506 and FKBP-12 together. "We know that it must work because Stuart's done it," he shrugged. "Anyway, its the logical next step."

"It's a serious blow," Murcko said, looking up painfully from his simulation. "Instead of going home tonight at midnight and having the first glimmerings of where 367 sits in the active site, we don't know. I have no idea when I'll know. Is it a week? A month? Three months? Six months?" He stopped himself, frustrated. After more than a year at Vertex, he still didn't have the information he needed to design with any real conviction new molecules, and the gap between Vertex's promise and its reality had begun to cause him to complain more than he liked. Science was tenuous enough without one's having to doubt whether one was a fraud. "I'll continue to do things to be useful," he murmured, "but in the absence of real data, anything I'm going to say is going to be speculative."

Boger was in command mode, managing a half dozen issues at once, another half dozen roiling at bay. He was jocular, intense, flippant, stern, unfiltered. He was moving from chessboard to chessboard with rifling speed. At his desk, the fingers of his right hand rolled over his turbomouse like a magician's over an eight ball as he stared imperturbably at the big screen of his Macintosh, his looking glass, his muse. He juggled phone calls—with Ken, with Holman, with Schmidt, with Kinsella—with round-robin visits from the scientists, Aldrich, and Nancy Stuart.

His ego, never restrained, was incendiary. To Ken he suggested calling the Harvard patent office "to put the fear of God in them." Vertex no longer had any claim on Schreiber, but now that it had pulled nearly even with him, Boger wanted to spray his path with tacks by suggesting that the timing of his agreement with Clardy might have violated his Vertex contract. Flipping through another company's red herring, he dug into an imaginary trench coat. " 'Pssst. Hot area. Inflammation,' " he sneered, then spat, "It's a technological Ponzi scheme written by investment sharks . . . It's Florida land that's land only six hours a day." Talking to Aldrich, he resolved to tell Burakoff: "We consider what he's doing a serious breach of contract and he's worthless to us as an SAB member as long as he's working in the same area." Taking pleasure in warning Kinsella that he might have to cancel a flattering *Wall Street Journal* article about him—a profile that Kinsella had eagerly coveted to promote his new company with Schreiber—he cackled when

Schmidt, on another line, couldn't remember Kinsella's name, referring to him as "that guy on the board who put in the least money."

Knowles, who had seen Boger this way ever since he was in graduate school, demurred in his call to Boger, "Perhaps I'd better be out of town for a few weeks."

Boger popped into the lunchroom. Ever since he'd started Vertex, he'd intended the area to be like his kitchen when he was growing up, a hub. It was here that the scientists, each reeling in his own orbit, took their pit stops. Scarfing some Ritz crackers and a juice, standing as always, Boger sidled toward Yamashita, who sat picking through the *Times*. Towering benignly, he told him the price of the chair he'd smashed was $280 and that it would be deducted from his paycheck.

"When I was a kid, each of my brothers and I had a spitting tree," Boger said. "When we were really mad, we could spit. I used to slime that tree all the time."

"I like to physically destroy metal," Yamashita said.

"You're too high-tech."

"Maybe. My lab chief in graduate school gave me an old terminal and a sledge hammer to smash it with," Yamashita said. "It was wonderful."

"You look relaxed now."

"And I didn't even have to destroy anything."

It was a loose, giddy moment, the first between them in weeks, and Yamashita seemed relieved to be able to joke about the violence that had only recently threatened to consume him. Indeed, nothing now was taboo. When Navia came over to congratulate him on his refinement, saying "These geometries are fabulous," Yamshita replied, "Or a fabrication."

With Boger providing the emotional cues, there followed an outbreak of verbal riffing that culminated, inevitably, with a mass skewering of Schreiber. The structure of FKBP-12, though it had been solved by Schreiber's and Karplus's graduate students and by Clardy's group at Cornell, was clearly going to be Schreiber's moment. By the rights and rules of science—credit accrues upward to the senior investigator; initiating an idea counts more than proving it—even the X-ray structure, which Schreiber had contributed to mostly by providing protein and synthesizing heavy-atom derivatives, was likely to be known as "Schreiber's structure." His string of path-breaking discoveries with FK-506 remained unbroken, and his crossover into biology was progressing even better than he'd hoped. And yet on the day of perhaps his most notable achievement, he was about to find himself sharing honors with an upstart company that had ingloriously fired him eight months earlier, a place where he was reviled. The scientists loved it, though clearly not as much as Boger and Aldrich, who gloated radiantly.

"Imagine Stu-Bob [Boger's and Aldrich's name for Schreiber] driving down Storrow Drive in his Porsche on his way to Harvard, hearing it on National Public Radio," Aldrich said. "He'll wrap it around a tree."

Navia picked up the theme. Humming the familiar introduction to "All Things Considered," he pretended to be driving self-satisfiedly. Suddenly his face contorted: "Screeech! Craaash!"

Only Thomson refused to join in. Sitting away from the group, glowering at Navia's clowning, Thomson nursed his conjunctivitis and a painful upper body rash that afflicted him during long sieges of cutting tissue and washing glassware. "Nobody appreciates what we have to go through," he said. "A lot of unpleasant, up-to-our-armpits-in-slop work that's caused a lot of ailments [he pronounced it "I'll-ments"] that if we were in an industrial setting would qualify us for workman's comp. For every one hundred milligrams of protein we produce, we have to process a fifty-five-gallon drum's worth of phosphylated thymus juice with a lot of nasties in it in one-pint chloroform jars. But we do it.

"I'd like to know," he muttered, "what some of the others do."

"Josh and I don't want to be quoted," Aldrich told the disembodied voice of Vertex's PR man on the speaker phone in his office. "Nothing about billion-dollar drugs. When calls come in, if they ask for me, the front desk is going to say I'm unavailable. If they ask for a scientist," he nodded to Moore, who had been corralled for the briefing, "it'll kick over to you."

Aldrich wasn't opposed to exaggerating claims on principle, but he knew that Vertex had to be exceedingly careful about how it handled press inquiries about Moore's paper. Moore sensed his caution. Though he thought the rehearsal unnecessary and disdained being coached by a PR person whom he couldn't see, he was eager to cooperate. Like a rookie pitcher in his first outing, he thought he'd play it safe. "I'll just say it's an important step," he volunteered.

"Facts, not suppositions," Aldrich advised. "You can say something like, 'This is an important step, but it'll be several years before there are drugs to show from it.' "

The PR man counseled, "You don't want to be in the position, Jon, of saying 'No, no, no. I can't answer that.' "

"I could just tell them research itself is never patentable."

"You don't want to say that," Aldrich interrupted, joking. "You never know what we may want to patent around here."

Aldrich and Nancy Stuart had prepared a backgrounder, explaining the importance of Moore's discovery and answering hypothetical questions. The PR man complimented them and, as was his job, intimated great things.

"We're going to put it on the wire so everyone and his grandmother will know about this," he said.

"That's OK with us," Nancy Stuart said, smiling. Aldrich rubbed his chin thoughtfully. Moore tried not to roll his eyes.

Sitting next to Navia in the X-ray lab at around 7:30 P.M., Yamashita hastily concluded that the solution to the authorship problem was to remove his own name from the paper. Quickly, he reconsidered. "Camus says it's stupid to try to make a point to the world by committing suicide," he said. "In a likewise fashion, if they give me first authorship and even if I might feel very strongly I shouldn't take it because other people are being screwed, does it help for me to say no?

"I may be unexpectedly ill tomorrow."

Navia leaned over from his microscope. "One of these moments," he said, "when you get a chance, I'd like you to see what a spectacular job you did."

He was scanning Yamashita's cocrystallization attempts with 367, not those from that afternoon, but others from earlier in the week.

In several of the wells, he saw incipient needlelike crystals growing in clear formation. Unlike a year ago, when Yamashita's first putative crystals of FKBP turned out to be salts, Navia had no doubt these were protein.

"Oh, man. That's crystalline."

"Do you think so?" Yamashita said, tempering himself. "We'll see tomorrow."

"No, Mason, I think you've got it."

Slapping Yamashita on the back, Navia left to rejoin the paper-writing effort. Obliterating all further thoughts of martyrdom, Yamashita shrugged.

"That's the way life goes," he sighed. "I've learned from being here, you just find the next experiment to do. That's what keeps us from destroying ourselves."

Boger charged from his office a half hour later with a first draft of the *Nature* abstract. A summation of no more than one hundred words, it was the most critical part of the paper, the "take-home message" most readers would scan as they foraged through the journal. Boger, who'd taken several stabs at it while Navia updated the text, handed it to Navia without a word.

"How far do we want to put out our necks?" Navia asked.

"What 'we,' Kemo Sabe?" Boger said. It was yet to be decided whether Boger's name would appear on the paper, though as a general rule, he, like Tishler, opposed being named except in those cases where he'd made a direct contribution. He was less interested in his résumé than in the larger prize of producing a drug. Credit, he thought, was the prerogative of those who did the hard intellectual work of eliciting data, not those who simply set it in motion or interpreted it comfortably afterward. A chemist, he applied the chemist's restrictive code for deciding whose name to put on a patent; either you drew the cartoon or you made the molecule. Everything else was superfluous.

It was after 9:30 when he, Murcko, and Navia stopped to eat. Tearing into containers of pan-fried dumplings, pan-fried vegetables, and egg rolls, they ripped with equal ferocity into the draft. It was the first Boger had eaten all day.

"Here's what I'm after," he said, legs swirled, pinching a dumpling with his chopsticks, stabbing the air for emphasis. "Stuart already owns the binding domain, effector region hypothesis, but I don't hear any enthusiasm around here for that hypothesis. I think the answer is going to be that something about it is right, but that it's not enough. I want us to stake out the alternative hypothesis. I want us to put forward the last great hope of biological explanations, a hypothesis through which you may be able to drive a truck-train later on but that stakes out new ground now."

Navia told him about Murcko's latest simulations in which he'd modeled one of Vertex's best inhibitors into the enzyme's active site. "It may be too small," he said. "The flap may be coming in and crushing it. We need a doorstop for the flap."

"A molecular doorstop," Boger said tantalizingly, as if finding the light switch in a dark room. "It's a larger hypothesis than Stuart's, but entirely consistent with it. Not only do you need an extended effector region, but you've got to hold the flap open just so." In other words, Boger speculated, it was not just the atoms that stuck from the bound conformation like a gaff that accounted for FK-506's biological activity, but the way that those atoms changed the outer shape of the protein, like a tongue pressing hard inside a cheek. It could be a critical observation for drug design as well as a direct challenge to Schreiber, and Boger, pleased, was comfortable with asserting it, even though Vertex hadn't proven it.

"The sin in science is not making mistakes," he said, hungrily spooning out more vegetables for himself. "It's overinterpreting your data. If you're disproved by a massive amount of new data, that's fine. What you don't want to be is disproven by a little more data. That means you went almost but not far enough."

Abandoning the mess in the lunchroom, Boger, Navia, and Murcko returned to their computers to justify their latest observation. With fourteen hours to go, they would do the next set of experiments.

"This is really enjoyable," Murcko said, knowing he would now be working through the night and right up to the next day's deadline, scrambling to produce simulated data to support a speculative theory that had been hatched spontaneously just moments before, "in a perverse sort of way."

• • •

In the chemistry lab, Roger Tung and Dave Deininger labored against less theoretical constraints. For weeks they had been working most nights until midnight, trying to make new molecules to inhibit HIV protease. Unable to sell the AIDS project in Japan, Boger had also been unable to expand it, and Tung and Deininger had chiefly borne the consequences. The company had now poured $2 million out of capital into HIV research, and the project was foundering in a kind of catch-22: They couldn't move ahead without more resources, and they couldn't recruit more resources without more progress. Tung, exhausted, was especially anxious and irate.

"We don't have a singular strategy. We don't have enzyme. We don't have crystals. We don't even have structure activity at this point," he said. "I've been sitting here saying 'Lord take me now. Let me open my veins.'

"We're banking that by incorporating some of the changes that other companies have used we'll be able to improve the overall biological activity of our molecules, but it's idiosyncratic. It changes from compound to compound. Mixing horses and donkeys doesn't necessarily get you mules."

If anything, Tung knew, the decision to go public would now increase the pressure. Immersed so thoroughly in his own work that he had not seen or talked with most of the other scientists in weeks, he was only dimly aware of how the requirements of doing business had come to dominate science elsewhere in the company. But he knew precisely the effect it was going to have on him and Deininger. They were going to have to produce with a late start and a fraction of the number of chemists engaged at Merck and elsewhere what no other company had been able to make: a convincing case that it was close to having a pill to help cure AIDS. To Tung, the very notion was an affront to science.

"We're in a situation where we can't afford to make mistakes," he moaned. "It's terribly foolish."

At 2 A.M. Navia, exhausted, again fell asleep in his chair. Boger woke him and told him to go home, then edited the manuscript himself until 5 before driving to Concord, showering, and grabbing a quick breakfast with Amy and the boys. He was back at his desk at 7:30. It was now Thursday morning, the day of the *Nature* deadline. Murcko, who'd gotten home at 4, was at his computer screen by 8. For a year, he'd tried to be careful about what he ate, but now he was chugging Diet Pepsi and scarfing the gnarled crumbs from a Dunkin' Donuts box. The room was littered with empty potato chip bags and pizza boxes. Murcko's briefcase, his floating office, overflowed at his elbow with paperwork—purchase orders, articles he'd been reviewing, candidate's résumés.

To test his hypothesis about the flap and to begin designing new drugs, Murcko modeled several inhibitors into Yamashita's finished structure. Boger wandered in, joined by some of the chemists. Collectively, they donned 3-D glasses and speculated on the molecular doorstop theory, whether what they were seeing was real, and what it might suggest.

"It's reaching in the dark," John Duffy, a bluff, red-headed chemist, said, looking at a tiny affector region sticking out from the wheel of the protein like a twig from a burl.

"There's not enough stuff out there to do what Stuart says," Boger agreed. "Something else has to occur. But our bigger problem is we need a factor of one hundred in activity. We need a break, a bunt single, something as dramatic as the first semaphor." The semaphor, a two-pronged subunit grafted onto the core of FK-506 like a miniature dousing stick, had been Vertex's first major success in chemistry, making its molecules as potent as cyclosporine.

The scientists began suggesting several ways to mimic FK-506, but Boger bluntly waved them off. "Nature didn't design FK too exquisitely," he said. "This is a molecule that lucked into a use. We shouldn't be too anal about mimicking it."

Murcko, taking the cue, disentangled an inhibitor from the protein and spun it around, breaking several crucial atomic bonds. "This is kind of extreme," he said, looking up from his keyboard. "Don't try this at home."

The others laughed. The process of discovery at Vertex, as collaborative as it was meant to be, had a way of shifting the pressure from one scientist to another, like a relay. It was now Murcko's turn, and he was mindful of how the experience had scalded Thomson, Moore, and Yamashita before him. Boger left, then Duffy, then the others, until just he and Yamashita were left staring into the reflective darkness of their screens.

"Vertex is entirely driven by hubris," Murcko muttered, half admiringly, half filled with free-floating dread. "We have no reason to expect that we can win, but we refuse to think anything else."

He was still uneasy about how speculative his work was, more so now that the company was depending heavily on him for its next critical breakthrough, now that its competition was no longer Schreiber but Merck. And yet he needed data. To do serious calculations he needed the coordinates for pinpointing exactly where and how FK-506 bound to FKBP-12—coordinates for which, at the moment, there was only one source: Schreiber and his collaborators. The previous Saturday Murcko had driven to Yale to hear Schreiber. Afterward, Schreiber told him he planned to release the coordinates as soon as his paper was published. Unhesitantly, Murcko called him now. The orgy of posturing, of Schreiber-bashing, was one thing, but this was science. Cordial as ever, Schreiber promised him the information.

Murcko marveled: "Stuart's really, really good at extracting what he wants from people and dumping them, which is what a lot of people say about Joshua."

At 11 A.M., after a final strategy session with Boger, Navia, Murcko, and Yamashita, and with the deadline less than an hour away, the X-ray paper was finally done. Moore had taken a couple of calls from the press, but the center of activity had shifted—and held—with the crystallographers. They had made their deadline. Working together, Yamashita and Navia had accomplished what they'd been unable to do apart. The emotional storm of two weeks ago had dissipated, replaced by an uneasy calm.

"Supervising someone else is like sailing a good boat," Navia reflected. "Everything's perfectly aligned. The sails are well trimmed, and all you occasionally have to do is gently touch the rudder to steer. With Mason, I've had to have my hand on the rudder full-time."

In another room Yamashita murmered bitterly, "Manuel is a compromised man."

Yamashita spent the morning agonizing over whom to name as authors. At 11:30, he, Navia, and Murcko wedged themselves uneasily into one of two small private writing rooms off the lunchroom to discuss the matter for the last time. The rooms, barely large enough for a desk, a computer, and a chair, were a concession by Boger to the scientists, a quiet place to think outside the relentlessly public channels of his social experiment. The scientists called them the "domes of science."

Determined to salvage a shred of idealism from the ruins of his experience with the manuscript, Yamashita had decided to make the decision employing "pure logic, even if we were obliterated by it." Rationally, that meant naming fewer authors rather than more as he'd previously tried. He now agreed with Boger that the struggle for credit that had erupted so venomously with his original draft of the paper was tearing the company apart and that it was in Vertex's best interest—in the best interest of science—to downplay such disputes by keeping the number of authors to a bare minimum, perhaps just those two or three who had done the actual work described in the manuscript. In the credit-driven world of academic science, there was an increasing tendency to include all those who, like one's collaborators, had done other work that had paved the way for a discovery or, like one's graduate students and postdocs, needed it for their résumés. Boger despised the practice as inflationary and dishonest, and Yamashita now agreed. Coming 180 degrees, he suggested in the dome of silence naming just himself, Navia, and Murcko.

It was a martyr's position, the most treacherous for Yamashita personally. Thomson and Moore would be—deservedly, Yamashita thought—furious. Moreover, he would look like a credit monger, though that wasn't his intent. Distraught once again at doing what he had deduced logically to be the right thing, he had already gone to Boger that morning to tell him of the decision, and Boger, anxious to have the scientists put aside their lust for credit, had promised to support him against the inevitable firestorm. Navia, too, now agreed with him. Only Murcko, hoping to lift some of the tension, balked. Telling Yamashita that he hadn't done enough to warrant credit, he asked him to take his name off of the paper, too.

Thus it was decided. Vertex would set an austere precedent on authorship, striking a blow against fatuous self-interest in research. The X-ray paper would bear just the names of the crystallographers, Yamashita and Navia.

However, what's logical isn't always practical, and Boger had designed Vertex above all else to work. Abhorring an anticlimax, he suddenly peeked in, uninvited, yet as if on cue. "I've changed my mind," he said. He told them he now favored crediting all those whose contributions, though not described centrally in the paper, were nonetheless vital. He urged Yamashita to reinstate Thomson, Fitzgibbon, and Moore. Smiling, he ducked out as swiftly and provocatively as he'd entered.

"We were flabbergasted," Yamashita would recall. "Shocked beyond belief. He said that under the rules of pure logic, it may have been wise for us to confront the problem of authorship now, once and for all. It was probably wiser to sacrifice me rather than to compromise. But he said he understood my pain and Manuel's pain and Mark's pain and that the issue wasn't worth hurting us any more than it already had. He said he'd made a mistake by not letting John publish right away. If he had, he said, the whole question of whether John should be credited on the structure paper would have been a non-issue."

Boger's reversal settled the matter, saving Yamashita from all further attempts at self-immolation. At 11:55, Boger faxed the X-ray paper to *Nature*, feeding the pages into the machine himself. Titled "X-Ray Structure of the Major Binding Protein for the Immunosuppressant FK-506, Solved by Molecular Replacement Using a Structural Fragment Derived from an NMR Study," it listed as authors Yamashita, Murcko, Boger, Moore, Thomson, Fitzgibbon, and Navia, in that order. Thomson and Moore agreed to be named as coauthors, though in the rush to get the paper out, neither of them saw it before it went. Miffed and annoyed, they refused to share in the overall rejoicing.

"What's a Judas goat?" Yamashita asked.

He, Navia, and Murcko were still scrambling late that afternoon to assemble photographs and diagrams in order to FedEx them to *Nature* by 5. The air was aflurry with self-congratulation, forgiveness, the telling back and forth of small lies meant, like an antiseptic, both to sting and heal.

Navia, wolfing a sub and scrambling to make a 6 o'clock flight to St. Louis, where he was due—another of his sidelines—to direct an NIH site visit, explained that it was a goat kept by stockyards to lure other goats to slaughter. A professional traitor, he said. The worst kind.

"Let's stop playing head games with Mason," Murcko laughed, pausing. "As long as Mason stops playing head games with us."

"I don't want to," Yamashita said. "How about Judas chickens? Judas cows? Judas . . ."

"Mason is totally awesome," Murcko said.

"I've been saying that since the day he got here," Navia said.

"His total awesomeness," Murcko exulted, "is exceeded only by his modesty."

If Boger's ability to set improbable goals and then meet them, seemingly against all odds, had not impressed the scientists before, it did now. He had steered the company all week through the sloppy conflation of business and science without so much as getting wet. Now, Friday, was to be his—and their—reward.

The media blitz had worked to perfection. The *Wall Street Journal*—the only paper that really mattered now, given Vertex's intention to go public—reported the discovery of the structure in that morning's paper as an unambiguous three-way tie between Schreiber (in collaboration with Karplus), Clardy (also with Schreiber), and Vertex. "Three teams of scientists reported they had uncovered a key molecular secret of a powerful drug that suppresses the immune system," the *Journal* reported prominently on the front page of the second section. Though Schreiber and Karplus were mentioned first, Vertex, being the only business, was identified in boldface, according to the *Journal*'s standard, busy-executives-don't-have-time-to-read concession to its readers. Investors looking for an item for their tickler files might not understand the importance of the discovery, but they now had a new name to associate with the "multibillion-dollar market" for new drugs prophesied by Clardy. Of the other papers reporting the story, only the partisan *Harvard Gazette* failed to mention Vertex prominently.

Schreiber, as Boger had hoped and predicted, was stunned. Recalling his conversation with Moore in late January, which left him, he said, "with the strong sense that they were miles away from having the structure," he'd also forgotten Moore's name, the younger, more junior Moore having joined the company after Schreiber became persona non grata and was no longer around. Schreiber had known for weeks that *Nature* would be reporting a tie, had even heard reference to "Moore et al.," but had failed to make the connection. "No, I was real surprised," he would later say, shaking his head.

What the confluence of papers most provided and what was practically forgotten in the stampede to get Vertex's out the door was affirmation. The ultimate test of whether a molecular structure is right, because it can't be seen even with the most powerful electron microscope, is whether other researchers, working independently, can duplicate it. That Moore's and Schreiber's NMR structures were practically identical and varied only slightly from Clardy's X-ray structure certified that they were probably both correct. Moore, especially, was relieved: Seeing a preprint of Schrieber's paper on Thursday, he had exaggeratedly crossed himself. He knew that the differences with Clardy were probably due to the presence of FK-506 in Clardy's structure. Much more troublesome was the fact that Yamashita's structure still varied from the other three in several key respects, particularly in the flexibility of the flap. None of the others had the flap swinging so freely. In the general euphoria, the differences between Moore's and Yamashita's structures were overlooked, but in private neither of them could release himself entirely from a deep sense of foreboding, of irreconcilability. The idea of a molecular doorstop depended wholly and utterly on the leniency in Yamashita's structure. Something had to give.

In the scientific community, eyebrows were raised about the latest example of what had become a suspect phenomenon: joint announcements of apparent ties. Under the subhead, "Race to Publish Results," the *Journal* noted *Nature*'s decision to break its news embargo. And in *Nature*'s News and Views, Dagmar Ringe, a crystallographer at Brandeis University, wrote scathingly that interest in immunophilins had invited a "plague" of publications about them. "More collaboration and less competition in this field would be a damn good thing," she wrote. For those who worried that biomedicine was being corrupted by blind competition; that in the race to publish, papers were being indiscriminately reviewed or, in certain instances, not reviewed at all; that publicizing discoveries in the media before they had been evaluated by other researchers led to errors, distortion, and fraud; that the need for businesses and academics alike to capitalize on their discoveries created an environment where meaningless redundancy was valued over rigor and innovation, there was now a new case to cite. Not surprisingly, Vertex, as the private entity in the affair, was hoised widely as the black hat.

Boger, of course, had no such qualms. His purposes were clear, and now that he achieved them, he was on to the next siege: going public. He, Aldrich, and Nancy Stuart left Vertex early Friday morning for Warner and Stackpole, the downtown Boston law firm where Ken was a senior partner and which had become a redoubt for Kidder and its lawyers, up from New York. The firm was in one of several overbuilt, underleased, marble-heavy holdovers from the construction boom of the mid- to late 1980s, though the firm itself was prospering. With its ovoid suite of offices encircling a postmodern rotunda, it was the antithesis of Vertex: polished, subdued, conspicuously stylish—old money with a new sheen.

The central question of offering stock in a company is how to value it. With an established business—one that makes money—the process is relatively direct: The underwriter simply mulitiplies earnings by some number reflecting the company's performance, then divides by the number of shares. With small, unprofitable companies years from having their first product, however, valuation is more mysterious. It depends on an amalgam of assumptions and guesses: When will the company start turning a profit? How much will it make? For how long? What rate of return will investors require, given what they can expect to earn elsewhere? With emerging biomedical companies like Vertex, the questions are particularly elusive. The FDA, not the company, determines when it can market a new product; a company may find out it has nothing to sell only after years of costly development. There are patent infringements, unanticipated shifts in technology, endless questions of product safety. The odds overwhelmingly favor complete, unutterable failure. Still, underwriters try, and the companies do all they can to help them, to put a value on these companies, using formulas and ratios and other more or less empirical measures of things that, like the objects of much scientific study, can't be seen and may be unknowable.

Boger had long since made his pact with these vagaries, but Kidder now wanted separate confirmation. It wanted to see, as Holman had put it, "if Josh was for real." Thus the task fell to others: chiefly Nancy Stuart as well as the scientists, who, dressed for the occasion, drove over from Vertex in shifts and cycled in and out of Warner and Stackpole's conference room like witnesses before a grand jury.

The lone woman in a roomful of men, Stuart led deftly through Vertex's timelines for making money. The company, she said, expected to identify a new immunosuppressant by the end of the year, launch clinical tests by early 1993, and begin marketing a drug in 1997—some six years away. With HIV, she offered an even more aggressive timeline: identification of a lead molecule by late summer, a drug in clinic the end of 1992, income in 1995. It was a spectacular claim: an AIDS drug ready and available for testing in people in just over eighteen months, profits in less than four years. Stuart tried ably to defend it. Vertex now had novel compounds that were active against HIV in cells, she told the lawyers. It had contracted with a California company to test them by midsummer in a new breed of genetically engineered laboratory mice that had most of the components of a human immune system and were infected with HIV. If the compounds worked, Vertex would be able to short-circuit the normal preclinical schedule and go directly into primates. Like Merck, it would then launch European clinical trials while compiling data for the FDA, which was desperate for new AIDS drugs and was pushing ahead furiously anything that looked the least bit promising.

Stuart worried that the Kidder men would consider the story too speculative, too filled with holes, too pat. "People have heard this a hundred times before: new technology, bright people.'Yeah, yeah,'" she interrupted herself, speaking afterward, "'so tell us about your clinical candidate.'" But Kidder was impressed. The biggest worry about small biomedical companies was that they would go on hemorrhaging money forever, but Vertex was obviously determined to make money soon. Indeed, it was the main reason it had ultimately gone into AIDS. True, dozens of other companies were fast-tracking potential AIDS drugs for entirely the same reason, but once again Holman found Vertex's story exceptional.

The scientists, on the other hand, were haunted. To make public their aggressive timelines simply added to their stress. "The HIV project is stretched outrageously thin," Tung complained. "What concerns me is that even in the most optimistic scenario, we may not be able to do what we've said we're going to do. It would be hard enough if we were fully staffed, but I don't see that happening now. We're running a lot more on luck than skill and that worries me. It worries me a lot."

Murcko, who also was interviewed, was even more outraged. He was disgusted by the intrusion. Yanking off his tie as he returned to the modeling room, he snapped: "A $500 lunch for fifteen people, and they billed us for every Coke we ordered."

Back and forth between the real world of scientific speculation and the illusory one of Wall Street Boger and the scientists swung. At noon, while Boger, Murcko, and Navia were lunching at Warner and Stackpole, a fax arrived from *Nature*. It said that in light of the similar X-ray structure of FKBP-12 appearing in that day's issue of *Science*, the journal had decided not to send Vertex's manuscript out for review. Yamashita, reading it, went cold. Twenty-four hours earlier, he, Navia, Murcko, and Boger had exulted not only about catching Schreiber, but about staking out an important new hypothesis about the biology of the world's hottest protein. They had driven themselves recklessly to meet a deadline that served *Nature* as much as it did them. Yet now their work wouldn't even be seen by those in a position to judge its scientific merit. Again, as it first had with Moore, *Nature* seemed to be rejecting a potentially vital piece of science for no reason but its own competition with a rival journal.

Boger took the letter immediately on his return and began composing his reply, an artful, exaggeratedly civil single-spaced three-page rant. Yet another occasion for Boger to rise to after a week of nearly continuous crises, it showed him at his most brilliant and most caustic.

Scientifically, he wrote, *Nature* had "compelling" reasons to reevaluate its decision. First, there was the significance of the structure itself: It was unique. Unlike Clardy's structure, it was unbound (it was the first X-ray structure of the enzyme by itself), yet it was distinct enough from both Moore's and Schreiber's NMR structures to raise the question of what the protein really looked like. At the same time, it posed a direct challenge to Clardy and Schreiber, who had observed that when the protein bound to FK-506 its overall conformation "changes little" from its native state. "As you will note from our data," Boger wrote, "this would appear to be obviously and significantly incorrect. . . . Rather significant changes in protein structure *must occur* [emphasis added] on binding FK-506, especially in the 'flap' region that guards the binding site." He also stressed the collaboration between Yamashita and Moore—"only the second case of NMR data being used to solve an X-ray structure" and "the first time" a fraction of an NMR structure had solved the problem of molecular replacement.

Any of these points might have been enough to warrant *Nature*'s reconsideration: Taken together, they made a convincing case. But Boger couldn't resist going further. He bluntly reminded *Nature*'s editors of the deal they had made in deciding to reconsider Moore's NMR structure. "We would not be frank if we did not express our disappointment at your abrupt reaction to our manuscript," he wrote near the end. "The favorable editorial decision in mid-April to send the NMR manuscript for review was made only after, *at your request*, we laid out in detail our X-ray observations and plans and committed ourselves to sending the as-detailed manuscript to you. We fully and completely met our side of what we cannot help but view as something akin to a bargain."

Boger had not come this far to be denied. He concluded the letter by giving *Nature* until the end of business Monday to reverse itself—one day. Otherwise, he said, Vertex would take its side of the FK-506/FKBP-12 story—"perhaps the 'hottest' subject in medically relevant biology," he reminded scoldingly—to a competing journal.

To the scientists, including Navia and Yamashita, whose names Boger affixed to the letter above his own, the letter was brash, impolitic, shocking, and potentially ruinous. They feared *Nature* would never again publish a manuscript with the company's name on it. Yet by now they had no choice but to go along. "We've already torched rule 1—'Never publish a paper we've already rejected,' " Navia said gamely. "There should be no problem torching rule 62—'We don't publish second X-ray structures.' " Still, underneath lay a jagged ribbon of fear. All they could do was wait and hope that Boger's hubris wouldn't lead them into an all-out Icarus-like plunge.

There was a five-hour time difference between London and Cambridge. *Nature* had until noon Monday, local time, to respond.

Nature's second reversal in a month came exactly twenty-seven minutes late, at 12:27. Boger accepted it munificently. It still meant that others—others more rigorous, closer to the situation—would have to approve the X-ray paper, but to Navia, the last great obstacle was now cleared. The stage was set for a full-scale confrontation with Schreiber—a confrontation he was every bit as confident as Boger that Vertex could win.

"Stuart has stuck his neck out," Navia said. "His work is a hypothesis driving a conclusion. He thinks FKBP is an inconvenient intervener in the process, that it's a socket for FK-506. But he doesn't understand biology. The *Nature* paper is going to rough him up a bit.

"If we get the complex," he said, almost as an afterthought, "we're going to chop his balls off."

CHAPTER EIGHTEEN

In 1974, when Boger was in graduate school, his mother divorced his father after thirty-four years of a marriage that had grown, for her, irreparable. Despondent, Charlie Boger arrived on a Friday morning in October at the big brick Georgian Colonial where Joshua and his brothers had grown up, the house his mother liked to say "lived so good . . . raised children so good." He put a gun to his head and pulled the trigger.

Joshua, at Harvard, was troubled but not shaken. He had not been close to his father for many years. The corollary of his deciding as a boy that he "controlled the world" was an ability to wall off those things he didn't or couldn't influence. Such had been his relationship with his father. Equally headstrong, they had clashed since he was a child: Once, on a fishing trip to nearby Lake Norman, he threw back a fish and Charlie, furious, had knocked him out of the boat. Drenched, impotent with rage, Joshua refused to talk to him for the rest of the afternoon and for days afterward. By high school, with his parents' marriage foundering, he had sealed his father out. When the call came that he had shot himself, Boger reacted cooly. "One of the consequences of deciding that you control your world is that you have to allow that everyone else does, too," he says. "I'd have been much more grief-stricken if a drunk driver had killed him or if he'd spent months in a hospital bed."

Now, with the IPO, Boger's mission funneled through that part of himself that derived chiefly from his father: his ability to sell. Not that he was acting out a latent, one-sided struggle for autonomy. But he was enough of a student of psychology to know that men, especially, become who and what they are by vanquishing their fathers and father figures. At Merck, Boger had risen swiftly in part by renouncing his first sponsor, a tempestuous Egyptian Jew named Joe Rokash, who had brought him from West Point to Rahway. "Like all good sons, I ended up stabbing my father, stabbing Joe, in the back," he says. Rokash had launched Boger's ascent from bench scientist to scientific leader—a rise that was now about to culminate with his becoming CEO of a public company. His ability to sell a major deal on Wall Street, even more than his ambition to revolutionize the drug industry, required him to expunge his father's failures as a businessman, to test how far he had come and where he would go.

It was ironic. Boger's chief ambitions had always been scientific; business was just an expediency, a necessity for enabling the science to occur. Success in business mattered to him only as a means to an end. Nor did he, like many in his position, glorify Wall Street as a proving ground. As head of Merck, Vagelos seemed to enjoy impressing investors. But long before his appearance at the Vista, twenty months earlier, Boger had concluded that Wall Street was hopelessly and irresponsibly capricious and thus beneath his talents intellectually, not to mention any discernible moral scope. More, though he was exceptional at it, he had little respect for the act of selling, particularly to an untutored audience. And yet Boger needed Wall Street infinitely more than it needed him. He would perform unhesitantly, single-mindedly, on its terms. He would do whatever it took to do an attention-getting deal.

Initially, that meant recasting Vertex's story to suit the rampant, overblown expectations of the stock market. As Aldrich put it, there was a lot of "product" now "cranking" to soak up the new money that was flooding into emerging biomedical companies. At the same time, the Regeneron debacle had made investors as skittish as cats. To Boger, naturally, this contradiction played all to Vertex's strengths.

Most of the companies going public were, strictly speaking, biotech companies: Like Amgen, their standard-bearer, they proposed to make drugs by manipulating genes. But unlike Amgen and the first generation of gene-splicers, they weren't simply talking about manufacturing proteins, which are notoriously hard to make, must be given as shots, break down quickly in the body, and are especially vulnerable to patent shakedowns. The new companies talked dazzlingly about "superdrugs": "smart" molecules that through more exquisite targeting or the application of advanced technologies or both would make conventional drugs—biotech and otherwise—obsolete. It was the difference between workstations and mainframes, pocket cellular phones and conventional desk models. There were stories about drugs that locked onto the sugar molecules that act like intercellular road signs, blocked them, and confounded inflammation: drugs that short-circuited DNA messages, thus defusing diseases like AIDS and cancer; genetically altered cells transplanted into organs to act as factories for churning out drugs against cancer and diabetes; turbocharged vaccines; kamikaze drugs that strapped antibodies, nature's own disease fighters, to little micromolecular rockets and sent them speeding into errant cells.

The major problem with these superdrugs, as Boger knew, was that no one had proven that they worked. There were major questions about the rationales behind them, about how to get them to their targets, about whether many of their targets were even active. Would the molecules survive in the body? What about side effects? No one knew because there were no data: The mechanisms were largely theoretical. Vertex, on the other hand, intended to make small molecules that inhibited known enzyme targets just as Merck and the other big drug companies had been doing for decades—"little pills in bottles with white cotton on top," Boger liked to say. Such drugs were relatively easy to make and to patent and were known to work. The markets for them were established and huge. Boger knew he had to satisfy Wall Street's hunger for superdrugs, but to the extent that he also could distance Vertex from the doubts about them, he knew he might also assuage Wall Street's uneasiness and position the company as a surer bet.

There were other structure-based design companies with claims equal to Vertex's, most notably Agouron Pharmaceuticals, a seven-year-old La Jolla firm that was the first company dedicated exclusively to using protein structures in drug design, but Boger scoffed at any comparison between them. In April, for instance, Agouron announced in *Science* that its crystallographers had solved the structure of a portion of the enzyme that enables the AIDS virus to hijack T cells. There had been a chorus of flattering press coverage, carefully orchestrated by the company, and its stock had soared on cue. A closer reading of the *Science* paper, however, revealed that the enzyme site, which the company identified as "clearly an important clinical target," in fact had "no detectable activity." "It's a dead target," Boger sneered incredulously. "Where's the drug?"

From a business standpoint, what differentiated Vertex's story—and what Boger now intended to flog most assiduously—were three things: Vertex's integrated approach, its first-among-equals attention to chemistry, and its lineage, which tracked impressively through the mainline of the pharmaceutical industry and Merck, then Wall Street's favorite company. "Smart ex-Merck guys making drugs with computers," Aldrich said, encapsulating. Agouron had exquisite crystallography, arguably the best in the world. But drug companies make their billions by selling molecules, not molecular structures. Unlike Vertex, Agouron seemed to commit itself fully to designing inhibitors only after a structure was solved, so that it would be unlikely, for instance, to produce a development like Vertex's 367, a promising drug lead that preceded having the structure of FKBP-12 by several months. Like Schreiber, Boger intended to show that the future of biology led through the synthesis of new and better compounds, through chemistry, and for that Vertex's bloodlines were impeccable. It had sprung from Merck. What else did investors need to know?

Throughout May Boger worked on his slide show. In the past, he had tailored it for venture capitalists, drug company executives, and other scientists, who needed little explanation. But Boger was now about to face the most unsophisticated audience he had ever addressed: institutional investors, portfolio managers, syndicate players, floor traders—people who had little concept of what he was trying to do and who bought and sold stocks according to criteria he disdained. As always, he tried to imagine their thinking. He juxtaposed on one slide, for instance, cartoon drawings of a hypodermic needle and a pill bottle. Next to the needle was the figure $5 billion, representing the total annual sales of all biotech drugs; next to the bottle, in much larger type, $160 billion, the total for small molecules. The slide deftly, if simplistically, combined several points. As Aldrich put it, "You don't see blockbuster drugs for acute situations. You see them for chronic cases, and that means ingestibles—pills. That's where biotech falls down." Boger was proud of the slide. Like a reference to "pond scum" in his explanation of natural products screening, it confirmed Vertex's place as a new paradigm in drug discovery, a Third Way, smarter and more rational than the big pharmaceutical companies, more lucrative and friendlier than biotechnology. It was something even the most uninitiated buyer could grasp.

Boger had learned from his father when to be explicit and when to back off, letting, as with his discussion of potential markets, buyers' reveries take over. But he also had to sell Vertex's science, a more exacting exercise. Boger agonized over this reductive process. He went around and around with Holman and Kidder's sales staff, who looked at him quizzically each time he used such fundamental nomenclature as *target*, *inhibitor*, even *binding*. After a morning with a Kidder delegation in which several of the salespeople urged him to use *stick to* instead of *bind* to describe the interfingering of molecules, Boger finally threw up his hands. "It's hard," he complained. "I've never gone down to this level before. I may be able to get away with *safety*, but *efficacy* I have to explain.

"I'm thinking," he sighed, "that this is all fine, but I better put it aside now and address the level of scientific education in America."

Ultimately Boger packaged Vertex's story into fewer than thirty slides, 50 percent less than usual. Many of them were stunning new computer images modeled by Murcko, who for several days designed and photographed them instead of doing science and ambivalently thought them, if anything, too facile. One series of stick-and-ball diagrams showed FKBP-12 first alone, then with FK-506 suspended over the active site, then with FK-506 and an unidentified Vertex proprietary compound docking with the precision of two satellites. The progression made it seem as if the company had indeed put to practice Aldrich's "connecting the dots"—inserting spacers between the atoms on FK-506 that were essential to its activity—in order to make better drugs. Even more tantalizing was a luminous ribbon diagram of HIV protease with an inhibitor nestled Venuslike in its maw. How simple, one thought. Elegant, sensual even.

Boger, an aficionado of slides and computer-generated molecular graphics alike, complimented Murcko profusely. "These are great slides," he said. "You can build all kinds of purposes into these slides." Aldrich, drifting in from his office next door, reprised the theme. Noting a slide stating that Vertex's scientists had forty years combined experience in aspartyl proteases, Dave Livingston joked to him: "And on the business side, how many years combined experience do we have selling smoke?" Aldrich smiled, "It's the intensity of the commitment."

Science and business together had driven Boger through the previous period of publicizing Moore's paper and handling *Nature*, but now he was in complete, unmitigated sell mode, thinking only about money. The rush to file a red herring quickly before the SEC, so that Vertex could go public before the investment "window" on Wall Street crashed shut, consumed all his effort and attention. Indeed, there were signs that he was already too late. By the third week in May, Regeneron was trading at $11.75, down from $22 just six weeks earlier. Isis, another glamour company with an intriguing story about blocking DNA, was said to be in trouble, and ImmuLogic, a Cambridge start-up specializing in peptide drugs, was forced to cut its initial offering price from a range of $14 to $16 down to $12. The business press had picked up on the jitters and was riding the story, particularly the *Journal*, which began publishing daily predicitions of an impending shakeout. An amnesiac market, which since the first of the year had somehow forgotten that IPOs were extraordinarily risky, that almost all new public companies traded down from their initial prices, that even Amgen had gone to $3 and stayed there before reviving, suddenly was shocked to wake up with a stunning hangover. "These people are just discovering compounds and they want a public market asking price," a prominent fund manager complained indignantly. "Why should I pay a public price for a venture company?"

Boger, typically, was undaunted; as always, he was "leaving things indeterminate" until the last moment when he would have to decide whether or not to take the offering forward. Aldrich just as typically blamed the *Journal* series as much as the overheated market itself for the current spate of whiplash. Writing all day, from 7 A.M. often until midnight, faxing draft after draft to Kidder and to Warner and Stackpole, where legal secretaries were now typing documents for the SEC around the clock, consulting hourly with Boger and the board members, he proceeded as if Vertex was moving inexorably toward going public, though more and more he was sure they were too late. "When I saw Icos melting down and I heard from my friend at Merrill Lynch how bad ImmuLogic was doing, I laughed, 'That's it. It's over,'" he said. "But you spend a day drafting. It draws you back in."

The red herring was finished on May 29, 1991, less than five weeks after Vertex decided to go public. It was an extraordinary document, not because it was different from the dozens of others new in registration with the SEC, but because it was like them—scant on substance (of which there was little to report), laden with risk factors.

Especially this last. The company was obligated to note, for instance, that structure-based drug design had yet to yield an approved drug; that Vertex planned to hemorrhage money indefinitely; that it had "no assurance" that it could find, develop, market, manufacture, or get approval for a drug much less profit from one; that many of its competitors had "substantially greater financial, technical, and human resources"; that it might not be able to survive without Boger, on whom it held a $2 million "key man" life insurance policy.

Given so little certainty, such narrowed odds, it was perhaps a wonder that Vertex or any other biomedical start-up would be sanctioned to sell shares publicly. But that presumed an orderliness and public trust that went beyond Wall Street's obligation or ken. Wall Street had only one cardinal rule, caveat emptor—buyer beware. By filing with the SEC, Vertex was fulfilling its debt to truth and the commonwealth. Whatever else, investors couldn't say they hadn't been warned.

Boger had never denied or dismissed any of the risk factors. In the pharmaceutical industry, failure is the norm, even more so in small companies. Indeed, his rationale for starting Vertex was to shave incrementally, one by one, the risks inherent in drug discovery, making the process more scientific, surer, more predictable. And yet now, with the risks openly acknowleged, Boger was free not to talk about them, to gloss them over, to sell Vertex's story as if they'd already been brought under his steadfast control. He was as buoyant and all knowing as ever. Vertex may in fact have been a small, desperately unprofitable company with no products, unproven technology, and no guarantee of success, going up against a mountain of nightmarish uncertainties, but so was every other small company. Compared to them, Boger liked to say, Vertex was "Amgen." Now that he would soon be talking to them directly, he was sure investors would have no choice but to agree.

Rolling, Boger entered the registration period—four to six weeks during which the SEC would review Vertex's filing while Kidder and the other underwriters premarketed the deal among the big fund managers who buy 60 to 80 percent of all IPOs—as fearless as if Vertex were reporting record profits. The company's initial asking price, worked out after endless rounds of talks with Holman and Kidder's sales force, was \$13 to \$15 per share—ambitious, but not overly so. Again Boger, starting far behind and deep in traffic, had pulled up even, only this time with Wall Street itself, and no one, not even he, could predict what Wall Street, in its infinite volatility, would do next.

The scientists, understandably, saw another view.

They saw unproven science, unmet expectations, uncertain goals—a story lunging out of touch with reality. They saw molecules that were still far from drugs, labs stretched to the limit, a Schreiber juggernaut. With immunophilins, they saw unanswered questions—How did FKBP-12 work? Was it the relevant target? How did it bind to FK-506? What accounted for the drug's increasingly cyclosporinelike side effects?—that left them unsure how to go about designing new drugs or even whether better drugs could be designed. With HIV, they saw sorely overtaxed chemists struggling to get from out behind other companies' patents, no protein for crystallography, and the brick wall of making a protease inhibitor to survive the gut. Perhaps worst of all they saw a growing lack of focus. It wasn't that they hadn't performed extraordinary acts of individual science, but drugmaking needs careful leadership, and without Boger—who was cloistered with Aldrich or editing his slides or flying off to New York to meet with the Kidder sales force, or squeezing in a last-ditch trip to Japan in early June to try to sell the HIV project to "our noodle buddies"—the project councils had begun to falter. Waiting for the emergence of self-actualized champions, Boger had deliberately not anointed lieutenants, leaving the scientists to chart through the myriad of scientific questions collectively. But the experiment was going badly, leaving the scientists frustrated, disappointed, resentful, spent, and, in a few cases, bereft.

"Joshua has to finish up talking to the money boys and get back and start stroking his people," Jon Moore explained. "Scientists need to be stroked. They need to be treated in a fatherly manner on a day-to-day basis."

Boger defended the paradox—going on the road to sell a scientific approach that required his leadership, which he was unable to provide because he was out selling it—as a necessary cost of doing business. He sounded like a father, like himself with his own kids or perhaps like Charlie Boger with him, explaining why he had to spend so much time away from home. Ultimately, he said, it would allow him to become more involved in the company's science, not less, since a big bolus of money would release him from the constant pressure of selling projects and chasing deals. He envisioned a day soon when he didn't have to go on every death march, when other companies would come to Vertex, when he would be more available to watch over things and make sure they worked as planned. The scientists, of course, didn't believe him. Watching him go off to Warner and Stackpole or to meet with institutional investors, they imagined he had found a new enjoyment, a new challenge, a new mistress of sorts. They began to wonder whether Boger cared as much about science as about climaxing his own rise—"being on the cover of *Business Week*," a few muttered.

The pricing of Vertex's stock had exacerbated their distrust. It was assumed, of course, that the IPO would make everyone connected with the company richer, perhaps spectacularly so, the cloning of millionaires being one of the most enduring—and appealing—clichés about small companies that go public. But because of the shaky market, Vertex and Kidder resorted to a mathematical sleight of hand in arriving at the price of individual shares. The maneuver, common among new publicly owned companies, is called a *compression* or *reverse split*. It works by reducing the number of shares in order to buoy the stock price. When Vertex had first decided on a $13 to $15 range, the market was still supporting such optimistic prices. However, as the window drew down, other companies like ImmuLogic were forced to cut theirs. Rather than reduce its price, which not only would diminish Vertex's cachet, but might be seen by investors as a sign of weakness, Boger and the board had opted for a 3 to 2 compression. The dollar value of each individual's equity was unchanged—three shares at two dollars equals two shares at three—but many of the scientists still felt cheated.

"It's like, 'What are you doing to me?'" explains Holman. "'I give you three, you give me two back? And then the stock price is lower than I thought it was going to be? It's like I've lost twice here.'"

In the labs, the compression was rumored for days before it was announced. Boger, harried and thinking it a non-issue, eventually disclosed it in a perfunctory, hastily called meeting in the lunchroom in which he seemed both impatient and irritated at having to answer for it. Pockets of anger suddenly exploded. Dave Armistead, lead chemist on the immunophilins project, was particularly appalled. He and the other scientists had come to Vertex in large part because of the stock incentives, but now, he said, Boger was changing the game. It was tantamount to fraud. Sounding out other scientists about possible legal action against the company, he flexed for several days before calming down. "As you can see, I'm not one of those people who's enchanted with the idea," he later fumed. "I'm not going to let it blow my year, but for a couple of days, I'll moan."

Throughout June, the episode refused to diminish, burning like a corrosive through the unanimity that had peaked with Moore's *Nature* paper. There was Boger's handling of it, which many of the scientists saw as a reflection of his and the company's new priorities. "It's Josh's bullshit mode," chemist Jeff Saunders complained. "He's not out to fuck anyone over, but at the same time he can be condescending. When he flips into business mode, you know he's doing something he doesn't want to talk about." There was the sense of collective impotence reflected by the fizzling of Armistead's insurrection, which even his closest friend and rival Saunders, not unhappily, likened to "background noise . . . mouse nuts."

Worst, perhaps, was the sudden and inevitable jealousy among the scientists themselves. Not only did the compression suggest a shrunken pie, but SEC requirements now made public several of their exact portions. Thus, even as Boger's 780,000 shares shrunk to 520,000, Navia's 103,000 to 67,000, and Aldrich's 87,500 to 59,000—holdings that, postcompression at $15 a share, would be worth $7.8 million, $1 million, and $900,000, respectively—resentment spiked. "It was like having your pants taken down in public," said Navia, who was doubly excoriated when Boger, needing to "get another Merck name," hurriedly made him an officer, so that Navia's $92,000 salary was also disclosed. Altogether, the aroma was tart, sulfuric. Said Saunders, "Club V is becoming a lot more Squibblike in its science management relations. I'm impressed by how much greed is showing through here."

Scientifically, it was another frustrating period. Despite receiving a preprint of an article by Schreiber detailing how to grow so-called cocrystals, Vertex still hadn't solved the structure of the FKBP-12/FK-506 complex on its own. Murcko particularly was distressed. "Merck has it. Abbott has it. Glaxo may have it," he said. "We don't have it. We're in a better position to understand the flexibility of the enzyme because we have the native structure as a starting point, but they've got a lot more computers and a lot more people. If the goal is to design drugs, they're ahead."

Far from being the straightforward challenge of blocking FKBP-12 with a well-designed molecular monkey wrench that Boger had first envisioned and sold so successfully to the scientists and Chugai, the immunophilins project had become a scientific tarpit. The scientists had no idea what they were trying to accomplish. Were they trying to mimic Schreiber's effector region? Hold open the swinging jaw of Yamashita's flap? Both? Neither? What did the exposed handle of FK-506 do inside the cell? Were the drug's side effects the result of binding to too many partners, or could they be trimmed away like so much fat and gristle? "This is one hundred times harder than anything I've ever worked on," said Murcko, "including HIV protease. It's a nightmare of terror. A year from now all of us are going to be a lot more religious."

Nature, with no reason to hurry now that Vertex's X-ray structure was second behind Clardy and Schreiber's (and perhaps retaliating for Boger's indiscretion), took its time reviewing Yamashita's paper. Navia, angered about having to "hang fire" in his imaginary war with Schreiber and stung by the reproach of the other scientists, fulminated. It took Yamashita, beyond caring, to stroke him and calm him down. "Manuel," Yamashita said, without a hint of irony, "is a very great man."

Without the FKBP-12/FK-506 complex, Vertex's two conflicting protein structures—Moore's and Yamashita's—offered little positive direction to the chemists, who were laboring unsuccessfully to make better compounds than 367. Both Moore and Yamashita assumed that the discrepancy, which was no longer relevant to publishing and unlikely to affect drug design, would be resolved imminently by having the structure of the complex. Thus they turned, in the absence of any clear direction and with the competition between them still unresolved, to other work. Boger was appalled. He was about to go around the world, trying to persuade wary investors that structure-based design at Vertex was already well underway, and its most marketable achievement was clouded with doubt.

"People have gotten lazy around here in the last few months," he snapped after a project council meeting at the end of June. "We've got two very different structures, and nobody seems concerned about it but me. These guys have short memories. Two months ago they pulled each other's cookies out of the fire and now they're not talking to each other. It's bad science."

As he had with the Chugai deal, Boger again straddled the core dilemma of his world: To do science you need money, but to raise money competitively you need to project illusions that are the antithesis of science. You need to scramble hard.

"Our strategy going in," says Holman, "was, Let's pull out all the stops. This is a market that has us all nervous, so let's at the end of the day not look back and say, 'Gosh, I wish we had done. . . .' " Kidder's plan called for a "full-blown, massive marketing in the United States, Japan, and Europe," a worldwide blitz that would start in Tokyo on June 27, 1991, gather speed through a five-day, five-city European whirlwind in early July, then peak with a two-pronged sprint across the United States—two teams each hitting two cities a day—climaxing in San Francisco and Seattle, Portland and Palo Alto, on July 12. By then investors would hopefully be in such a state that the deal would be oversubscribed two or even three to one, enabling Vertex to get its asking price or, as Regeneron had, a higher one. Still, Holman was taking no chances. Normally such road shows target only institutional investors, but he insisted on a second team consisting of Navia and Vertex's new comptroller, Keith Ehrlich, to meet with retail buyers—individuals—who weren't as "price sensitive" as the fund managers and might be induced to bid up Vertex's stock.

"Vertex is the kind of story where we were going to get the deal done regardless," Holman says. "But the market was overheating. It was the classic example: too many deals, deals being priced poorly, and everybody getting hurt by it. At the time we were finalizing the road show, Kidder alone had twenty-two deals in registration, with eight or nine of them on the road. The portfolio manager we would call, say, in Boston, would say, 'Hey, I've got twelve invitations today for lunch.' We had to do an enormous premarketing of our institutional people. We brought Josh to a hotel here and got people in from all over the country—insiders—to get them so excited that they would tell their client, 'I know you've got twelve, but of the twelve this is the one you've got to come to.' "

"The problem," said Aldrich on the eve of the road show, "is that a lot of investors have seen issue after issue go out and tank, so they say to themselves, 'Why should I buy when I can wait a month and the stock'll be half the price?' It means we have to express total confidence in the offer. Even with my friends I'm saying 'This is gonna go, there's no question, so get some—if you can.' "

Stoking the atmosphere of a hot deal, Kidder's plan was to present Vertex as the most tantalizing story in a season of stories—in Holman's phrase, the crème de la crème. Boger, who always conceived of Vertex as being at the head of any class it was in, gladly obliged. He walled off those aspects of the story that made it seem less than absolutely convincing, absolutely true. Reality would not, could not, intrude. Indeed, his final instruction to Nancy Stuart and the scientists as he left for Japan was not to let him know what was happening in the labs while he was gone. Ostensibly to preserve the accuracy of his presentation, the edict also allowed him a fig leaf of deniability in case of a setback. If there was bad news, he didn't want to know it lest it detract from the picture he was perfecting, that he had to perfect, before the fact, of clear and orderly success. As if to reinforce this open-faced image—and perhaps another, of himself as the ex–Harvard/Merck wunderkind—Boger shaved his beard shortly before his departure.

But reality did intrude. On June 24, the day before Boger and the others were scheduled to leave for Tokyo, a drug industry tip sheet, *F-D-C Reports*, known commonly as the "Pink Sheet," published a story touting "Vertex Pharmaceutical's Orally Active HIV Protease Inhibitor." Cribbed from the red herring with no independent follow-up, the article reported that Vertex had three small molecular compounds that blocked HIV protease in cells and that it planned to begin testing a drug in human patients by the end of 1992. In fact, the prospectus had said only that the company was "seeking" to design an oral AIDS drug for asymptomatic—healthy—carriers of the virus. Vertex had no such miracle pill, nor did it claim to. Still, as Boger and the others took to the road, there existed now the impression that they would be selling an inside position in AIDS, an impression that didn't displease Boger, who was seeking to position the company, but which worried several of the scientists.

Deeply fatigued from a year of failing to do what the company now seemed to be saying he had already done, Tung especially was concerned. Like Boger, he thought science was like crossing a minefield, things constantly blowing up in your face. That was the thrill of it. By himself, Tung used to hop freight trains in California and Oregon when he was a student at Reed College, a school known for cultivating intellectually gifted but unconventional students. Drug discovery had that same atmosphere of risk, of being alone, on the edge. It was also messy, convoluted, unpredictable, inexact, and generally pursued under the most breakneck conditions. Tung worried that by oversimplifying the process, Boger was creating the false impression that making drugs was like making pizzas. "I don't think we've reached that point," he moaned, "but I don't know."

In fact, within days he and Livingston would learn that Vertex had made compounds that blocked HIV-infected T cells from reproducing and were "bioavailable": Fed orally to mice, they were getting to the bloodstream. The problem was, they were different compounds—"horses and donkeys," as Tung had put it. Livingston was torn. Boger could use such "data points" as ballast during his road show, but they were ambiguous. One obvious solution was to have the orally available compounds tested quickly in cells to see if they stopped the virus from spreading. But Vertex didn't have its own cell assay, and Boger hadn't authorized anyone else to contract for one. Standing in front of Boger's office, Livingston, Murcko, and Tung puzzled over what to do. If they told Boger about the new data, they would contaminate his story. If they didn't, they would be delaying vital experiments while maintaining the fiction perpetuated by the Pink Sheet.

"Tell him we have to make a deal but don't tell him why," Tung suggested. "Tell him he doesn't want to know. Tell him that the people who are most informed believe it's necessary and leave it at that."

Murcko shrugged. "He's a clever fellow. He'll figure it out."

They decided to tell him nothing.

As Holman had predicted, the road show was gathering momentum, playing to receptive out-of-town audiences. "In Japan," he says, "we had eighty-five people for lunch. It was the largest crowd I'd ever seen. In Europe, the crowds everywhere we went—with the exception with one or two cities—were enormous. People wanted to hear the story." Speaking in London, Zurich, Geneva, Stockholm, and Paris on successive days, Boger had no indication whether the deal would sell—only 500,000 of the 3 million shares the company intended to offer would be available outside the United States, and it was too early for buyers to commit orders. Still, he returned to Boston on late Friday night, July 5, strongly encouraged.

Boger brimmed with restless energy. He had traveled so frequently in the past two years that he had trained himself, as a necessary efficiency of family life, to reorient himself quickly when he got home. Thus while Navia, Aldrich, and Ehrlich spent the weekend getting their clothes to the dry cleaners and trying to sleep off their jet lag before the next leg of the trip, Boger, a competitive if rusty athlete, shot baskets for two and a half hours Sunday in his driveway in ninety-five-degree heat, then drank a half gallon of water. Throughout the weekend, his youngest son, Sam, who'd seen little of him in recent months and at first hadn't recognized him without his beard, said "Bye-bye, Daddy," whenever he left the room.

Monday morning was Boston, as Vertex's home city and a locus of big mutual funds, second in importance only to New York. The market was holding—barely. At 7:30 A.M., a chauffeured limousine collected Boger and Aldrich at Vertex and swept them to an already broiling financial district for a day of meetings: 8:00, one-on-one with Fidelity Management and Research; 9:30, one-on-one with Massachusetts Financial Services; 11, one-on-one, State Street Research; noon, luncheon at the Meridian Hotel, where three weeks earlier Holman had shepherded another Kidder client, Cambridge Neuroscience, through a road show appearance attended by only some fifteen people. "Part of getting a happy client is making sure you control their expectations," says Holman, who told Boger and Aldrich to expect a crowd of perhaps thirty while privately hoping for thirty-five.

Fifty-five came. Inured by months of storytelling, most seemed impressed, if not entirely sold. "This is either the greatest thing since sliced bread," one institutional investor muttered on his way out, "or a total scam."

Barred from attending, no Vertex scientists heard the remark. Had they, they might have despaired even more deeply for themselves and for the precarious validity of the science they were so desperately summoning in the labs.

Modeled after the Palazzo della Cancelleria in Rome, the Villard Houses of the Helmsley Palace Hotel on Madison Avenue and 50th

Street in the heart of Manhattan are not simply an advertising shtick for Harry and Leona Helmsley, but a full-blown American Renaissance palace, just opposite the great granite apse of St. Patrick's Cathedral. It swims in European opulence: vaulted gilt ceilings, lavish bas reliefs, towering marble fireplaces. Holman picked it for Vertex's New York road show appearance over any number of other spaces nearer to Wall Street both for its grandeur and to emphasize Kidder's faith in Vertex. "The last time I used a room in New York that size for that many people was Genzyme in 1986," he says. "The book was oversubscribed. We had 3 million shares to sell and 30 million in orders."

How far Boger had come from the "meat market" at the Vista, twenty-one months earlier, few of the nearly eighty people who crowded to hear him now could suspect. Boger, who had trudged through more fruitless sales talks than he cared to remember, didn't begrudge them their erratic interest. Still, he seemed slightly unnerved, as New York intends of even its most confident out-of-town guests.

He began slowly, less assuredly than in Boston, though he quickly modulated himself. By now he had given this particular slide show perhaps a dozen times, and its rhythms were familiar. As the clanking of silverware on dishes subsided, he resorted to stock assertions—"We have compounds moving toward clinical trials that have been designed by using this breakthrough approach," he said—that both oriented his audience and let them know quickly where Vertex stood in areas that were central to them both.

As elsewhere on the road show, Boger stayed within the lowest-common-denominator script he had devised with Kidder. Showing, for instance, Murcko's FKBP-12/FK-506 slides, he lectured: "We can see which parts of this drug actually touch the protein target. Only the atoms, or the bits, of this drug that actively touch the protein are necessary for its action. . . . Think of it as building a scaffold, or as we often call it in the lab, 'connecting the dots.' "

Explaining how the company selected drug targets, he said: "We do not go into projects where the biology is uncertain. Vertex will never bring a compound into the clinic to test someone's biological hypothesis. That's too risky; we leave that to the NIH. . . . We look for the biology to be well understood. We look for the chemistry and the biophysics to be doable in a short period of time."

Flipping ahead, he continued: "This information is used and has been used at Vertex to design much smaller compounds than FK-506 that bind to this protein in the body. . . . These compounds are much simpler and much more specific than FK-506 for this binding site. . . . We have shown these compounds have the expected biological activity in human cells."

Boger was assiduous about not simplifying his story so much that it was patently false or misleading. But it was equally clear that the dictates of selling necessitated a high sheen, the impression of a sure thing, perfect control. Parsing his speech, one could not help but question whether his story—of orderly, rational discovery—had anything to do with the harrowing fits and starts back in Cambridge.

It was true, for instance, that Vertex had "compounds moving toward clinical trials." But had they been "designed" through a "breakthrough approach"—an approach that had only just solved, with enormous difficulty, the structure of its first protein and had yet to yield any vital information about how it stuck to FK-506 or anything else? And was it true that "only the atoms, or bits, of [FK-506] that actively touch the protein are necessary for its action"? A compelling story, but in fact it was still unknown which bits of the molecule accounted for its activity; and given Schreiber's increasingly viable assertion of a gafflike "effector" domain, it was less and less likely that it was only those atoms that bound to FKBP-12. Whatever Boger was describing, it was not the "nightmare of terror" that now entangled Murcko and the others.

There were other zealous assertions that probably would have exasperated the scientists had they heard them. Did Vertex really know the "essential parts" of FK-506? Did anyone in the labs but Aldrich ever actually talk about "connecting the dots"? True, Vertex knew that if one gave an immunosuppressant to an organ recipient, it helped block rejection, but not whether FKBP-12, the object of all their structural efforts, was biologically relevant. It was still possible that the enzyme was a poor target or, worse, that the "smaller and more specific compounds" designed at Vertex, while active, would be useless as drugs. Was this biology that was "well understood"? How could Boger say Vertex had used structure-based drug design when it didn't know what its target did or how it did it?

It was a measure of Boger that even as he painted the gritty reality of science in its most simplified and positive light, he defended himself by saying that compared to most others in his situation he had been a paragon of restraint. It scarcely mattered. Wall Street had become so intoxicated with stories, so self-delusional, that any attempt at ethical salesmanship was like a reed in a flood. The investors would see what they wanted to see, hear what they wanted to hear.

And what they heard seemed to enthrall them. If Vertex wasn't at the point of designing drugs, Boger had clearly positioned the company to exploit those breakthroughs that would soon make it possible to do so. How soon was still a question, but Boger was a pioneer; he wouldn't get trampled in the stampede. He knew the right moves, how to sell. Satisfied, the investors lobbed him a few easy questions, mingled determinedly, looked at their watches, and filtered outside and back downtown, there to add Vertex's name—a name many of them had only first heard—to the roster of hot new companies, to ignite a buzz, to send it swirling into Wall Street's raging cacaphony.

Aldrich pressed the phone receiver morosely to his ear.

"So you're calling in your chits, right?" he muttered. His sunken eyes flashed gravely, beseechingly, as if he had a migraine, his voice a thin well of barely concealed abuse. "Does Banker's Trust have any chits?"

There was a pause. "What did Alkermes trade for today?"

Another pause. "So did they trade up?"

Aldrich rubbed his tired face. It was almost 8 P.M., Thursday, June 13, the day after the road show. Aldrich was in his office, talking with Holman from New York, trying to establish with as much diplomacy as he could why the deal still hadn't come together. He was querulous, incredulous, wrung out, like an absent son insistently trying to get a doctor to explain over the phone why the treatment he'd advised for his father wasn't working, why the old man had collapsed and now, suddenly and unexpectedly, teetered near death.

"Is the sales force still pushing this," he asked, "or are they looking at it as a dead duck?"

He shook his head, dissatisfied with Holman's reassurances. "How many more do you think you need before you get the rest to pile on?" He stared blankly. "Yeah, you don't want to call too often. It's a psychological thing. Once they smell weakness in it, forget it."

Hanging up, Aldrich was fuming. "Everyone did everything they had to except for the guy selling," he said. "The guy had us going out at $13 to $15, talked us into making a big drop, and still hasn't delivered one order. That's what really burns me."

It had been an unfathomable week. Buoyed, if exhausted, by the road show ("I wouldn't wish it on a dog," Navia moaned), they had returned expecting that Kidder would start racking up orders at once: A few institutions would take 100,000 shares apiece, Kidder's salespeople would put out the word out that the deal was starting to gel, then a stampede. The illusion of scarcity: That was what they had all discussed. But Kidder had inexplicably failed to land a single account. Between them, Aldrich and Boger had gotten commitments for 400,000 shares, more than all three of their underwriters combined. The previous afternoon, desperate, they had agreed to drop the offering price from $13 to $15 to $9 to $11—a 30 percent discount—but still no takers. The scientists, predictably, were incensed. "The ball," Boger said wryly, "mysteriously disappeared from the court."

To Holman, Vertex had been caught in a fatal cross-shear, a convergence of overwhelming events. "By July, most institutions were looking at year-to-date increases in their portfolios of 25 to 30 percent," he would recall. "That's a home run hit. What these guys were now interested in doing was preserving that for the rest of the year, because they'd get a nice year-end bonus for having a 30 percent return, and they were already there. What they didn't want was to lose any money.

"Everyone moved to the sidelines. Six of the seven medical deals that were brought to market directly prior to Vertex were priced at the bottom of or below their filing range, and all but one of the seven traded down within two weeks after the offering. So institutions were saying, 'Market's overheated. We're getting burnt royally. I don't care if you've got IBM, I don't care if you've got Microsoft at the IPO, we ain't buying.' There was just this incredible fence-sitting going on.

"Vertex," he says, "became the Company from Heaven, Deal from Hell. And it was the Deal from Hell not because of anything in our control. The only thing you can do in that situation is go back and tell them they have to lower their price."

Aldrich, in his anger, refused to believe that Kidder couldn't simply muscle a few of its regular customers into buying Vertex stock, that like the remote, recalcitrant doctor on the phone, it couldn't resort to heroic measures. But as one of a number of less than dominant companies in the field and with its recent history of regulatory and business difficulties, Kidder was wary of too hard a sell. "Drexel used to have the ability in the junk bond market to tell Vernon Savings, 'You buy this goddamn bond, or I'm never trading with you again,' " Holman says, "and guess what happened to Drexel. . . . Our style as a firm is not to work that way."

Boger, typically, wasn't bitter about the sudden downdraft in Vertex's fortunes; factoring everything, he still thought the company would go out at the upper end of its new range. But he was furious about the wanton cupidity it reflected. "I'm not depressed about Vertex," he said after Aldrich told him about his conversation with Holman. "I'm depressed about the world.

"I've had investors say to me, 'I never buy a company that's not in clinic for more than $100 million.' They're checklist people. It's voodoo. I want to say to these guys, 'Lighten up. All you've got to do is beat the S&P by one point and you're a hero. Think a little longer term. If the stock goes down 20 percent after the IPO, big deal. It's completely nuts to be spending your time trying to catch the last dollar.' "

One irony of going public was that Boger, who exalted control, was now powerless. The offering was entirely in the hands of Kidder and the other underwriters—in a larger sense, in the indifferent hands of the market itself. He and Aldrich waited throughout the next morning and early afternoon for Holman to call. When he didn't, they responded reflexively. Aldrich, the emergent dealmaker with only Chugai to his credit after a chain of bitter disappointments, ate his liver; Boger, the scientist, scrounged for data, checked and rechecked his hypotheses, groped for explanations. He checked the activity of seventeen biotech stocks every thirty minutes on his computer. The Dow was up thirty-two points overall by midafternoon, but there was little movement in the sector. It was frozen. Boger began to believe that that was Vertex's problem. Institutions accustomed to trading biotech stocks didn't know how to value a small company determined to take on huge markets against mammoth competitors. "We're too hard to pigeonhole," he said brightly. "The analysts who are telling these guys what to do are all biotech guys. You'd have to have been around since Syntex [prior to Amgen, the last full-fledged pharmaceutical company to emerge, in the late 1940s] to understand what we're doing."

Once again, as during the darkest days of the Chugai deal, Boger and Aldrich resembled the faces of Janus. Aldrich—glowering, pessimistic, funereal—detected a "blood in the water situation." Every passing day that Vertex's underwriters couldn't close the deal told investors that the company was in trouble, making them wary and killing their incentive to buy. "We may not be able to sell this deal at any price," he fretted. He was still furious at the underwriters but also blamed himself and Boger for being "naive."

"This is an acute demonstration of what happens when you go public, which is that what you feel about yourself is what others are willing to pay for you," he said. "A couple of months ago we thought we were this great management team. Now we're saying, 'What the fuck are we doing?' This is going to defuse a lot of hubris around here." He was talking, it seemed, about Boger.

Boger, conversely, was upbeat, philisophical. He, too, was self-critical but mostly for listening to Kidder in the first place, for relinquishing control on crucial decisions. "It's completely clear to me that if we'd set our range at $15 to $17, as we first wanted to, we'd have had to come down the same 30 percent. It's a cat and mouse game with a single operant rule for the buyer: 'Whatever the initial price is, I want it down 30 percent.' Listening to Kidder and Cowen on that cost us $6 to $10 million. Now that we know that it's an auction and has nothing to do with real value, we'll take our lumps and get down the road." To Dave Livingston, one of the few scientists who dropped in repeatedly throughout the day to gauge how things were going, he predicted, "We'll go out in the middle of our reduced range, with orders for about 4.5 million shares." That would give Vertex $30 to $35 million, far less than the $50 million Boger had told the scientists to expect.

"That's the nightmare scenario," Livingston said afterward, "that we raise only enough money to get us through the next couple of years and there's not enough time to do what we've told the world we're going to do. That could get very ugly."

By the following Tuesday morning, July 22, Vertex's new price range had "jarred a couple of people," said Holman, but still too few to assemble a deal. It was four trading days since Aldrich's conversation with Holman—an eternity in such a volatile market—and Aldrich's dark prophesy taunted them all. If they didn't sell the offering soon, very soon, the market would simply turn against it, leave it for carrion. Vertex and the underwriters would be stigmatized: a putatively hot start-up that couldn't make it out during one of the richest speculative bubbles in history. Canceling an IPO would make it much harder, perhaps fatally harder, to raise more money in the future. What corporate partner, what investors, would want to touch Vertex after it had been scorned by a market that had happily imbibed such dizzying stories as Regeneron? Aldrich imagined having to return defeated to the board, at whose pleasure he and Boger ran the company. If the board was forced to dig deeply to cover Vertex's onerous burn rate, how likely, he wondered, would that pleasure be to continue? It wasn't too late for the board to bring in bona fide management—senior people who, by implication, would have been able to take the company public where he and Boger hadn't. Boger had no such worries—he was still confident that the deal would succeed and confident of his own indispensability—but he, too, was reluctant to arouse the board.

At Kidder, the stakes were equally high, the mood as steep. Memories were fresh of its meltdown in the late 1980s. A failure by Holman, its top New York dealmaker, to pull together perhaps its most prestigious offering in years would have dire implications. Holman's banking staff decamped to the sales floor, doubling up on account calls that started before 7 A.M. and finished after midnight. They patched in Boger at home. Chortled Boger, "I don't think Kidder can afford to have this one go down."

With a "book" of 3 million shares, Vertex needed commitments of something more than that before it could set a final price and ask the SEC and NASDAQ, the over-the-counter stock market, to approve trading. As the number of commitments crept up Tuesday night and early Wednesday, Kidder thought it might be able to sell a smaller offering, but not all 3 million shares. Holman quietly began "putting pressure on the system." "Many people in the firm, many senior people," he says, "now bought stock to get the deal done. You don't like to go to your partners and say, 'We need you to put up some of your cash in your personal account.' But a number of people did that. We were looking for any order we could get."

Finally, at 3 P.M. Kidder's top managers met internally to price the deal. They calculated that at $9 a share, the bottom of Vertex's range, they had scraped together enough subscriptions to sell 2.75 million shares. The meeting was tense, urgent: They were running out of time. Another day and the deal would likely unravel. Holman phoned the other underwriters, who wanted an even smaller offering. Subscriptions had reached barely 3.25 million shares. Without an "aftermarket"—enough interested buyers so that those who bought at the IPO would have someone left to sell to—Vertex's stock was headed inevitably down. At 4:15 Holman called Boger, who all along had accepted Kidder's recommendations. Now, Boger told him he was willing to yield on price but not an iota on the size of the deal.

"Our line in the sand," Boger recalls, "was, we do 3 million shares or we do nothing.

"I could hear Al's voice breaking as he saw his next three bonuses crumbling to sand."

It may be, as Aldrich believed, that Boger's resolve was pragmatic: During the pricing meeting, Kidder's lawyers had warned that if they cut the size of the deal, the SEC might require the underwriters to recirculate it. "The deal was too fragile," Aldrich said. "Recirculation would have killed it." But Boger was determined for other reasons. Having come to regard the underwriters, who he thought had done a fine job up to and including the road show, as ineffective, he resolved to salvage the final marketing of the deal himself. "We got no good predictive advice about this deal from any of them," he said. "When I realized they didn't know any more than I did, I decided to take over." Oddly, he justified his hard line with a perverse populism. "I think its important to have those shares out there. There are worse things than making a lot of stupid investors rich, but if I'm going to do it, I'm going to make a lot of them rich. If we only sold 2 million shares, that wouldn't be enough investors. I want to make a lot of people happy."

Boger had not wanted to try the board's faith in him, but by now he had no choice: He needed to sell another 250,000 shares, fast. Even if its members agreed, Vertex still would have only a hair-thin margin, a so-called gross book of 3.25 million shares on a 3-million share offering. "We couldn't afford to drop out Aunt Ruth," he joked about the precarious ratio. At 5 P.M. and for the next three and a half hours, he became a whirlwind, phoning board members nonstop. He got Schmidt in New York, Kinsella in his car. Frank Bonsal, a board member from Maryland, was vacationing in Wyoming, miles from the nearest phone. Insistently, Boger scared up a driver with a jeep and a portable phone to track him down. Nothing throughout the IPO had been so distasteful or so fraught. The last time the original investors had bought Vertex stock, a year earlier, it had cost them $2.50 a share; now Boger was asking them to reach down and buy blocks of up to 80,000 shares at $9. Remarkably, they all agreed. "They were troopers," Boger said, "but I had to spend capital I didn't want to spend. Benno had to put up $720,000. It was a crime."

Holman, a bachelor, was hosting a dinner party that night at his apartment for twenty-five Kidder interns, but he stayed in his office till nine, awaiting Boger's call. Satisfied now that they could sell all 3 million shares, he and Boger agreed to price the deal at $9. It had been an interminable scrape: from the $18 to $20 that Boger and Schmidt first imagined they could get after the Regeneron offering to the $15 to $17 Boger proposed at the outset of the IPO in early May to the shrunken $13 to $15 of the postcompression period to last week's discount of $9 to $11 and, finally, to the bottom of that range. Other companies had gone lower, to $7 in some cases. Taken together, the compression and the plummeting price had devalued by two-thirds the instant riches that the scientists first envisioned when Boger began selling them on the idea of going public less than three months earlier. Still, it was $27 million for a two-and-a-half-year-old company with no products, no profits, and no assurances—other than Boger's overwhelming confidence—that it would ever have either. By absorbing the last measure of risk themselves, Kidder and Vertex had assembled a deal that they—and Wall Street—could live with.

Holman, exhausted, caught up with the party at his house after 9. "My job was basically over," he said. Boger, attending to the piles of paper on his desk, got home shortly before 12.

But the deal was not easily retired. Throughout the night and deep into the next afternoon, a succession of endlessly unnerving legal and regulatory snarls held up final SEC and NASDAQ approval. Boger and Holman spoke by phone four more times between midnight and 3:30 A.M., then increasingly, alarmingly, throughout the day on Wednesday. Aldrich was grim, beside himself, fearing that the deal wouldn't be closed that afternoon and that skittish investors, again smelling blood, would back away. Finally, at 3:59 P.M., an hour before closing and after the company had threatened to take the deal to the rival American Stock Exchange, NASDAQ approved the sale of Vertex stock. It finished the day unchanged.

It was over. From a dead stop, Boger had taken Vertex in less than three months from a cash-poor, closely held start-up to a public corporation with $30 million in the bank, he and Aldrich selling more than 600,000 shares, a fifth of the offering, themselves. He had turned himself inside out to do it, from a scientist cum entrepreneur to a public showman and administrator who also was in charge of the company's science. Now, it was that science, struggling and unfocused, that needed his attention most. Boger turned to it full-faced. Whether it would receive him back as unambivalently was another matter.

"The problem with the IPO was that it raised people's expectations," Dave Armistead said. "People thought we were going to go out at $15 to $20—precompression—and started figuring in their own minds what it would take to become a millionaire. Then the compression came and knocked them down a third, then the final price discounted them by a third again. . . . Then there was the prospectus, which showed who'd gotten how many shares. Chemistry had a pretty good year last year. Crystallography did nothing. But crystallography as a group got more bonus shares than chemistry, even though there are more chemists. It's like Merck. People began to see that it wasn't just performance that determined compensation. That's dangerous."

Among the scientists, Boger's absence had been interpreted as neglect, his selling as betrayal. As he returned to the day-to-day operation of the labs, he knew he had to allay the damage of his being away, but how much and with what intensity he didn't know and resented having to consider. Having done his part, he found himself more and more intolerant when the scientists didn't do theirs. It was not, he acknowledged, a healthy attitude: a leader impatient with those he's leading. Boger had drawn his ideas on leadership from Hannah Arendt, who said that power devolved directly from the confidence of the governed. But the IPO had exposed a rift between the dictates of his position and theirs. Now, for the first time, the scientists, the original buyers of his story, had come to distrust him, some acutely. Boger knew the feeling. He had had it often himself about others above him.

CHAPTER NINETEEN

Though it was paid for by Fujisawa, the First International Congress on FK-506 was ineluctably Tom Starzl's show. He was conference president, host and lightning rod, the reason 1200 researchers were drawn to Pittsburgh for a week in late August, and why they came anyhow. Summer is when science meets, but Fujisawa, the drug's manufacturer, was rumored not to want this meeting. Its much-delayed European and U.S. clinical trials were going badly. Frustrated transplanters were unable to replicate Starzl's work. The company wanted time, a lower visibility. But Starzl had insisted. He had not slowed down in the year since his heart operation and was determined to spare no amount of effort to advance the drug. Not wanting to shut down his transplant service for the third time in two years in order to take his case to the world, he used his position to bring the world to him. Gaunt and erect in a white or red turtleneck, he had no formal speaking role but plied the cavernous conference center like a specter, dropping in on friends and competitors, occasionally approaching one of the floor mikes to ask a barbed question or lash back at one in his pursed, quietly apocalyptic low-plains drawl.

It seemed a rueful valedictory, this stalking. Starzl still believed that FK-506 was an extraordinary drug. He defended it stoutly, unequivocally. But the molecule had never been for him an end in itself. He had pursued it solely as a tool for overcoming the biological barrier to transplantation. In experimenting with it, he had become perhaps one of the world's foremost clinical immunologists. He had taken from it what he could, and now, as always, was rushing toward the next Armageddon. "This is the last thing I'm going to work on [with FK] minus two, maybe minus one," he said wearily on the eve of the meeting.

But Starzl couldn't easily extricate himself. Even before the conference began, there were harsh signals that he would be called severely to account for the drug's erratic performance, much more so than in Barcelona or San Francisco. More than a dozen medical centers in the United States and Europe now had the drug, and none of them had come close to matching his results. The discrepancy blasted away the cordial veneer that distanced him from his peers, rendering Pittsburgh once again a bunker. Not only was his data widely suspect, but many considered his promotion of the drug improper and self-serving. "It's not the drug that we were led to believe it was," a transplanter indignantly told a confidential marketing survey around this time. Snapped another, "I shouldn't have to get my information about drugs from the *New York Times*." Starzl had long been scorned as a heretic for his methods. Now, an ever-growing list of critics was assailing his credibility and ethics. Though he was past them now, past FK-506, perhaps even past the great span of his life's work in transplantation, he was forced to double back to defend his flanks.

His central claim was that FK-506 was a *qualitatively* better drug than cyclosporine. As long as no one else had the molecule, few could argue, not that many hadn't tried. Sandoz, maker of cyclosporine, had furiously campaigned sub rosa to discredit FK-506, dubbing it "Fujitoxin" and plying transplanters with data suggesting that the Japanese molecule, far from being unique, was what one of them called "a turbocharged cyclosporine—both good and bad." Starzl still insisted it was more than that. Sandoz hadn't needed to send its own researchers to the conference to rebut him. Because virtually all transplanters prescribed cyclosporine and most had taken substantial help from Sandoz, from free samples to draughts of research money, the company had no shortage of surrogates.

Leading them was Sir Roy Calne. A strutting bantam of a man,

Calne, of Cambridge University in England, was Starzl's chief and perhaps only rival. He had pioneered the clinical use of cyclosporine and had paved the way for Starzl's rescue of FK-506 by deeming it too toxic for humans. They were publicly friendly, lavish in their mutual homages, but there seemed to be little love lost between them.

At a panel discussion the first morning, Calne confessed to being "surprised and amazed" by Starzl's success with FK-506. But his iciest testimony concerned the drug. Noting its "almost spooky similarity" to cyclosporine, Calne predicted FK-506 would be approved but as a backup either for those who couldn't tolerate cyclosporine or to alleviate specific side effects, such as "the very unpleasant hirsutism" among children. "It could be," Calne sniffed, sounding at once congratulatory and dismissive, "that children will be the greatest beneficiaries."

Neither Starzl nor Fujisawa, of course, saw FK-506 as a "crossover" drug. To them, FK-506 was a revolutionary molecule, the kind of drug, as Starzl had put it, "that comes along once in a lifetime." Yet they were almost useless to each other as allies, so divergent were their aims, so toxic their recent relations. During the more than two years since Starzl's initial triumph with FK-506 in liver "rescues," Fujisawa had tried at times forcibly to wrest control of the drug's development back from him. Fearing that his outraged refusal to compare FK-506 and cyclosporine in randomized trials would slow or even derail the drug's approval, it had tried to cut him out of critical meetings with the FDA, above the agency's objections. It had refused to recommend his protocols to other transplanters, urging drastically higher doses that Starzl and his people considered criminal. Strategically, Fujisawa planned to grow into a worldwide pharmaceutical company on the billions of dollars in revenues it anticipated from FK-506. The drug was to be its flagship in the United States, where it was the first Japanese company with what appeared to be a major new therapy, and its efforts to establish itself had been costly and, so far, disappointing. It was not about to yield its future to a fractious and unyielding American maverick like Tom Starzl.

Fujisawa kept a determinedly low profile at the congress. Several of its scientists delivered talks, but mainly its contingent remained ensconced in a food- and smoke-filled lounge one flight up from the main hall. All but invisible to the mass of scientists, they met in scattered groups among themselves and with a few invited business guests. "Scientifically, we are very confident that the drug works," said a senior research executive, "but whether this can be a good drug or not from a marketing point of view we are still uncertain. We have to accumulate data to show superiority to cyclosporine."

Starzl's people, conversely, were ubiquitous, embattled, contentious, and indignant. Having dominated the meeting's organization and with the most experience with the drug, they gave the most talks, sat on the most panels. They swarmed in and out, occasionally substituting for one another so that they could race across town to the larger drama of the transplant wards. Throughout the week, they led intense, nonstop parallel lives—one minute immersed, steel-eyed, in a maelstrom of medical center, the next engaging in scientific brinkmanship in the conference hall. As with Starzl, the exigencies of the first gave rise to a profound disrespect for the second.

Their tribulations were extreme. For instance, within an hour of staving off criticism during a discussion on blood monitoring, Andreas "Andy" Tzakis, a soft-spoken Greek-born surgeon with bulging, forlorn eyes and a Harpo Marx–like insouciance, one of Starzl's closest lieutenants, was back at the liver clinic at Pittsburgh's Children's Hospital ministering to Mary Arthur, a pretty seventeen-year-old from Louisville, and to Arthur's family.

Nineteen months earlier, in January 1990, she had had "everything out"—her entire lower viscera—as the first recipient of an experimental "cluster" transplant, wherein the insulin-producing cells in the pancreas are infused into a newly grafted liver. Her surgery, insulin production, and immunosuppression had been uniformly successful—a breakthrough on several fronts—and she had returned home to an active life. Then, in January 1991, a metastatic cancer reappeared. Despite intensive chemotherapy, a tumor materialized in her jaw and had grown virulently, tripling in size in the past two weeks. Sitting with her anxious parents in a noisy waiting room, fingering a small gold cross on a chain around her neck, her cheek delicately puffed, she told Tsakis she wanted to do "whatever will get rid of the cancer." For the next two hours, he and several other doctors—oncologists, radiologists—wearily discussed what to do. Their desperation about recommending radical surgery that might destroy her appearance while only buying her a few months' time was plain, particularly Tsakis's, whose exhaustion and anguish seemed to weigh on him like irons.

It was a posture of interminably beset moral urgency that many of Starzl's people shared and that was represented most vehemently at the conference by Dave Van Thiel, Starzl's acerbic, 300-plus-pound field manager and intellectual second-in-command. "They're overdosing their patients," groaned Van Thiel, Pittsburgh's chief of gastroenterology, during a panel on European and Asian clinical testing. "It's remarkable," he said after Calne's left-handed remarks, destroying a Styrofoam coffee cup by inches. "Despite the fact that they're poisoning people, they still think it's a good drug."

That FK-506's value as a drug was anything but foregone was a medical issue: It would be resolved in the clinic. But Starzl, determined as ever to cull whatever he could, also recognized the pivotal role of basic research in answering practical questions about the drug. Like Boger and Schreiber, he had long since stepped beyond the narrow limits of his field. Only by probing the cellular and molecular biology, he knew, would the drug's maximal value be revealed.

Of course, those questions invited their own rivalries, and Starzl was forced to become involved in them as well. Schreiber, by virtue of his continuing priority in the field, had been given along with his friend and collaborator, Gerry Crabtree of Stanford, the longest presentation—each had a thirty-five-minute plenary talk, early on the second morning. It was more than repayment for good work or because he had promised big news. Both as a member of the committee that reviewed abstracts and as cochair of the session on cellular biology, Schreiber had known for months what everyone else would be presenting. In at least one case, he apparently influenced the scheduling, requesting that a Merck scientist named John Siekierka, one of the codiscoverers of FKBP-12 and a presumptive competitor of his in the search for new partner proteins, be moved to a less featured slot.

"Siekierka is pissed purple because we put him in a different position," acknowleged Starzl, who smoothed things over by inviting

Siekierka back onto the second day's program, though only after, and with less time than, Crabtree and Schreiber. With Vertex SAB member Steve Burakoff, Schreiber's other main collaborator in immunology, introducing the morning session, Schreiber was guaranteed that he wouldn't be upstaged. Indeed, it hardly mattered. By now Starzl had come to see in Schreiber a key ally and had taken—unsolicited—to praising his work lavishly. Even without Starzl's interceding, Schreiber was a made man.

Vertex, as Boger had promised, sent a large group: Thomson, Harding, Livingston, Moore, Peattie, Nelson, Stuart, and several others. However, none of them—Boger suspected an illicit boycott by Schreiber—had been invited to give a major talk. They would, in the pyramidal hierarchy of many such meetings, deliver so-called open papers—ten minutes from the podium—or posters, standing next to blowups of their work in the lobby during designated hours, like vendors at a trade fair. In fact, a few of the calls were questionable: Moore, whose *Nature* paper was both well known and highly regarded, was only given a poster. But on the whole, the assignments seemed to reflect what the scientists already knew: that their research so far had been important but hardly earth-shattering and of less than general interest to a medical audience.

Boger had talked for months about coming himself but, conceding that Schreiber "owned the meeting," finally decided against it. "This could have been an important meeting for Vertex," he'd said, "but it won't be because of Stuart. When Vertex can flash a slide and say, 'This is a small molecule inhibitor of FKBP that can be made by a masters-level chemist in his or her kitchen in an afternoon,' that's when we bring down the house at an FK-506 meeting.

"The *second annual* FK-506 meeting," he vowed.

Navia also changed his plans. Invited only to give a poster, he decided first to fly in for just twenty-four hours, then canceled abruptly at the last minute. It had been a grim time for him professionally. In July, after months of deliberating, *Nature* finally rejected his and Yamashita's structure paper. The structure itself was still wrong, and he and Yamashita humiliatingly had to correct it using Schreiber's and Clardy's coordinates. Meanwhile, his "baby," using cross-linked enzyme crystals as supercatalysts, was rejected by *Science*, and he still hadn't produced structures showing Vertex compounds inhibiting HIV protease. Navia had come to Vertex boasting that his group would be a powerhouse, but it had faltered repeatedly while he had been increasingly absent and, it seemed, consumed by his own ambitions. Consummately concerned about his place among crystallographers, he hadn't published any work in more than two years, since he left Merck.

What with the Yamashita psychodrama and Boger's decision—widely resented within Vertex—to make him an officer, Navia had become alienated, morose, unsure of himself since the road show, and prone increasingly to sudden rages alternating with acts of contrition. The thought of standing by a poster of a protein structure that he had needed Schreiber's help to solve while Schreiber himself dominated the congress from the stage would have been unbearable. He still believed Schreiber's effector hypothesis was too simplistic, but without data or an adequate place to stand, he was not about to challenge him on it in Pittsburgh. Latching on plaintively to Hurricane Bob, then blowing off the Atlantic coast and playing havoc with flight schedules, he asked Harding and Moore to put up a sign at his poster station saying Hurricane Victim. "Yeah," sneered Thomson sardonically, "Hurricane Stuart."

If Navia was a casualty, he was a casualty of the need of all scientists to be first and of his putting his neck out too far. He had misjudged—himself, the competition, the questions he was trying to answer. But even those who had succeeded and were now struggling for priority were at risk of being subsumed. FK-506, if anything, had proven more mysterious and powerful than any of them had imagined. They had all striven to know the molecule: its shape, how it interacted, what it did, what it was capable of doing, how to manage and control its extraordinary potential as a drug, what to *do* with it. They had dared one another and themselves to know more, do more, explore more deeply. And yet even the most Promethean of them—Starzl—had met with inevitable dissatisfaction.

FK-506 had pushed Starzl relentlessly beyond himself, beyond even his own murderous strictures for reaching past what was possible and for cheating death. Once islet cell transplants like Mary Arthur's proved successful, for instance, he immediately pushed his group to the next logical step: the intestines. Without the ability to absorb food, islet cell recipients continued to suffer greatly. They were either tied to feeding tubes or, like Arthur, nibbled down protein shakes and bagels, unable to gain even a few ounces. And yet the intestines were, as Starzl observed, "the forbidden organ"—filled with deadly bacteria that if they broke into the bloodstream caused massive, often fatal infection. Still, of the first five patients to receive new bowels, all survived, eating and—after a grueling recovery—digesting food.

With each new round, the attacks on Starzl grew. "I read . . . the report . . . with mounting horror," a British doctor wrote after reading Tzakis's published account of Mary Arthur's operation. "I thought that this sort of mutilating surgery had ceased long ago. How many more cruel and inhumane operations will be done in the name of advancement of science?"

Starzl had always been obsesssed, but his urgency and resistance to criticism had been case-hardened by age and ill health. At sixty-six he finally seemed to be acknowledging his mortality. Ever youthful and irreverent, he had long ago given up any religious hopes for himself. But transplanters cannot but be changed by their patients, many of whom are sustained only by a profound and abiding faith. A hand-written note on the wall next to a young woman who'd just had a liver transplant quoted Romans 8:18: "For I reckon that the sufferings of this present time are not worthy to be compared with the glory which shall be revealed to us."

So it now seemed with Starzl. For thirty years, he had not stopped working. He had performed thousands of grueling and macabre operations despite being tormented each time he picked up a scalpel by the horror of what he was about to do. In the process, he had imperiled himself—overcoming hepatitis twice, an ulcer, temporary blindness after an operating room accident with a laser, and his recent bypass surgery—and destroyed his first marriage. Meanwhile, he had become the most prolific scientific author in the United States, publishing 503 scholarly articles between 1981 and 1990, many in prestige journals and all, if not dictated personally, written with his unstinting—many of his staff would say bullying and heavy-handed—editorial assistance.

And yet the ultimate prize had eluded him, not only a key biological discovery that moved medicine irremediably ahead, but the honor that science reserves for those who account for such breakthroughs. Schreiber, whose own ambitions were clear, recalled talking to Starzl by phone on the day the previous fall when two other transplanters shared the Nobel Prize for medicine, all but foreclosing another Nobel for transplantation anytime soon. Schreiber marveled at Starzl's steely equanimity. He had not gone into seclusion, as one of Schreiber's Harvard colleagues was known annually to do, but kept up a full eighteen-hour work schedule throughout the day and in the days ahead.

Starzl knew his dominance was coming to an end. He had already decided to devote the rest of his career to research, having stopped operating earlier in the year, and was now facing the turbulent conclusions drawn from all his restless endeavors: that the future of transplantation, as it always had, lay ultimately in the ability to control rejection and that even that would never be enough. Mary Arthur's resurgent cancer, like so many relapses before it, reminded Starzl and his people dramatically of the ultimate futility shadowing even their greatest successes. Transplantation went only so far.

This was the great imperative of Starzl's life: to go further. And it explained his zeal in trying to ensure his legacy. Bursting headlong into several new areas at once, brimming with his customary resolve, Starzl also had recently begun campaigning more actively for recognition, an effort that included flying off more and more frequently to receive awards and prizes he once would have spurned, aggressively cultivating the media, and publicizing his achievements. The Pittsburgh conference itself, however important medically, was equally an effort by Starzl to orchestrate the attention of several worlds on *his* molecule, even as its cachet began to fade.

"There is a strange thing about the dimming vision of aging eyes," he would soon write in his memoirs—another of those self-promotional acts to which Starzl now committed himself utterly—which he wrote in the breathtaking time of just three months. "What cannot be seen clearly, the mind fills in more vividly than reality." Thus it had become with immunosuppression. Peering into the heart of immunology, Starzl had begun to perfect his own understanding of it. What he lacked in data, he more than made up from an unparalleled breadth of experience, of vision. No one else, certainly no surgeon, had descended the ontological ladder so completely, from organism to organ, from organ to cell, from cell to molecule, bearing such powerful witness at each stage.

Ironically, he began thinking more and more about multidrug therapies that not only abridged the need for the kind of radical surgery he'd spent his life advancing, but might eliminate it. "Transplantation," he began to say, "may just be a footnote to this entire story."

This was the message Starzl intended to press at the conference. However, there were problems. His best data had been developed by an Italian liver expert, Antonio Francavilla. Starzl's team had always been a polyglot group, with Japanese doctors struggling to communicate with Greeks, Swedes with Italians, and all of them attempting to overcome—with greater and lesser success—deficient English. Yet even in that mix, Francavilla's accent was persistently torturous. ("His English was worse than imperfect," Starzl would recall twenty years after their first meeting. "I could not understand a word he said.")

The problem had surfaced acutely two days earlier on the eve of the conference. As Starzl and nearly one hundred members of his regular Monday night FK-506 group gathered to rehearse their presentations, he began to doubt whether Francavilla, who pronounced the drug's name as "effa-kaya-fiva-oah-seex," wouldn't simply garble his.

"This piece is a companion piece to Schreiber's," Starzl had begun encouragingly as Francavilla stepped up to practice his talk. "We have to very critical about this. If it's not precise, our message will be lost."

Van Thiel, sensing that the one real piece of "earthshaking" science that the Pittsburgh group would be presenting was being strangled, was less patient. "I think we should scrap the first two conclusions. They're somebody else's," he cut in after Francavilla showed his first several slides. After another incomprehensible explanation, Van Thiel wrung his huge hands. "I think it would be great if Tony spoke perfect English," he muttered. "I don't think you want Tony to defend that slide. I wouldn't want to defend it in Italian."

"Can we practice again?" Starzl looked up, exasperated. Francavilla, apologetic and perplexed, said he would try.

Throughout, the strains escalated. Invoking Schreiber time and again, Starzl exhorted his researchers one after another to push their conclusions, particularly with autoimmunity. It was here, on the clinical side, that Pittsburgh still maintained an edge, and Starzl was determined to wring every advantage he could. He had wanted the FDA, for instance, to approve a clinical study with FK-506 for patients with MS. But the agency had resisted authorizing treatment for a disease of the nervous system with a drug of known neurological toxicity. Now, when a neurologist named Benjamin Eidelman reported that three patients with MS—two transplant recipients and one other—had been receiving FK-506 and had shown dramatic improvement, Starzl was ebullient. He noted: "I saw the third patient today and she looks ravishing. She's flakier than hell, but she's ravishing."

Eidelman, a senior researcher in his own right and not prone to excessive claims, was cautious. He dourly reminded Starzl that MS is notorious for spontaneous remissions, but Starzl pressed on. "You're sitting on your hip pockets," he said. "You're telling a great what-a-big-boy-I-am story, but you're being too conservative. Jesus Christ. With three MS patients, this is really big news."

"Neurotoxic means neuroactive," Van Thiel reminded them. "It's a matter of dose."

Eidelman again resisted, but by now Starzl was on him. "I would deemphasize the what-a-big-boy-am-I character," he said tautly. "Who gives a shit? The important thing is that the drug's been used now for three patients with MS. It's almost the biggest news of the whole damn thing. The public needs to know this. This is the big glitch with the FDA. This is where they need to feel some pressure to come along."

Eidelman stood his ground. He would not make any claims for the drug, only report his findings. Starzl, smiling through gritted teeth, accepted his decision.

"One of the things about discovery," he lectured, "is that you've got to know you've discovered something. You don't want to be afraid all the time."

It was an oddly fateful reprise of the scene at Vertex months earlier when Boger and the others were preparing Yamashita's crystallography paper, driving till the last minute to make the most of their data, risking overstatement to build a more magnificent bridge.

But by now it seemed too late. By relegating himself to the role of arbiter and coach, Starzl could only, as he put it, "engineer . . . make sure of quality." The larger lessons of the drug's action no longer favored those who made their observations in the operating room and the clinic, but in the lab. Despite his high-handed treatment of Eidelman, Starzl seemed resigned to the fact. Indeed, in a sense, he'd already ceded it. The next morning, after working through dozens more papers and staying up much of the night, he advised reporters at an opening press conference what he thought their story would be: "It's unfair to focus on anyone, but I think that Schreiber's group in Boston has big news. . . . I think you ought to talk to Schreiber."

Schreiber wandered through the conference at an airy distance from the proceedings. Not only did he know very few of the participants—he and Starzl had only just met face-to-face—he was practically incognito. Most of the surgeons and other doctors had probably not talked to a synthetic chemist since they were undergraduates, and the molecular and cell biologists, even if they knew Schreiber's name from his articles, were unlikely to put it together with the loping, mildly self-conscious figure casually perambulating through their midst. Schreiber was like a sightseer, famous in his own country, but now abroad, waiting for someone to notice him. Like Starzl, he had crossed several intellectual time zones to get to this place, yet was still something of a stranger. Even the oversized video monitor at the entrance to the hall showing a computer model of the complex structure of FKBP-12 and FK-506 (Clardy's structure "courtesy of the Harvard Chemistry Department and S. L. Schreiber") elicited near total indifference. Most of the conferees didn't know what they were looking at.

Those who did know him, of course, were his former colleagues from Vertex, and Schreiber and several of them sought each other out. Always irrepressible, Schreiber was dying to talk about his new work. After restlessly sampling the crowded poster session after lunch on the first day, he finally tracked down Moore and Harding, who, hearing the rumors of his impending announcement, had begun looking for him. Not just curiosity, but time drove their search. The sooner they knew what Schreiber had, the sooner they could respond. Merck, if it hadn't been already, would soon be all over him. Boger had once said of the chase for data: "The average third-year revenue on a new drug is $300 million, so three months now could be worth $75 million later on." Every day, every hour, had a price. Eagerly, the three of them huddled by the coffee and picked-over danish near Moore's poster.

"We've got it," Schreiber blurted. "We've got the immediate receptor for both complexes. It's going to blow everybody away."

Harding was stunned. It was literally, as Livingston would say, "a transcendent discovery"; as Calne would put it, "a revelation." Since the discovery first of cyclosporine, then of FK-506, there had been a rising tide of questions about the drugs' actions: How do they work? Why are they immunosuppressive? Are their side effects intrinsic or superfluous? Are there partner proteins? What are the relevant portions of the molecules for drug design? Why are they so strikingly similar? Together the questions had formed one of the most tantalizing mysteries of modern biochemistry. Now, incredibly, exasperatingly, Schreiber appeared to have solved them all, affirming in the process almost everything he had predicted.

In a single, brilliant set of experiments, he, his graduate students, and collaborators at Stanford had shown how both drugs worked. Once in cells, FK-506 bound to FKBP, and cyclosporine to cyclophilin—that much was never in doubt. But the drugs themselves didn't do anything. Only upon half-burrowing into their hosts and presenting their remaining atoms to, in each case, a third molecule did they initiate a biological response. Schreiber and his collaborators identified that common molecular partner as calcineurin, an abundant enzyme found in all cells.

"Of course you always hope for the simplest possible scenario, which is that the immediate target is the same," Schreiber said. "And it is. It is!"

Schreiber was smiling wildly, a Roman candle: "My God. It's unbelievable. They bind competitively—two different drugs, two different structures, two different immunophilins, and they bind to the same site. How did that ever happen?

"I mean," he continued, "nature evolves these two microbial products, and they're *glue*. They're molecular *glue*.

"These drugs," he grinned fabulously, telegraphing his punchline, "aren't drugs at all."

Harding maintained a dispassionate front. He asked the obvious questions: How big a protein was calcineurin? What did it do? What were *its* partners? Was there anything else, either a natural product in the body or another drug, that inhibited it? Yet behind his resolve lay an angry snarl of emotions.

Harding was prey to a seething envy: He had wanted badly to discover the relevant biological target for the drug himself, to explain how it worked, to maintain his priority, and had been routed by his brilliantly successful ex-collaborator and friend. As he often did, he also now felt resentful, betrayed, as if by some combination of his own accommodating character and the overpowering forcefulness of those around him, he had been oppressed, pushed aside, rebuked. He felt defeated, a victim. He had come to Vertex two years earlier at the peak of his promise, codiscoverer of cyclophilin and FKBP-12, to collaborate with Schreiber: two young conquerors overtaking the world. But Schreiber had now gone on beyond him alone. Worse, Schreiber's new work had instantly and thoroughly eclipsed their earlier accomplishments together. A protein's cachet—and its discoverers'—rose and fell strictly according to its biological significance, what it did. Science, like fashion, could be exceptionally cruel as it moved ahead. And here was Harding, codiscoverer of two proteins that in the end didn't do anything but present atoms, like servile eunuchs, to the real molecular kingpin, the real "protein of interest." Because inhibiting FKBP-12 was apparently no longer what mattered for immunosuppression or drug design, so did the protein become instantly marginalized, passé, a footnote. Harding could feel himself going with it. (It was more than spontaneous diminution or paranoia: The next day, Schreiber put up a slide during his speech that credited "Schreiber et al." with the discovery of FKBP-12. Harding, first author of the *Nature* paper, was never mentioned.)

The conversation broke up, with Harding and Schreiber chatting perfunctorily and making plans to get together. Moore, barely concealing his astonishment and admiration, returned to his poster; Schreiber, to the meeting. Harding, immediately, instinctively, headed through the skywalk and up to his room on the eleventh floor of the adjoining hotel, a Vista. He was going to call Boger.

In the elevator, all his resentments from the past two years hewed to the surface like a sudden, violent rash. To Harding, Schreiber's triumph and the corollary of his own defeat were a referendum not on which of them was the better scientist, but on which was the better system for doing science, Schreiber's or Boger's. In his mind there had been no contest: Schreiber had won every round. And he and the other scientists had had to pay for it with their careers, their invisibility.

Angrily, his face reddening, his eyes more mournful than usual, he began to blame Boger. Boger's supreme maxim was, Do the experiment. But doing the experiment at Vertex had become harder and harder, at times next to impossible, he thought. There were the incessant pressures to produce, run assays, test compounds. There was the chronic understaffing, the shortage of reagents, Boger's absences, the vagaries of the project councils, the clashes of egos, the distrust of outside collaborators, the niggling distractions of having to help sell the company's story to bankers and lawyers and assorted financiers. There was no time or place to think. Schreiber had won, Harding concluded, because he was utterly devoted to winning. He, on the other hand, had lost because of a handicap. He was a victim of priorities that were as unfocused and at odds as Schreiber's were singular and acute. "I'm a technician with a Ph.D.," he muttered. "I was hired to be a scientist, but all Josh expects me to do is come in day after day and flog away at the bench. That's not science.

"It's heartbreaking for me, because if I was back, bitching and moaning, working at Yale transplants, I'd be one of the world's most highly recognized scientists now."

It was understandable that Harding's bitterness would be shared by others in the Vertex camp as news spread of Schreiber's bonanza: They had been killing themselves over the wrong protein. But the outpouring was swift, violent, total. Boger had no defenders in Pittsburgh. Whatever their individual response to Schreiber's victory, it occasioned for each an opportunity to air long-festering grievances that they felt Boger had made it impossible to discuss. For some of them, the mood would flash over loudly into desperation.

Thomson, not surprisingly, was the most upbeat. A connoisseur of proteins, good science, and the Nietzschean overtones of Boger's and Schreiber's rivalry, he couldn't help but admire Schreiber's hat trick, even though it meant his own achievement with FKBP-12 was now as diminished as Harding's; even though he would soon be up to his armpits in calf brains, extracting and purifying calcineurin. The next morning, before his talk, he would say, "I feel like getting up and saying, 'I don't want to waste your time. I give my ten minutes to Stuart so he can talk some more.' " From the podium, simulcast in his bomber jacket at ten times his size on the projection screen behind him, he repeated the panegyric by congratulating Schreiber publicly—an unusual gesture. He was dismayed again over the loss of Schrieber as a potential collaborator, moaning, after a researcher from Smith-Kline acknowledged Schreiber's generosity in providing materials, "Is there *anyone* here who doesn't have reagents from Dr. Schreiber other than the company he was scientific advisor to?" But he failed to blame Boger for either the rift or the outcome. He reveled in Schreiber's achievement and was politic enough to pay tribute, though not before calling Vertex and telling Matt Fitzgibbon to order from Sigma, a major protein supplier, its entire stock of calcineurin so that no one else would be able to get it.

The biologists were not as forgiving. Dave Livingston argued that Vertex should shut down chemistry completely for six months while biology confirmed whether or not Schreiber was right and came up with a reliable assay for testing new compounds. What good was it, he asked, to keep making molecules to inhibit FKBP-12? It was ridiculous. He, Debra Peattie, Harding, even Patsi Nelson, perhaps Vertex's truest believer and one of its most uncomplaining scientists, all bemoaned Boger's decision to pick a project where the biology was so poorly understood. Unanimously, they agreed that his lack of formal biology led him repeatedly to underestimate it and them. They were horrified, irate, that Vertex was claiming to do structure-based drug design while it still wasn't clear what they were designing molecules *to do*. "So do you guys have a drug yet, or do you just have a target?" a nephrologist from Tampa would ask Harding on Wednesday at a communal lunch table. "Until a half hour ago we had one of each," he grumbled, "but now we have three more targets."

As in the ritual slaying of a king, Boger was blamed for everything. One group, led by Thomson and Moore, complained that he hadn't been directive enough, hadn't simply told them what to do and made sure they'd done it; another, spearheaded by Livingston, Peattie, and Harding, countered that he had been *too* directive, making major decisions unilaterally. Livingston theorized that he was, in fact, both, each at the worst time. For example, he said, Boger had promised Chugai, without consulting his senior scientists, that Vertex would deliver on two major priorities—small molecule inhibitors of FKBP-12 and the structure of the protein—*before* it had determined whether those goals were relevant. He then let the scientists decide largely how to meet them. It was perverse, Livingston said. It should be the other way around. Boger, he argued, should consult his senior people before setting goals and priorties, then use his authority to execute decisions on his own. To Livingston, the inversion reflected Boger's arrogance, the failure of which now seemed egregiously apparent to those who worked for him.

There was little joy or vindication in the analysis: Livingston, like the others, was disappointed, bitterly frustrated, dismayed. He dreaded confronting Boger, knowing how Boger turned things around. Even if what he said was true—that by his pose of being smartest in the class, Boger had created a cult of personality that was bad for the company's science—he also knew better than the others that Boger's approach had been the right one. Boger had no choice but to be arrogant. He had no choice but to make bold promises based on incomplete data. Vertex would be dead otherwise. There was no way Boger could have raised $70 million in two and a half years, packed the labs with world-class scientists and equipment, put them in the middle of the hottest areas and made them competitive, given them stock that was now worth hundreds of thousands of dollars and jobs that were secure for at least several years, no way that he could have done any of this without flaring the great calipers of his ego.

Had Boger been there to defend himself, he'd have said what Livingston and the others knew to be true: that Vertex was a drug company; that it was not in business to make important biological discoveries but to use them; that the competition with Schreiber was a sideshow in which Schreiber had all the advantages. He'd have pointed out that Schreiber, supported by Harvard and the federal government, hadn't had to tap-dance his way around the world several times a year to scrape together capital. He'd have noted provocatively that Schreiber had at his disposal dozens of the world's smartest, most aggressive graduate students and postdocs with no more complicated goals than to please him, not a contentious, ego-driven group of journeymen researchers with their own oversized ambitions (who, he'd have added, he thought were acting like children). Schreiber didn't have to run a fast-growing company, which either grew or collapsed; he did. Schreiber didn't have to make anyone rich; he did. Schreiber didn't have to satisfy Wall Street, only himself. Indeed, it was best to let academics like Schreiber sort out the biology while Vertex chased the supreme prize: designing molecules.

But Boger wasn't there, which, to the scientists, was just the point. By running the company's business, by isolating himself out of necessity but also self-righteousness, he had also isolated them, stranding them in a world that thrives at the forefront on access and collaboration. Merck, Starzl's group—they were big enough to go it alone; Vertex wasn't, and the flaws of hubristically thinking that it was now etched themselves in its defeat.

To Schreiber, this also was the crux of his breakup with Boger. "A characteristic quality of Vertex that is highly unusual—and many people recognize it—is that they are absolutely determined to do everything completely themselves," he said. "I really don't think they want any input from the outside.

"That's the real issue between myself and Vertex. The other stuff," Boger's concern about what he perceived as Schreiber's utter lack of ethics, credit mongering, and apparent contempt for the company's need for secrecy, "is a smoke screen.

"I have a certain visibility in this field, and from that point of view, they take a certain risk, in interacting with me, that the perception may exist that I have had a lot of influence over what they're doing and therefore should get some credit for what they've achieved. They have to evaluate that concern. It's not to say that they can't solve every one of their problems by themselves, except that the more activity they're made aware of, the quicker they will learn."

Symbolically, Pittsburgh had captured the fundamental difference between them. In their sense of their own superiority, in their drive to be great, Boger and Schreiber had drawn sharply different conclusions about the need for other people. Schreiber inhaled other people, their ideas, sublimating them effortlessly, without pangs, into his own ambitions. They were his instruments. He expected the same of them. Boger, just as solipsistic, treated others as he treated himself. He believed they were as omnipotent as he was and thus unneeding of anything more than opportunity. He thought they *should* do the right thing, intuitively, on their own because it was less fun and efficient not to. Those under him he let make their own mistakes.

Throughout the conference, the scientists would become more and more frustrated with Boger's view, even if grudgingly they agreed with it. It might be right for Vertex, for business, for the drug industry, for making better molecules, even for leading an ethical life, as Boger insistently maintained. But it was wrong for the world of winner-take-all science for which the conference was a stand-in. Here Schreiber was everyone's darling, a figure of immense respectability and esteem, having a status distinctly at odds with their own anonymity. They had done well, but he had *won.* At a gala reception in the Vista ballroom on the night after his talk, Schreiber glided in an almost ethereal light. He was surrounded by people. If Boger's laughing confidently with Holman and Aldrich outside a waiting limousine had become the defining, if ambivalent, image of his newfound success, this was Schreiber's: to be at the center of a group of well-dressed senior researchers, all leaders in their fields and with some of the most powerful labs and connections in the world at their disposal, heading off brightly, their drinks clinking in the powering din, to have dinner together, to talk about, as Starzl put it, "our strategy . . . what we're going to do next." Starzl was in the group. So were Burakoff and Barbara Bierer, still a Vertex consultant. And sitting off to the side of this shining tableau was Vertex—desultory, remote, cursing Boger loutishly for their fortunes. Harding, drinking heavily, would spend the night in a fever of abject jealousy.

Harding got off the elevator, slid the card key into the lock, and entered his hotel room. The room, bright, had an air of pleasant abandonment, like a classroom during vacation. It had just been cleaned. It was that hour approaching midafternoon when the halls of convention hotels are empty as tombs, with the guests gone and the maids having finished up. He dialed Vertex.

"Stuart's identified the critical partner protein," Harding told Boger. He was purposely breezy, as if he was reporting a dramatic car wreck in which no one had been hurt. "It's calcineurin."

"I don't know what that is," Boger said. His voice was cool, restrained: Harding could see him scrolling through the contents of his computer screen. Then, softly: "How does he keep doing this?"

They spoke for several minutes, Harding doing almost all the talking. Boger, assumably taking notes, asked basic questions that, on the one hand, he needed answered to begin his own data collection and, on the other, showed how completely and stunningly original Schreiber had been.

"How do you spell calcineurin?" he asked.

Harding told him.

"How big is it?"

Boger wanted a rough measure of how hard it would be to solve the structure of the protein. Harding told him. "Good," he said, finding out that it had a molecular weight of about 55,000. "We don't have to worry about Stuart beating us with NMR. It's too big."

Boger asked several more perfunctory questions. Then, business done, he permitted himself a terse groan.

"How does he keep doing this?"

"You mean why is he so lucky? It's because he sits all day, every day, thinking about what he wants to do and how he's going to get it done, and he's got an army of people who are willing to work hard."

"Supposedly we do, too." Boger sounded more exasperated than acid, though clearly he was annoyed.

"He's blessed," Harding volunteered.

Silence.

"Why," Boger asked, answering his own question, "does it always have to be Stuart?"

CHAPTER TWENTY

Best in a crisis, Boger averted one now by breezily refocusing the debate. Forget Schreiber, he counseled the scientists returning from Pittsburgh. All that matters, all that ever matters, is data, data that now was emerging spectacularly, data that was Vertex's, exquisitely if not uniquely, to exploit.

It was a powerful balm—redemption in the next experiment—and Boger administered it staunchly, without recrimination or doubt. He reminded them that this was the moment they'd worked for. Thanks to Schreiber, who as promised had quickly released the computer coordinates for his and Clardy's X-ray structure, Vertex finally had sufficient information to make a first substantive pass at designing drugs. It had detailed blueprints both of the native protein and of the protein in bound conformation with FK-506. Meanwhile, Navia, moiling penitently at the bench while the others were in Pittsburgh, had grown earring-sized cocrystals of FKBP-12 and Vertex's lead semaphor compound, 367, and had passed them to Yamashita, who, revivified and cramming for his medical school boards, was now within days of solving that structure as well. And there was, whatever one thought of it, Schreiber's calcineurin bonanza. Excusing the source of much of the information and the significant new problems it presented for drug design, Boger couldn't have been more pleased. Exuberant, he left the scientists few excuses to mope.

He himself overcame the loss to Schreiber as he often had: by minimizing Schreiber's role, writing it off to a character flaw, damning the academic star system that allowed Schreiber to take credit for it, and then, to ensure his point, ridiculing the discovery. If Schreiber now considered Boger too self-reliant for his own good, Boger dismissed him as a black hole, a view freshly supported by the stream of publications following the Pittsburgh conference.

According to this paper trail, the first evidence of a specific recepter for an immunophilin complex came not from Schreiber's lab but from that of a Stanford immunologist named Irving Weissman. Weissman was world renowned, a leader in the molecular biology of T cells, far more accomplished—and famous—within cell biology than Schreiber. Five months earlier, one of his graduate students, Jeff Friedman, had discovered that cyclophilin and cyclosporine together bound to an unidentified protein with a molecular weight of about 55,000. Weissman called Schreiber, who he knew was looking for a similar receptor for the FKBP-12/FK-506 complex. "It is clear to all of us that there had to be a common mediator, and this seemed to be a candidate," Friedman says. Schreiber flew to Palo Alto. He took Friedman to dinner at an Italian restaurant, where Friedman agreed to share his data. The work, obviously of great promise, was the essence of his doctoral thesis.

Back in Cambridge, Jun Liu, a postdoc in Schreiber's lab, quickly identified Friedman's protein as calcineurin. Working further, he found that it was a common binder for both complexes and determined that two other small proteins were involved as chaperones. Thus, as the publications now indicated, Schreiber's lab had broken the story, but not without being given the basic facts by Weissman and Friedman.

Yet in Pittsburgh, Schreiber had credited Liu strenuously but had scarcely acknowledged the contribution of the Stanford group. To the world, calcineurin—like FKBP-12 and its structure—was Stu Schreiber's discovery. Schreiber had used his prerogative in announcing the work to assume overall credit for it. Said Friedman, "My view of Stuart is that he's pretty much a megalomaniac. I will never work with the guy again."

To Boger, who scarcely needed persuading, this was yet another example of Schreiber's persistent self-aggrandizing. Schreiber had seen the whole picture, had put himself in a position to enable it to emerge. It was a brilliant example of scientific leadership. But Schreiber apparently had not had the initial key insight himself. Typically, Boger dismissed not only the accomplishment but the implications, particularly Schreiber's assertion that cyclosporine and FK-506 were "molecular glue." Scoffing at the idea that the effector domains of cyclosporine and FK-506 were strong enough to tat together two much larger proteins on their own, he said: "This is Uri Geller bending spoons. Don't show me the evidence. It's impossible."

Boger's cleverness and deft moral certainty blunted most of the attacks on his leadership but not the scientists' despondency over what was now expected of them. Schreiber's assertion that FK-506 wasn't a drug at all but an extraordinarily fortuitous dollop of molecular adhesive, in fact, had calamitous implications for everyone in the field, particularly Vertex. Indeed, the impact could hardly have been worse.

Boger notwithstanding, there was little question that Vertex's rationale for improving on FK-506 had taken a severe, if not mortal, turn. In less than two years, the project's degree of difficulty had increased exponentially. What had started with a simple enzyme inhibition problem—to insert a tighter binder than FK-506 into the active site of FKBP-12—had been replaced by a quagmire. If Schreiber was right, and what was necessary for immunosuppression was to inhibit not FKBP but calcineurin, everyone in the field now had to change horses. They faced the immeasurably harder task of mimicking a molecular architecture that was largely unknown, that involved as many as five entities that changed shape, perhaps dramatically, on contact, and that would require the bound structure of at least several, if not all, of them to visualize. It was like designing pieces for a sloshing, scissoring three-dimensional jigsaw puzzle, blindfolded, with outmoded templates.

Boger, as ever, was confident such an effort was possible. "The idea that FK-506 is the best it can be is ridiculous," he said. His faith that drugs discovered through screening are accidents of nature and thus, by definition, flawed was unshaken. He dismissed recent reports that Merck, despite hundreds of man-years, had been unable to make a single change to FK-506 without sharply reducing its biological activity. Typically, and with notably less success than usual, he exhorted the scientists to join him.

There were solid reasons for the scientists' reluctance. It was, after all, extraordinary. However unintentionally, Schreiber's tying of FK-506 and cyclosporine to a single partner protein legitimized the view, promoted most assiduously by Fujisawa, that the two drugs were, in fact, not accidents at all. According to this interpretation, the molecules were evolved specifically by nature to do exactly what they seemed to be doing in T cells: tacking FKBP-12 and cyclosporine to calcineurin (all of which are found in all cells and in all organisms) as part of some more universal biochemical interplay. Such a view would help explain the drugs' almost mystical activity. Sandoz, for instance, with nearly a decade head start on Merck, had reportedly made 1200 cyclosporine look-alikes, virtually all of which were less potent. From the standpoint of simply tatting together relevant proteins, FK-506 and cyclosporine were beginning to look more and more like what Boger and the scientists could only find ruinous, inconceivable: They were beginning to look perfect.

Boger, unsurprisingly, rejected this view. He was especially caustic regarding the implication that because the molecules had defied improvement by conventional medicinal chemistry, they were therefore unbeatable as drugs. Even if they had evolved to perfection inside their respective microbes to glue proteins together, he reasoned compellingly, they had not evolved to become perfectly bioavailable in humans cells, to survive the gut, to be optimized for reducing side effects. In science, Boger knew, God was in the details, and 4 billion years of microbial evolution, however perfect in itself, was still no basis for designing drugs. That the universality of their targets implied increasingly that the drugs' side effects were inextricably bound to their activity was a point Boger prudently failed to raise.

It was here—their therapeutic profiles—that FK-506 and cyclosporine were still most vulnerable and where Boger was sure Vertex could still win. However, it remained a "religious question," as he might have noted, whether Vertex could make a better molecule in its present circumstances, that is, before it ran out of money.

Aldrich, especially, worried about this. He continued to believe Boger's predictions but was unwilling to dismiss Merck's and Sandoz's failures as mere wrongheadedness. "If the big boys can't do it," he said, "it makes you wonder." As so often before, Boger's fearlessness, determination, energy, and focus were a rallying point within the company. But ultimately his success depended on his ability to convince those who worked for him that they could do what many of them were beginning to think impossible. "Josh drives us faster than he should," said Tung, "but not as many people are as ambitious as he is, and not many people get as much done with as little as he does."

Boger came to the first project council meeting after Pittsburgh wry, supportive, and, as he often was when he thought the scientists were becoming too self-involved, deeply provocative. The councils, which had gone on unevenly without Boger during the IPO and with increasing tension ever since, had become exhibit A in the indolent mutiny of the past few weeks. Privately, some of the scientists hoped that Boger now would simply disband them, put himself in charge, name department heads and project managers, and tell them what to do.

Nothing was less likely. Far from causing him to renounce his social experiment, the moment occasioned for Boger its first real test. To retrofit a project spontaneously in the wake of new information—to move quickly where big companies like Merck couldn't—this was Boger's ideal. The councils, like Lenin's Soviets, were his sword for permanent revolution, for doing things the way he'd determined was best. He was hardly about to lay it down. Besides, he confessed disarmingly, he didn't know what to do, not specifically. He would learn from them, from the results of their experiments. When he had more information, he'd decide.

In the short term, he was more concerned about resources. He strolled to the whiteboard and began scribbling numbers. Vertex had thirty-five researchers in immunophilins: five in chemistry, nine in biophysics, twenty-one in biology. As an index of the company's priorities, the numbers implicitly rebuked the biologists, who had complained the loudest in Pittsburgh and who now swallowed their tongues. He then followed each number with a question mark under the heading Future. Insisting that the overall number wouldn't grow, he told them to prioritize their experiments and return within a few weeks with new staff adjustments.

Though he'd have preferred not to be considering a complete overhaul in the company's lead project just two months after telling the world it was on the verge of having a drug candidate, Boger loved these moments, loved to rile things up. He compared them to "free climbing," where climbers scale mountain walls without ropes. Throughout September, he reveled as the scientists struggled to assemble a response. Certain experiments were obvious. Before they reoriented the project toward designing inhibitors of calcineurin—a challenge so daunting that if it were to come up in the New Project Council, Vertex's long-range planning group, it would have been howled down at once—they had to test Liu's and Friedman's data. Schreiber and his allies could, after all, have been wrong. Vertex needed quantities of calcineurin and an assay for testing its inhibition. Meanwhile, they would keep making and testing molecules against FKBP-12. Livingston's proposal for shutting down chemistry notwithstanding, Vertex had already generated small molecular inhibitors of the protein that were both immunosuppressive and orally available *without* knowing the ultimate receptor. It would be foolish not to continue.

Attracted by Boger's rough optimism, goaded by his energy and enthusiasm, most of the scientists affected a speedy convalescence. They sloughed off the last of the postcalcineurin panic and, by the end of September, were moving forward again. They had another incentive: Vertex's stock. As disconnected as ever from events within the company, it had begun climbing in August and by September 26 was at $15 a share—a 66 percent increase. It was now at the precise level that Vertex had first proposed four months earlier and had been forced to slash during the IPO.

"Two months ago we couldn't sell it for $10," Aldrich groaned absurdly on a day that Vertex's stock jumped 7 percent. He spent the afternoon fending off reporters wanting to know why the company suddenly was so hot. The run-up, Aldrich understood, had little to do with Vertex per se. As if to confirm that the company had gone public at the worst possible moment, Wall Street's brutal fence-sitting in June and July had resolved itself in a second manic romp almost as euphoric as the first. Again biotech boomed. Again the market hallucinated. A start-up called MedImmune, years away from making immune system modulators for treating AIDS, had gone out at $9.25 in May; by the end of August it was at $27. Another company, Somatogen, counting on a questionable demand for artificial blood substitutes, jumped from $19 to $36—in a month. Story stocks once again had a kind of herd immunity: They were inured, even against their own shaky internals, by the artificial robustness of those around them.

It was caveat emptor all over again. Nothing—or at least nothing the public knew about—had changed to warrant the increase in Vertex's price. Indeed, the very moment its paper worth was soaring by two-thirds, some of the scientists were despairing of it ever doing what it said. It was hard, under such irreconcilable circumstances, not to be cynical, and Boger moved swiftly to remind the researchers that Wall Street's fickleness should not be confused with the reality of their progress or their worth. If the year on Wall Street had proven anything, it was that investors didn't—couldn't, perhaps *shouldn't*—know what they were buying. Discovering drugs was a hairy, uncertain business, freighted with precipitate moments like this one when everything simply and stunningly, to use Holman's phrase, "fell out of bed." Small unprofitable companies were nightmares of disorder and discord. Like sausage making and writing laws in Churchill's famous apothegm, it was probably best not to know too intimately what went on with them. And yet an industry that lived by the story died by the story. The point again came home chillingly in October when a lackluster company called Anergen saw its stock rocket 400 percent, then crash as a result of widely misinterpreted news accounts about another company promoting a similar technology. Like a look-alike in a mystery, Anergen was hijacked and taken for a white-knuckle ride by mistake. The company stopped answering its phones.

Boger discouraged any competitive rejoicing. Anergen was working in autoimmune diseases. There was no telling what the toll would be for other small companies doing the same when irate investors realized how they'd allowed themselves to be deluded. "It's like a Sunday school talking point," Boger said. "Yes, there'll be a day of reckoning, but that doesn't mean the damage is going to be inflicted justly. Weren't there any good people in Sodom and Gomorrah, or were they all wicked? I'm not worried about us. I'm worried about others falling on us."

The stock run-up was a safety valve for the disheartened, especially Yamashita, who planned to use the proceeds to put himself through medical school—to escape. Others, particularly Murcko, were annoyed. "I wish it would go to $4," he said unironically. He was thinking about future stock options, where lower prices would be more valuable.

In mid-September, Yamashita completed the complex structure of FKBP-12 and 367, giving Vertex its first detailed look at the bound anatomy of one of its own compounds. Murcko inhaled the data like a drowning skater suddenly discovering a pocket of air under the ice. At last, he would be doing more than modeling; no more bald speculation, no qualifications, no apologies. Within minutes he had the new structure up on the screen, side by side with Schreiber's and Clardy's structure of FKBP-12/FK-506. He would show, once and for all, why FK-506 was an immensely potent drug and 367, a weak imitator. He would see, *see*, at last what he had to do.

The juxtaposition was stark, suggestive. FK-506, as Schreiber had predicted, had a protruding effector domain that reached out from the portion of the molecule that bound to FKBP-12. The right-hand flag of 367's semaphor, meanwhile, lay crumpled in the active site, also reaching, but not as far. Together the structures looked like before and after pictures in a shaving ad: FK-506 sticking hairlike from the follicle of the binding pocket; 367, same full follicle, but a shorn stump, cut clean. Even without knowing how the molecule fit with calcineurin, Murcko knew Vertex was going to have to put something into the region beyond the protein's undulant surface, a hook of some kind. Admiringly, he noted that Schreiber's effector hypothesis was substantially right, though not necessarily his assertions about calcineurin. "It's one percent of the protein in the brain," he said. "I don't think you want to just knock it out."

And yet Schreiber's hypothesis didn't explain everything. Looking at the two structures three dimensionally, Murcko noted the relative insignificance of the effector. It was like a nail jutting from a warped plank: enough, perhaps, to join it to another piece of wood, but not if one wanted them to stick. For that, one wanted more surface area contact, a coat of glue. Whether or not he needed a molecular doorstop—a group of atoms to prop open the flap region—Murcko now doubted more than ever that the effector region alone accounted for the drug's action. The atomic configuration around the active site, as Boger had speculated, seemed also to be involved.

Murcko pinned Vertex's computer network with dozens of studies comparing the binding energy of the two complexes. He had always talked about designing drugs as an "iterative process," a kind of smart person's trial and error. Not that he considered himself smarter than the chemists, but by modeling different atomic configurations, then calculating their efficiency, he could predict which ones would be most potent. He could see how moving a few atoms an angstrom or two to fill an empty pocket might make a compound bind more tightly. He could juggle electron clouds, substitute charges, make thermodynamic adjustments measured in millionths of calories. The key was to take the information he was generating and wrap it into specific suggestions that Armistead, Saunders, and the other chemists would listen to, suggestions that were not only sound, but easily tested. No chemist wanted to hear about a potentially great molecule that might take a month to synthesize. Who would bother? Murcko's degree of confidence—a sore point for all modelers, and particularly Murcko, who had grown more and more anxious to prove both the validity of his methods and his own worth and whose frustrations had long been huge—depended not only on the quality of his predictions, but on his ability to persuade the chemists to make what he designed.

Weighing everything—the obvious need to build out into the effector region, the differences between 367 and FK-506, the unchartable but apparently significant changes in the surface—Murcko concentrated on the semaphor. Drawn on paper, its arms lay outstretched in a Y. Three dimensionally, however, he could see that they crumpled on binding, collapsing like a fighter's arms in the clenches. Here, possibly, was an opportunity. Harding had once compared 367 to a skater falling through a hole in the ice. As long as its arms were loose, it would slide through. Rigidify them somehow, though, and they might catch. The molecule might save itself.

Murcko modeled several new molecules with arrays of atoms inserted to keep the semaphor pointed outward, outward against the flap, outward toward the effector region. Calculating a substantial improvement in binding, he interpolated which ones might most improve the drug's overall activity while being easist to make, then suggested them to the chemists.

There is a presumption among nonscientists that research breakthroughs are inherently dramatic—great eureka truths, blazing revelations, thunderous insights, shouts of joy, the combined emotional stimulation of a great college basketball rivalry and a soaring aria. More often, the opposite is true. Some small deed is made. An adjustment. A scientist, frustrated, half rises with an idea just different enough to elicit a subtly better result that reinforces the conviction to go on. Always, the seed is the moment when one experiment, one reagent, one method is chosen over another, a small, but critical, branch point.

The chemists welcomed Murcko's suggestions, but ambivalently, noncommitally. Such communications between modelers and bench chemists are inherently awkward, since a decision by a medicinal chemist to make someone else's compound, besides requiring a commitment of time and energy, is equally a choice not to pursue some favored idea of one's own. And yet the chemists were stuck. For months, Armistead, Saunders, and the others had been making derivatives of 367 that bound ever more tightly but with no corresponding gain in immunosuppression. They, too, had concluded that the molecules weren't extending far enough and had faulted the semaphor. The problem, from a production standpoint, was how to shore it up without making it so hard to construct that it became worthless. Said Armistead, "The bottom line for us was that we knew it took Stuart's group six to eight months to synthesize 506BD. We didn't want to do that."

Working in series, Armistead and Saunders concentrated on one of Murcko's ideas: inserting a small planar ring of six carbon and six hydrogen atoms into the semaphor's crotch, as a chock, to pinion its appendages. Six compounds later, Saunders made a molecule—563—that was three times more active in cells than Vertex's previous best hit. Within weeks, Patsi Nelson noted a commensurate gain in activity in mice; the molecule was immunosuppressive in animals. It was the most potent Vertex had made yet.

The chemists, their confidence rebounding, immediately began scaling up production: for more assays; for toxicological studies; for Navia and Yamashita, who began at once to try to cocrystallize it with FKBP-12 so that Murcko could compare that complex with 367; especially for shipment to Chugai, to whom Boger had again promised just such a gain in molecular activity.

It was a bravura leap, an exceptional piece of science, not a great theoretical breakthrough, but a great practical one. Gains in cellular activity tend to come slowly, unpredictably. Yet another one like this and Vertex would have a drug candidate. Suddenly, Murcko's "feedback loop" was installed for one brief, exhilarating cycle and had worked perfectly, just as Boger had predicted.

Boger was electric, infused. The process was unambiguous proof of his concept for drug design, a prototype moment, as momentous in its way as Schrieber's triumph with calcineurin. "It wasn't the obvious, next medicinal chemistry thing to do," he told the project council, uncharacteristically understating the achievement. Vertex had deliberately made a more biologically active molecule. It had predicted, on the basis of structural information, an incremental gain in activity, thereby designing a better compound. For an instant, it had turned on a switch illuminating the mystery of molecular binding and introduced choice into a process that for fifty years had swung between random sampling and brute force, between frustration and luck.

Fifty years earlier, Tishler and the other pioneers of scientific drug discovery launched the process of screening systematically for new drugs that placed the drug industry in its modern-day arc. They had found extraordinary drugs in epochal quantities and had learned to reassemble their atoms to make them even better. But they hadn't been able to improve them at will. Rationalizing the process of discovery, making it fabulously productive and profitable, they'd fallen short of the ultimate prize: to control it, to drive it ahead on the gales of their own brilliant imagining.

That had always been Boger's goal, the hard, irreducible thing inside that impelled him. He had come to discovering drugs not, as Tishler had, as a moral act, but as an intellectual one, and now it was in its first, primitive fruition at Vertex. He was there by the standard that measured most, his own intellectual satisfaction. He was controlling the process, its architect, its avatar. Boger had hoped that Tishler had understood this, but it seems equally likely that he died feeling hurt by Boger's defection from Merck and believing that it was meant to be so.

By the norms of science, Vertex's new information loop was an invisible conquest. It wasn't publishable. There would be no talks about it. One chemist's deliberations in choosing to make certain molecules over others was far too slender a data point for making any public assertions on structure-based drug design or anything else.

And yet ironically, for he had gone to great strains to make mountains out of far less, Boger seemed not to care. Schreiber was right about him: Boger's fundamental connection to the outside world was different from his, different from most scientists'. They thrived on recognition for the sake of their careers if nothing else. Boger, too, needed attention, but he needed the idea of himself more, of his freedom, of being able to accomplish what he set out to on his own because he had figured out what to do and how to do it. And what he had set himself to do was to perfect a system of using proprietary information in the advantageous design of new drugs. Better molecules, he was absolutely certain, would follow. Someday, the drug industry's pipelines would be choked with them, and the people who made them first would, one hoped, go into teaching, as Boger often said he still wanted someday to do.

His competition with Schreiber again had come down to a siblinglike match over parity. Schreiber had made himself a beacon of synthetic chemistry in biology to illuminate its secrets. He was a discoverer. Boger aimed at the next rung: to control biological activity, to impose himself on it in order to change it. This explained their life choices, the speed at which each flung himself ahead and the difference in their respective views. Science exalts conceptual breakthroughs; business, practical ones. Thus although Schreiber appeared at times, such as in Pittsburgh, miles ahead of Boger, they were still straining neck-and-neck for the larger prize: to make a lasting mark on chemistry, science, the world. Each one's work informed, completed, made whole, and was mutually essential to the other's.

The scientists, understandably, regarded the situation more ambiguously. Five-sixty-three was not a drug. It had undergone none of the critical tests that make a molecule an approved therapeutic. It could be blindingly toxic, cause hallucinations, hypertension, strokes. Vertex's iteration, its one turn of the crank, was just that: a single isolated success. As the technology was brought forward and improved, it would take perhaps dozens of such increments to demonstrate that the concept worked well enough to be used with any degree of confidence. Even then, it might be useless for reducing the overall side effects associated with FK-506 and cyclosporine, for these side effects now, with Schreiber's and Starzl's work, appeared more than ever to be linked inextricably with the drug's action rather than with extraneous atoms that could simply be cut away and discarded. There were enormous problems, chiefly with certain kinds of toxicity, that lay forever beyond the promised land of structure data.

Nor was 563 a strict display of the lock-and-key, we-like-to-be-sure-of-the-biology paradigm long promoted by Boger. The scientists still had no sure understanding of the biological events they were trying to control. Was calcineurin the ultimate target? If so, how did the complexes bind to it? What of the other FKBPs? Murcko had built his better molecule by examining keys only. Of the tumblers he was trying to hit, he knew depressingly little, less than that. He knew nothing at all.

Murcko, curmudgeonly as ever, reproved Boger's euphoria. "I think all the arguments that we make, Abbott can make the same," he said, referring to the large drugmaker known for having some of the drug industry's best biophysicists. "On what basis can we claim to be uniquely qualified to be doing rational drug design?"

The question lingered like an unresolved chord throughout the fall as the scientists shook off the last effects of Schreiber's victory and dove, one by one, back into their work. It was more than rhetorical. As Boger had warned, the great risk in science wasn't being wrong, it was overstating one's data. In truth, it was more than that. It was deciding what to conceive possible in the first place. Scientists like Boger were deeply, immutably contradictory by nature. They were shamans. Either they were right or they weren't. Eventually, with data, they would know. Everyone would. Choosing to believe them in the meantime could be like stepping off the edge of the world.

Could Vertex design drugs? Could better drug molecules be designed? Boger was as certain as ever.

Who else thought so failed to interest him.

"*I* know," he smiled impetuously, serenely. "Everyone who matters already knows."

EPILOGUE:

FINDING THE PONY

April 28, 1993

Boston's World Trade Center, unlike New York's, is not a beacon of wealth and power but a refitted waterfront mercantile mart. It sits on a pier asquat the city's famously ruined harbor and has a postcard view of downtown, which, after the harshest winter in decades, shimmered this brilliant Wednesday morning like an aurora borealis. Two months earlier, the World Trade Center in Manhattan rumbled and shook as a car bomb blasted a crater in the underground parking garage of the Vista Hotel, where in 1989 Boger had taken his manifesto to Wall Street. Now, in Boston, he faced 250 worried members of the Massachusetts Biotechnology Council, waiting to proclaim—not unpleasurably, for he had predicted this, too, and was confident Vertex would prevail—a day of reckoning.

Seldom had an industry's fortunes fallen so fast. In just eighteen months since the biotech bubble of 1991 finally burst amid a string of spectacular flameouts, all drug stocks, big and small, had lost almost half their value. Worse—far worse, many in the room believed—the drug industry's long-standing expectation of limitless profits had become a ripe political target. The Clinton administration, as one drug-industry lobbyist put it, had calculated that the best way to win its health care reforms was to "go to war" with the drug industry. Hillary Rodham Clinton's task force was talking ominously about cutting at its heart, with price controls. Without the specter of enormous profits, stocks could only go lower, investors would flee, capital would dry up, innovation would flag, companies would die. Despite its extraordinary successes, high-risk research, the presumptive justification for all those profits, was beseiged, demonized, as perhaps no other time in its history.

Saying he expected "to see frogs raining down from the sky sometime soon," Boger told the assembled executives that only the smartest, fleetest, most adaptable companies—companies like Vertex that were burning tens, hundreds of millions of dollars in pursuit of fast, novel drugs for big markets—would compete successfully in the new period. Why only them? "We're more motivated," Boger said, "and the fear of death and God is closer to us." Since all drug firms were likely to be deprived of two of their three main avenues for making money—raising prices and developing copycat "me-too" drugs—there was little denying Boger's analysis. Highly focused innovation, backed by a feral organizational leanness, was now key. Gregariously, Boger ended with a slide of a cartoon showing a drug company executive talking to a scientist. "I like it," the caption read. "Find a disease for it."

Boger could afford this bit of light heresy. Despite the hard uncertainties besetting the industry, Vertex's outlook couldn't have been much brighter. Two weeks earlier, the company had signed a $20 million deal with Kissei Pharmaceuticals, the fastest growing drug company in Japan, to develop and market an AIDS drug in Japan and the People's Republic of China—a drug that Vertex now was within weeks of selecting from several strong candidates. (Kissei's ceremonial gift was a framed photograph of Mount Fuji, just as Chugai's had been.) Another leading company had declared a "special urgency" to license the compound in the United States and Europe, and hadn't balked when Boger and Aldrich floated a trial balloon for a $100 million deal. Meanwhile, barely twenty months after being blindsided by calcineurin, Vertex had two other near-clinical candidates rapidly under development, one of them a stepchild of the hard-pressed immunophilins program. Within 110 people, nearly $50 million in the bank, a new subsidiary based on Navia's enzyme work, and a relatively buoyant stock price, Vertex had outshone even Boger's most aggressive predictions.

The conference itself, formally an annual meeting of a statewide trade group, affirmed Boger's new power and influence. Not only had he been asked to provide the industry's perspective but Vertex's stamp was everywhere on the proceedings. Brochure covers featured Moore's FKBP-12 structure and the company's crystal structures of HIV protease done up in Dali-esque portrayal. The meeting's major scientific section was on structure-based design and resembled a Vertex road show. Chemistry and biophysics were whispered about with new reverence. In the aggregate, the meeting seemed a coronation, a concession that the future lay not in drugs that were proteins but in those that inhibited them—just as Boger had long said.

Of course, respect within the biotech industry had never been Boger's main goal or even a major one; he was out to shake up the big drug companies, particularly Merck. Here, too, a reversal of fortune gave him ready satisfaction. The keynote speaker was Ed Scolnick, Merck's brilliant and tempestuous director of research and Boger's chief sponsor at Rahway. It had been Scolnick who had told a Harvard Medical School audience three years earlier that having the structure of HIV protease had given the company "some help, not dramatic help" in designing drugs. Now, with Murcko, who had talked dazzlingly about Vertex's design program in HIV, seated next to him on the dais, Scolnick issued a staunch and—given the audience—surprising defense of Proscar, Merck's drug for shrinking prostates. With Merck's market capitalization off a staggering $20 billion in the past year due in no small part to Proscar's disappointing debut, Scolnick's investment-style sales talk indicated that even the industry's titans were now having to stoop. Boger thought the talk "bizarre." He beamed when a few days later he received a hand-written letter from Scolnick, congratulating him on his apparent success.

Roaring into adolescence, Vertex had in four years come of age by every measure but the one that mattered most; it still had no drug.

In fact, it was tantalizingly close to having as many as three. As was typical of Boger and science, luck—and the ability to capitalize on it aggressively—had been instrumental in at least a couple of them. Hair-raising reversals had been the norm.

After the initial demonstration of structure-based design with its first semaphore compound, 367, Vertex's immunophilins project bogged down in the implications of Schreiber's and Weissman's calcineurin work. Nothing the chemists made came close to FK-506, which was proving in European clinical trials to be as effective as but significantly more toxic than Starzl had said and which Merck, Sandoz, and other companies were finding impossible to improve simply by altering its structure. Then, remarkably, chemist Jeff Saunders made a compound that was highly immunosuppressive, bound tightly to FKBP-12, yet seemed to work through another channel, avoiding the quagmire of calcineurin entirely. For seven months the chemists toiled almost exclusively at optimizing Saunders's molecule only to discover it a false miracle, the result of what Boger would call (with uncharacteristic melodrama) a "diabolically misleading" cell assay that had been born out by animal experiments that turned "mysteriously negative" soon after the initial assay was found to be in error. Distraught, beleaguered, they returned in the summer of 1992 to Murcko's "nightmare of terror"—trying to block calcineurin, for which they still had no structure and little hope of getting one soon.

It was a grim time. Yamashita, making good on his promise, left Vertex in July 1991 to go to medical school in Hawaii. He moved in with his parents, leaving Vertex briefly with half its bench effort in crystallography and with many of the scientists believing he'd have stayed if Boger had instructed Navia to give him his own project. Boger refused to do so. Navia, meanwhile, began traveling nearly full time to promote his enzyme technology, which would soon engender new labs and a new entity, Altus Biologics, Inc., Vertex's first subsidiary. Murcko, deprived of data both in immunophilins and HIV protease, again became agitated, complaining bitterly about what he considered the company's faltering commitment to structure-based design. Morale sank as the project councils stumbled and lost their footing.

Vertex had raised another $25 million by selling more stock during the final days of the bubble the previous fall, but it had been nearly two years since the deal with Chugai, which was becoming impatient. The company's burn rate had climbed to $7 million per year, and Boger had begun preparing the board for "double digit losses"—$10 million or more. Where it would turn for funding next, no one, not even Boger, could say. Of Wall Street's renewed iciness, he said, "You couldn't sell hair tonic that gave eternal life right now."

For a time Vertex appeared to have a promising lead in HIV protease, but at a conference in August 1992, chemist Roger Tung learned that the compound class was stuck squarely behind another company's patent. Tung spent the next twenty-four hours in his hotel room, despairing and furiously sketching new molecules. By fall, Vertex seemed for the first time in its brief history to be adrift, with no clear path to the next stage.

Boger, as ever, remained confident, but he also knew that changes were necessary. Vertex had gotten too big for him to be all places at once, all things to all people. Quietly, he began recruiting Vicki Sato, head of research at Biogen, Aldrich's first firm, to oversee the company's science. It was an admission that the demands of chasing money had overwhelmed his ability to do active science and that his ultimate role, in a world dictated by business, was (to use an overused word) in "growing" the company. Boger retained final control of the projects and continued to be their driving force, but Sato, a quick, street-smart no-nonsense former biology professor at Harvard would be the company's field commander, the person responsible for keeping it on track and on time. Boger hired her, he said, in part to make peace with the company's biologists, many of whom still distrusted him over calcineurin.

Within months, Sato reorganized Vertex's project councils. She named project heads—Dave Armistead in immunophilins, Roger Tung in HIV protease, Dave Livingston in a new project in inflammation, and Matt Harding in a new anticancer program. Because there weren't enough of the positions to go around, Boger simultaneously named Thomson, Harding, Murcko, Livingston, and Debra Peattie as senior scientists, a rank previously held only by Navia. Repudiating the flat organizational structure of Boger's social experiment, Boger and Sato felt the moves were necessary to keep several people from quitting the company. Boger's first hires had now been with him almost four years. Their founders' stock would soon be fully vested, and they would be free of the "golden handcuffs" that risk-taking young companies use to keep people from leaving. Headhunters were calling the scientists more and more frequently. Reluctantly—regretfully, it seemed—Boger acknowledged that the scientists would need conventional rewards like titles and status to satisfy their ambitions and the imperatives of their careers.

With hierarchy came new stresses—Saunders now reported, for instance, to his friend, nemesis, and former labmate Armistead, who no longer ran reactions but spent his days in meetings, riding herd. Immunologist Patsi Nelson left the company rather than report to Harding, with whom she considered herself professionally equal. Competitive preening among the scientists exposed jealousies previously without outlet. Thomson, whose group arguably had produced more value for the company than any other, resented being promoted with people whom he considered less productive and committed to the company than himself. "Like graduating with the class," he called it.

Still, few longed for a return to the open-ended egalitarianism of Boger's social experiment. Boger considered that experiment a success, for it had produced the champions, the leaders, who would become Vertex's new scientific cadre. But most others welcomed the change. Quietly, the company made the transit to a more conventional management structure without smothering itself, as other brash start-ups often did. By the end of the year, Vertex resembled more closely other companies at its stage, a place with a ladder and individuals elbowing to climb it at every rung.

Boger was less disappointed than he seemed. He had gotten what he wanted—self-selected leaders and a sense among almost all of the scientists that their ideas were valued and would be heard. As with most concepts, egalitarianism was for him a tool, not a goal. What he believed in was equal opportunity, not equality of status. The ranks had held. Among senior researchers, only Patsi Nelson had left.

On June 28, 1992, doctors in Pittsburgh transplanted a baboon liver into a thirty-five-year-old man dying from hepatitis B. There were twenty-seven people in the operating room, including ten surgeons. Starzl didn't direct the experiment himself but had authorized it to address a desperate shortage of human organs: a shortage, ironically, that resulted from Starzl's success with making liver grafts standard therapy and that had left Pittsburgh with half as many livers as patients and a dangerous amount of what economists call "surplus capacity." Using FK-506, doctors kept the man alive for seventy days. He died of a stroke and massive infection, his unrejected liver still functioning. Later in the year, the group transplanted five organs—pancreas, liver, stomach, large and small intestines—into a four-year-old girl. In January, they repeated the baboon experiment, then cancelled the program.

Starzl, dauntless as ever, pressed ahead with the urgency of someone half his age. His transition to immunologist was now all but complete, if not yet accepted by those who studied the body's defenses at their atomic level. But he was to be no gray eminence. Typically, he was determined to prove them wrong. In 1992, he began publishing a series of articles—a unified theory, so to speak—of graft acceptance. He now believed that when a body received an organ graft it became a chimera, a mixed being, a hybrid of host and donor. Cells from the transplanted organ migrated throughout the body while those of the recipient invaded the alien tissue. What immunosuppressive drugs did, he hypothesized, was protect the cells of each from the other, causing a "biological truce," an equilibrium of crossed souls. Unproven, the theory would ultimately explain both transplantation and autoimmune disease, he insisted. Coming upon it had been the apotheosis of his long career. "A glimpse of eternity," he rhapsodized, and a "fair trade for the thirty-five years of work preceeding it."

Proportionately, Starzl's contributions with FK-506 waned as the field around him exploded. He still flooded the FDA with requests to launch new clinical trials, especially with autoimmune diseases, but there were others now with more experience and insight to advance the drug's human experimentation. Brain researchers, for instance, now thought the drug might treat strokes. Weissman's and Schreiber's calcineurin discovery had invited new realms of thought, new interpretations of old mechanisms, and the interest of the world-class researchers throughout medicine. By mid-1993, it appeared that FK-506 would finally be approved by the FDA, probably by the end of the year. Starzl, who had salvaged the drug seven years earlier when it was rotting the guts out of dogs, was systematically evaluating new compounds at the time, looking ahead.

Mary Arthur, the Louisville teenager who was the first patient to receive transplanted islet cells and whose cancer preoccupied Pittsburgh surgeon Andy Tzakis during the FK-506 conference in Pittsburgh, survived the surgery to her jaw without disfigurement. Cancer-free three and a half years after her transplant, she was still producing her own insulin. She planned to get married. After considering a career as a chef, she switched her college major to pharmacy.

Schreiber charged ahead brilliantly, if less ecstatically. Determined to control the burgeoning field of immunophilins research, he continued to work seven days a week. But as his influence and visibility grew, so did his burdens. Still shy of his fortieth birthday, he began complaining of the relentlessness of science, how it drove you to be first, then pitilessly gave no pause before the next race. He sounded at times like a man strapped to a mast. "The world would be a lot better place," he said mordantly in the fall of 1992, "if there were no scientific prizes."

He had become "almost a caricature" of the obsession with individual credit that was both science's driving force and the source of much of its bitterness, a former collaborator said. Yet his determination remained immense. He was reshaping science, just as he planned. Biologists still dismissed him as a revisionist chemist, chemists as an apostate and amateur biologist. But they couldn't dismiss what he had done. Perhaps as much as any of them, he and his colleagues had elucidated the basic mechanisms of how cells communicate internally. By May of 1993, he had stopped referring to the "black-box" of signal transduction. He believed the problem had been cracked.

The signal transduction company that he and Kevin Kinsella

had originated, Ariad Pharmaceuticals, opened six blocks from Vertex in an MIT-sponsored Research and Development park, and Schreiber consulted there approximately one day a week at a starting salary of $75,000. Meanwhile, he amassed a freezerful of natural molecules that, like cyclosporine and FK-506, had intriguing biological effects, banking them like so many sperm samples. With enough compounds in the bank to seed "more research projects than I'll be able to study in my lifetime," he swung headlong toward the next frontier—gene therapy—developing a method for switching transfected genes on and off. By summer, he was negotiating with venture capitalists to start a new company.

Vertex's problems in immunophilins taunted the scientists acutely. Thanks to Yamashita, who determined the right crystallization conditions in the weeks before he left, the company was now generating every few weeks an X-ray structure of FKBP-12 with a different Vertex compound bound inside the protein. But without a calcineurin structure, they remained lost. When it turned out ultimately that they and Schreiber had both been right—that FK-506 worked through an effector region *and* by changing the outer shape of FKBP-12—it only reminded them how forbidding a task they faced. Not only did they need to move too many atoms in too many places, but even if they did, they would still be trying to block an enzyme so biologically vital that it might seem a nightmarish target for a drug.

More and more, the chemists felt they were beating a dead horse. They were doing structure-based drug design but on a project for which that still might not be enough.

It was Boger who found the pony.

Cyclosporine had been known to switch off a mechanism in cells that enables them to expunge toxins. Like a micromolecular bilge pump, the process, known as multidrug resistance (MDR), had long frustrated oncologists, who watched helplessly as they infused powerful cell-killing agents into their patients only to see them flushed into the blood and carried off.

Boger speculated that Vertex's semaphore compound, 367, because it bound to FKBP-12 but wasn't strongly immunosuppressive, might block MDR as effectively as cyclosporine but without cyclosporine's prohibitive side effects. Presumably, it would let doctors get more cytotoxins to more cells and thus save more lives. Boger suggested the idea to Harding in the winter of 1991, and it was soon given further credence by an article suggesting that FK-506 also blocked MDR, though very weakly.

Harding sent 367 and several other compounds to Yale for testing. The molecule looked extremely promising. Suddenly Vertex had a lead in a virgin market worth up to $500 million. By early 1993, less than eighteen months after Boger's initial insight, the company had chosen a derivative of the molecule to begin developing for human experiments.

Far from the orderly, data-based process of iteration and reiteration, the selection of Vertex's first clinical candidate illustrated another Vertex dictum: The ultimate reward for research may turn up elsewhere than intended. The key is to draw the right lessons, be astute, act decisively, do the proper experiments.

Within weeks, Vertex was pursuing a deal to license the rights to a possible second clinical candidate—a promising treatment for sickle cell anemia—snatching it from under the nose of a top European drug company.

May 19, 1993

Gratification suffused the Vertex conference room like a glow. In an unambiguous display of the speed and effectiveness of the company's "feedback loop," as Murcko called it, crystallographer Eunice Kim was presenting a new structure of a strikingly potent HIV protease inhibitor. For months, as the chemists had churned out better and better molecules, becoming steadily more excited, Kim had been reducing her turnaround times for feeding them detailed images of how the compounds bound atom by atom within the enzyme. This one, 328, had taken her exactly five days from the time she got the molecule. (Yamashita's ill-fated structure of FKBP-12, by comparison, had taken more than a year.)

"What took you so long?" Boger said admiringly.

"Where's 330?" joked Sato. She was referring to a molecule submitted within the last week that was five times less potent but that had looked spectacular in its ability to survive the gut and remain intact within the blood. It was the inability to last in the body long enough to work, of course, that doomed almost all of the best protease blockers, and 330's bioavailability had been deliberately engineered into the molecule using Kim's structures. The compound differed from an earlier one in its series by the repositioning of two carbon atoms within a binding pocket measured in billionths of meters. Vertex could spare the loss of potency: for sheer binding, its most potent molecules were now so perfectly designed that they were off-scale in the company's assays.

Six compounds in all were still in contention as Vertex now set about choosing which one to scale up for human testing—an enormously complex and expensive process requiring quantum jumps in activity at every level of the company—but as Boger had been pressuring for months, the project council was running out of time. It needed to "pull the trigger." Tung, cautious as always, importuned for more data. In three weeks he would be leading a contingent of six Vertex scientists to the massive annual AIDS conference in Berlin, there to disclose the company's work publicly. He wanted to tell a complete story.

Boger grew impatient. "I'm all for stalking horses," he said, "but when the real horse is ready, I don't want to keep him in the paddock because the stalking horse is at the last turn and we want to let him finish so that he doesn't feel bad. I want to go out and shoot him."

To Boger there was only one issue: getting the FDA to approve the testing of Vertex's compound in humans before Merck's protease blocker, announced in February, was licensed. He had been surprised that Merck's compound, which he conceded looked impressive, wasn't a "killer molecule" and had guessed that the company had rushed it into the clinic so that it might be on the market by the time CEO Roy Vagelos retired in November 1994. Boger believed that Vertex had the better drug—smaller, easier to synthesize, more likely to get into the brain, which is protected by a chemical filter and which harbors the virus. But if Merck's drug was already on the market, all bets were off. No one dominated competitive markets better than Merck. Worse, Boger half feared that Merck would simply give its drug away, as it had with its cure for African river blindness, in return for a huge tax break and a mountain of glowing press. At the very least, Merck's drug would set the standard for approval for everyone else.

"The rest of the world is rushing headlong," Navia said. "The other guy can have an order of magnitude shittier structure, but if you're number 2, you're fucked."

"Throw a dart," Sato urged.

After a few minutes, the decision was made. Vertex would proceed with the development of 330, changing course only if new data arose to indicate that it had something better.

It was done. Navia's "hallucination" and Tung's embrace of it to escape Armistead's dominance in chemistry; Boger's searing antialtruism and obsession with beating Merck; Vertex's calculation to go into AIDS because it was the quickest route to profitability; its use of the program to sell the IPO—more deliberate origins and higher motives were easily imaginable. And yet not to Boger. For him, lofty motives were infinitely less powerful, less trustworthy, less *useful*, than pure ones. Science was too difficult for people to engage in solely because, as he had written at age thirteen, they wanted "to help rid man of the burden of disease . . . and to help man get along with man." They did it because they were absolutely certain it could be done, and to prove to themselves and the world that they could do it first. They did it to bash their competitors, to think themselves divine, to win, and to avoid the terrible, deathly anguish of losing. Backbreaking science and unblemished greed and raw fear, not moral correctness, would conquer AIDS. Boger was absolutely sure of that. He didn't want to save the world. He wanted to control it; he believed he always had. Now the world would see the fruits of that fierce presumption.

He wasted no time. Dispatching most of the scientists from the room with a hail of laughter that was part congratulation, part triumph, and all, in its unalloyed arrogance, roisterously appealing and charismatic, he huddled quickly with Tung, Sato, and Livingston to attack the next set of experiments—toxicological studies, animal studies, multidrug studies with other compounds, one-on-one comparison studies, formulation studies to determine how to deliver the drug, blood assays, ultra-pure large-scale preparations of the molecule; studies of the compound in several kinds of animals and every type of cell, at every temperature and pH imaginable, over short periods and long periods, in duplicate, triplicate, and with every conceivable risk and ambiguity addressed. By Vertex's timelines, the molecule had to be available for human experimentation in AIDS patients by the end of the year or during the first quarter of 1994. An immense amount had to be known about it before that time; more, immeasurably more, after.

As one Merck veteran much admired by Boger would say: "There are those who make the case that finding lead molecules is the easy part. Making drugs is hard."

The following day, May 20, John Thomson flew home to Australia for his first visit in four years. His parents were ill, he was eager to see his children, and he had to apply in person for a visa to go to Berlin. It had been Thomson who had again, through sheer force of will, enabled Vertex, after struggling for almost two years, to produce so much HIV-protease that the need for crystals of it was no longer an impediment. He had also isolated and crystallized a highly scarce and extremely competitive protein that was to be Vertex's next target.

His labors, anointed with these and other successes, had taken him through a passage. He was still driven, but no longer by anger. His self-respect had returned and with it his respect for others. His drinking and smoking were down to levels appropriate to someone who imagined for himself a future. His group had grown to eleven people, and with more pride than he would concede, he accepted—because he needed it, he emphasized, not because he wanted it—a small, windowless office. On the day before he left, he wrote a check for a terrifyingly sleek new motorcycle. "A rocket," he called it, laughing self-knowingly in a way that made it seem as if what interested him was simply the pleasure of going faster than before.

December 16, 1993

It was Vertex's work on HIV that rescued the company from the vagaries of immunosuppression and propelled it to the next tier.

Climaxing two years of maneuvering—two years during which the immunophilins project foundered and when hopes for conquering AIDS looked increasingly dim—Vertex announced that it would develop its lead protease inhibitor jointly with Burroughs Wellcome, the British-owned drugmaker and manufacturer of AZT. The molecule was not 330 but a second-generation compound, VX-478. The deal, which would eventually bring Vertex $42 million, was in fact worth several times that since Wellcome would pay the full cost of development—perhaps as much as $200 million. Vertex's stock rose $2 on the news, to $17.50.

It had been Wellcome, of course, that had led the drug industry into AIDS, and Boger was mindful of what had followed. Ever since the company first introduced AZT during the mid-1980s, it had been besieged and vilified. AIDS activists had organized around-the-clock picketing at the company's North Carolina headquarters and had breached security at the New York Stock Exchange to chant "Sell Wellcome!" and "Fuck drug profiteers!" For nearly a decade, the firm had been mired in costly lawsuits that threatened its patent position and, ultimately, more than $500 million a year in revenues. So onerous was the specter of such well-publicized controversies to a small firm that Boger and Aldrich had initially resisted going into AIDS research largely because of it.

Boger was pleased to be allied with the drug company that knew more about AIDS than any other, yet in such a way (Wellcome now owned the North American and European rights to the drug and would pay royalties to Vertex) that sharply diminished Vertex's own exposure. "I'm glad our money comes off the top," he said.

Boger's satisfaction was equalled, clearly, by Wellcome's own relief at having lined up a potential successor to AZT. Six months earlier, a study in Europe showed that the drug did nothing to prolong the lives of people with AIDS, despite slightly delaying the onset of symptoms. More recently, a Harvard study concluded that even the drug's minimal benefits were often cancelled out by its side effects. Accompanying nausea, vomiting, and fatigue made taking AZT not worth the trade-off of delaying the advent of full-blown AIDS for many patients. AZT did have therapeutic value; it seemed to protect the fetuses of women infected with HIV and apparently kept many people with AIDS from becoming demented. But the consensus, as the *Times* reported, was that it was a "moderately useful drug that can slow the course of AIDS in some patients for a limited period of time." Nevertheless, with AZT accounting for almost 15 percent of Wellcome's income, the company was hardly going to walk away from the one truly lucrative franchise in the AIDS market.

AZT, by default, remained the drug of choice for people infected with HIV, but frustrations with the compound—and with the vast scientific effort that had failed for nearly eight years to replace it with something better—had grown huge. Thirteen years into the epidemic, hope seemed to be running out. In November, prospects seemed to reach a new low when an entire battery of experimental vaccines—vaccines that had been highly effective against laboratory strains of HIV—was wiped out in tests against strains of the virus taken from people. Not one showed the slightest efficacy.

Boger, typically, was undismayed by the failures. He considered them less a result of complex biology and a wily virus than of misguided and misapplied science. Much of his confidence, as always, was based on Vertex's choice of target. Several other companies had now shown that blocking HIV protease remained the best hope for stopping the virus from spreading. The latest, and perhaps most enthusiastic, of these was Merck. For several months, blood tests of four patients who were given Merck's protease blocker showed dramatically reduced levels of virus. In this one admittedly limited field trial, the compound seemed to slow the spread of HIV better than any drug yet devised.

Not only was Boger confident that Vertex could beat Merck's drug, he was sure it could generate even better ones through structure-based design. A week earlier he'd sought to prove the point at an AIDS meeting in Washington. For months, Vertex had been trying to get enough of Merck's compound to begin comparison tests with VX-478—a competitive necessity for any company hoping to position a new drug. Having finally made enough of the molecule to determine how it bound to enzyme—how it worked atom by atom—Vertex chemist Dave Deininger had given a sample to crystallographer Eunice Kim, who promptly solved the structure of the complex. At the meeting, after Merck's scientists conceded they didn't know how their compound worked specifically at a molecular level, Boger, at the podium, concluded his own talk with a slide showing how Merck's drug sat in the active site of the protease, in effect answering the question for them. (Without a guarantee of patent protection, Boger refused to disclose the structure of Vertex's own molecule.)

Several leading drug companies were now developing protease blockers—Merck, Roche, Abbott, and Merck-DuPont. Agouron, the early structure-based design company long dismissed by Boger, had such a molecule, as did Searle, the Monsanto subsidiary whose patent disclosure had temporarily derailed Vertex's chemistry effort when Tung had learned about it in 1992.

Most of the others were ahead of Vertex and Burroughs in getting their drugs into the clinic, and the competition—for attention on Wall Street, for scientific recognition, for clinical investigators, even for patients—was extreme.

None of this stayed Boger's enthusiasm, which was typically unbridled. First and most encouragingly, he had Vertex's compound, which, while not as potent as some of the others, had several extraordinary properties. Unlike most protease inhibitors, it wasn't easily cleaved; even when delivered orally, it remained in the bloodstream at high concentrations for several hours, meaning that it would provide what many of the other drugs couldn't—constant protection at acceptable doses. In test animals ranging from rats to primates, it was practically nontoxic, so much so that the only side effect that the company had been able to induce with megadoses of the drug in rats was to clog their intestines. And it was cheap and easy to synthesize, taking roughly seven chemical steps compared to Merck's twenty-one. As Boger would point out, in the calculus of drugmaking, where yields decrease, often dramatically, with each step, VX-478 was "much more than three times" easier to make than Merck's compound.

Without clinical trials to prove that the drug worked, these advantages were speculative at best. But as Vertex moved toward human testing, the next and most critical phase of its development, it had a formidable partner in Burroughs, the world's pioneering antiviral company. This was the other cornerstone of Boger's optimism. It had become by now an article of faith that HIV was too mutable to be stopped by a single compound. Multidrug therapies were now considered essential; indeed, the hottest story of the year in AIDS research involved the use of three drugs to cripple the virus. Given that any multidrug regimen must include, at least for the time being, AZT, and that no company knew how to work with AZT and other antiviral agents as well as Wellcome, the alliance gave Vertex a critical edge.

(This headiness persisted throughout the coming months, even after Merck was forced to announce in March that it was discontinuing its clinical trial. The company had found that viral levels in those patients taking its compound had eventually returned more or less to where they had been before treatment—suggesting that HIV had mutated yet again into a new drug-resistant strain. AIDS patients, the scientific community, and Wall Street were dismayed by the news. Institutional investors, interpreting Merck's difficulties as a failure of protease blockers in general, punished those companies, including Vertex, that had a stake in them. Boger and Wellcome scoffed at Merck's disappointment. They attributed it to weaknesses both in its compound, which wasn't as long lasting as VX-478 and thus gave the virus a period of time to regroup and evade, and its clinical trial, which sought to test the drug alone rather than in concert with other drugs.)

Boger and Aldrich had always believed that the time to raise money on Wall Street was not when you had to but when you could. Clearly now was such a time.The day after the announcement, on Dec. 17, Vertex filed a fast-track stock offering that would yield, six weeks later, another $62 million. Though it was a volatile period for drug and biotech stocks, the offering built from start to finish. The company's stock traded at $16 on the day in mid-January when Boger and Aldrich started their road show in Europe and $18 when they returned to the East Coast two weeks later. It still would be years before Vertex might have a drug. But with three deals in 1993 alone, it was now in the black, showing a year-end profit of more than $2 million. It had salved its burn rate and put $120 million in the bank. By the measure that distinguishes most new companies—penury—Vertex at age five was no longer a start-up.

Internally, the passage was marked in several ways. The company added two senior drug industry executives to the venture capitalists on its board of directors: Donald Conklin, president of Schering Plough Pharmaceuticals, and Barry Bloom, an MIT-trained chemist and recently retired head of research and development at Pfizer. It issued hefty stock options to most of the original scientists to keep them from defecting to other companies. Aldrich was promoted to Senior Vice President, by title and in fact the second most important person in the company. In a press release announcing the move, Boger credited him generously with playing a "major role in Vertex's success."

July 1, 1994

With six projects and 135 employees sprawled among five buildings, Vertex no longer resembled its fragmentary beginnings. It was sturdier now, more evolved. The original SAB, with the exception of Jeremy Knowles (now dean of faculty at Harvard) and Steve Burakoff, was long gone. Of the ten scientists first hired by Boger, all but Debra Peattie, who'd left to attend Harvard Business School and to have a baby, remained, although fewer and fewer of them at the bench. Harding, Tung, Armistead, and Livingston were now project heads, Vertex's equivalent of middle managers. So was Thomson, who had been tapped to direct the company's efforts against Hepatitis C yet who occasionally still worked through the night purifying protein and, it would seem, himself. Ironically, only Navia, recently becoming a father for the first time and refocusing some of his priorities, spent more time in the lab than he used to.

In mid-April, after more than five years of human experimentation, the FDA finally approved Fujisawa's application to market FK-506 in the United States. Although licensing of the drug, now called Prograf, was relatively quick following a unanimous recommendation by an advisory panel, the clinical consensus was that it was equivalent to cyclosporine but no safer or more effective. Indeed, Starzl's early claims notwithstanding, and as Schreiber's calcineurin discovery would suggest, the drugs are more alike than they are different. The point was reinforced two months later when a 15-year-old liver transplant patient, complaining of severe headaches and leg and back pain, made news by declaring that he would rather die than suffer anymore from Prograf's side effects.

Long anticipated, the approval confronted Vertex with a painful choice. The company had tuned itself up on FK-506. It had mastered the processes of structure-based design by mimicking every part of the molecule. But it was now at least five years behind a drug it could no longer say, with any reasonable certainty, that it could beat. Meanwhile, times had changed: as Boger put it, "The moment has passed when an improved FK-506 is what the world's been waiting for." There were now other experimental immunosuppressants, with other targets, that looked more promising. Sticking with its original strategy could only take Vertex further out of the race.

If it had been up to Vertex alone, it is likely that the company would have dropped FKBP-12 and calcineurin in favor of other, more productive targets; it would have conceded the problems in immunophilins and gone on. But Chugai had, somewhat inexplicably, became even more enthusiastic about the project the longer it went on. This left Vertex in an obvious quandary. Boger was too irrepressible an optimist to believe that the program had been anything less than a success. "But for biology," he said, "we did a fantastic job." But it was a success that was draining the company. Not wanting to displease Chugai, which had been more than patient, he and the scientists resolved to keep going at least until Chugai's financial support ran out in early 1995.

All research-based companies, of course, encounter such blind alleys. One fortunate consequence of being secure financially while having several projects in development was that Vertex was now in a position to cut its losses sooner rather than later; it could afford to be truer both to its science and to investors. To Aldrich, this meant not having to prop up less-than-desirable drug candidates. More and more biotech companies, confronted simultaneously with onerous burn rates, disappointing clinical results, slender portfolios, and an ongoing drought on Wall Street, now seemed to be pursuing questionable therapies longer and longer in the face of ambiguous or even negative clinical data. Such desperation, Aldrich thought, inflated expectations ruinously, making the fall that much harder when it came.

Indeed, if there was a recurring image now within the industry, one that made investors fittingly wary of *all* biotech stocks, it was the slow-motion crash-and-burn of the formerly hot start-up. Such a well-publicized crack-up had recently taken place, and it reminded Aldrich and everyone else at Vertex of the price that all drug companies, particularly unprofitable ones, pay for sticking too long with a loser. Regeneron, whose $99 million IPO had signalled the crest of the 1991 biotech frenzy when Vertex went public, had gone into a death spiral with its lead drug candidate, a therapy for Lou Gehrig's disease. The company had known the drug had serious side effects since it first began testing it in mice years earlier. When rumors of similar toxicity in humans finally forced it to revise its clinical trials in March, its stock sank immediately by a third, to $8.75. (It had gone out at $22). Two months later, when Regeneron announced it was finally abandoning the trial, its stock plunged again to just over $4.

Whether such a fall awaited Vertex only time would tell. One of the luxuries of not having a drug in clinic was that a company's expectations couldn't be sullied by impartial evaluation. For that reason, Boger and Aldrich took no joy in Regeneron's unraveling, though they'd long predicted it. Indeed, their own attempts to manage expectations were about to grow much dicier.

Vertex now had four compounds in or near the clinic. Two of those were for blood diseases—sickle cell anemia and beta thalassemia. The molecules hadn't been invented by the company's scientists but rather had been licensed in, which, while not inconsistent with Boger and Aldrich's original business strategy ("Vertex didn't get founded to prove a principle," Boger would say, "we're trying to put drugs on the market"), did little to advance Vertex's claim as an avatar of structure-based design. A third treatment, for cancer, was the found pony of the immunophilins project. The company's entry in the suddenly hot field of attacking multidrug resistance—a similar molecule was now the major product hope in Sandoz's pipeline—it too reflected perhaps more of Vertex's savvy opportunism and marketing gamesmanship than its well-advertised scientific prowess. And while the drug appeared promising, it didn't have the superior profile that Boger had long predicted for the company's molecules. "The therapeutic index [the ratio of efficacy to toxicity that is the primary measure of a drug's effectiveness] is not what I want it to be," Boger said. "It's not VX-478, where you can serve it up on hamburger buns."

That left VX-478 to shoulder Vertex's claims. "We know what we want to do," Aldrich said in June, "We want to build the greatest drug discovery company in the world, we want to develop innovative therapies, and we want to make a lot of money for ourselves and our shareholders." Now, in the summer of 1994, it was within the once-taboo realm of AIDS that those aspirations were finally to be tested.

For more than a year Vertex had withheld the compound's structure, meaning that the scientific world had heard a familiar, and decidedly partial, story—great pre-clinical data, phantom information on how and why the molecule worked. Deservedly, the posture invited much skepticism. In Berlin, where the public rollout of Vertex's AIDS program had first begun, Tung was chastised for the omission. "At a meeting like [the international AIDS conference,]" a frustrated researcher with connections to a rival company had complained, "it's useless and a waste of time without a chemical structure."

Now Vertex could afford no such secrecy. In June, kicking off what Boger would call "a real major scientific publicity blitz," Tung traveled again to an AIDS conference in Europe—the first of several such appearances scheduled throughout the summer and into the fall. Boger knew that before Vertex and Wellcome could sell an AIDS drug to people infected with HIV, they first had to sell it to those clinical investigators whose patients are the first to take experimental drugs and whose support is thus essential. Like Starzl, these doctors care less how a drug works than that it *does* work and that they have the opportunity, exclusive if possible, to prove it. Disclosing the chemical structure of VX-478 for the first time to such a group in Nice, Tung was met, inevitably, with the first real criticism of the compound.

Reactions seemed to range from cautious optimism to mild disappointment. Presented at last with an atom-by-atom accounting of the drug's activity, no one felt that Vertex had been blowing smoke, but neither was anyone immediately convinced that VX-478 was all that the company had claimed. There were questions which, though easily washed away if the drug proved to be effective, now seemed to undercut Boger's glowing predictions. Was the molecule unobtrusive enough to pass the so-called "blood-brain barrier" and enter the central nervous system, something Searle's compound, though similar, had not been able to do? What of possible allergic reactions? VX-478 had chemical groups similar to those thought to cause some people to react violently to Bactrim, a widely used antibiotic. Though the issue of allergies in immunosuppressed, HIV-infected people seems minor, perhaps even moot, it nonetheless raised the still quite real specter of unanticipated side effects as the drug entered human trials.

Even more troubling perhaps from a business standpoint was Vertex's patent position. Boger had delayed revealing the molecule's structure until after the company's European patent application was made public. But no patent on the molecule had yet been issued nor would one be for some time. Because the compound's chemical core was tantalizingly similar to Searle's, despite distinct differences in its overall structures, it remained possible that Vertex's strategy for inhibiting HIV protease was, in fact, blocked. "I'm not sure their patent is as secure as they think it is," said Dr. Carl W. Dieffenbach, chief of developmental therapeutics at the National Institute of Allergies and Infectious Diseases and the federal government's point man for assessing new AIDS drugs.

It is a measure of Boger that he had anticipated these problems —particularly this last—and had done much to alleviate them with the Wellcome deal. As part of its due diligence, Wellcome had no doubt long ago satisfied itself that Vertex was the sole and rightful sole owner of VX-478 and its derivatives. Just as clearly, Wellcome's patent lawyers, who had won—and kept—the rights to AZT through 10 years of intense legal strife, provided Vertex with an intimidating ally. To Boger, drug development, like war, was a strategic engagement, won or lost through attrition along multiple fronts. With little else remaining before the fall's final "ramping" of VX-478 into patients—"There'll still be leaves on the trees," he predicted, though he didn't say where—he was secure in the knowledge that Vertex was positioned to prove itself at last.

As always, he believed the future would answer for itself.

ACKNOWLEDGMENTS

A great many people have generously assisted in making this book possible, and to all of them I wish to express my sincerest thanks. I am especially indebted to Josh Boger, Tom Starzl, Stu Schreiber, and everyone at Vertex. Without their willingness to have me around, I'd have had no story.

At Vertex, I owe many thanks to Rich Aldrich, David and Sharon Armistead, Mike Badia, Cathy Beechinor, Dave Deininger, John Duffy, Laura Engle, Matt Fitzgibbon, Matt Harding, Jeremy Knowles, Grace Lee, Chris Lepre, Judy Lippke, Dave Livingston, Hal Meyers, Jon Moore, Mark Murcko, Manuel Navia, Patsi Nelson, Steve Park, Dave Pearlman, Debra Peattie, Brian Perry, Govinda Rao, Sergio Rotstein, Vicki Sato, Jeff Saunders, Nancy St. Clair, Nancy Stuart, John Thomson, Roger Tung, Al Vaz, and Mason Yamashita, all of whom spent considerable time with me, and to the original board of directors—Frank Bonsal, Bill Helman, Dan Gregory, Kevin Kinsella, and Benno Schmidt—who approved my being there. I am indebted especially to Mason Yamashita, who lent me his spare room on my visits to Cambridge.

In Pittsburgh, John Fung, Andy Tzakis, and Dave Van Thiel generously provided access to patient data and interceded for me with graft recipients and their families, who usually agreed to speak to me on short notice and in times of great distress, and were always helpful and inspiring.

Libraries and archives that provided me with invaluable resources were the Neilson and science libraries at Smith College, the Frost Library at Amherst, the Hampshire College and Mount Holyoke libraries, Forbes Library in Northampton, the University of Massachusetts central library, the University of Pennsylvania archives (where Gail Pietrzyk was especially helpful), the Center for the History of Chemistry, the Carnegie Institution, and the Library of Congress.

Steve Burakoff of Harvard Medical School, who introduced me to Josh Boger, was vital in helping me to get started. Dick Todd, former editor of *New England Monthly*, supported the project in its early stages, and his predecessor, Dan Okrent, was encouraging and full as ever of useful advice. I am also indebted to Bruce Weber, Katherine Bouton, and Jim Atlas of *The New York Times Magazine*. At Simon & Schuster, my editor Bob Bender has been a patient and sympathetic ally. My thanks also to his assistant, Johanna Li. My research assistant Portia Keating consistently did more than I asked. I am especially indebted to my agent, Amanda Urban, whose judgment and commitment sustained me throughout.

Finally, I am blessed with a fine and generous group of family and friends. I am immeasurably indebted to my parents, Hilda and Herb Werth; my sister, Susan Werth; and to my in-laws, Muriel and Phil Goos. Alan and Anita Sosne and Fred Eisenstein provided inspiration, moral support, and clips, as did Bill Newman. Kathy Whittemore and Stella Schwartz put me up. Bill McFeely, Jon Harr, and Anthony Giardina gave writers' succor, encouragement, and an experienced ear. Joe Nocera did much work on early drafts of my manuscript and provided wise professional advice. My nine-year-old daughter, Emily, and six-year-old son, Alex, cheered me on with good nature, faith, and acceptance, more than I often was able to summon in return, and I am indebted for their understanding attempts to keep it down. As always, my wife, Kathy Goos, made—and makes—everything possible. Her suggestions were invaluable and her patience, often sorely tested by my travel and other, less explicable lapses, unstinting.

I regret that there may be people I've forgotten, or omissions. Whatever shortcomings they reflect are, of course, my own.

SOURCES

Obviously, this project could not have been undertaken without access to the people and events described in the narrative. As a condition of my having nearly complete freedom to come and go at Vertex for almost four years while I researched and wrote this book, I agreed to let the company review my manuscript for factual accuracy and proprietary disclosures. Also, all those individuals quoted directly in connection with Vertex (including Stu Schreiber) were allowed to review their quotations, again to ensure accuracy and confidentiality on scientific matters. For scenes involving discussions with individuals outside the company from which I was excluded, I was able to reconstruct dialogue extensively due to the cooperation of the participants. In many cases, that meant talking with everyone who was there. Only in a few cases were people unwilling to talk with me. In those instances, in addition to corroboration from Vertex, I confirmed with at least one person who was present at the discussion on behalf of the other party.

For sections of the book that fall outside of the narrative, I relied on a variety of sources. I am indebted to Mary Snead Boger for helping me to reconstruct her son's family background and Ken Boger and Amy Boger for their thoughtful perspectives. Director Don Cummings of the Stonewall Jackson Training School helped me with Boger family history. Bill Holder of the Wesleyan University News Office facilitated my search through back issues of the *Argus*, the student newspaper. Other helpful sources regarding the Bogers and life in Concord were the Concord *Tribune*, on microfilm at the Concord Memorial Library, and *North Carolina: Through Four Centuries* by William S. Powell (1989, Chapel Hill, NC: University of North Carolina Press).

I relied heavily on the scientists whom I interviewed to tutor me in their fields. I owe a particular debt of gratitude to Jim Mullins of Stanford University and Mark Feinberg of the University of California at San Francisco, neither of whom I've written about here, for their patient explanations of molecular biology and of the world and culture of high-stakes biomedical research. Among print sources, there are two that tower as the most comprehensive: *The Molecular Biology of the Cell* by Alberts et al. (1989, New York: Garland Publishing), the standard undergraduate text; and *The Eighth Day of Creation* by Horace Freeland Judson (1979, New York: Simon and Schuster), which chronicles the revolution in biology. I also relied on the following articles and books:

Angier, N. 1988. *Natural Obsessions: Striving to Unlock the Deepest Secrets of the Cancer Cell.* Boston: Houghton Mifflin.

Bishop, J. E., and M. Waldholz. 1990. *Genome: The Story of the Most Astonishing Scientific Adventure of Our Time—The Attempt to Map All the Genes in the Human Body*. New York: Simon and Schuster.

Borek, E. 1961. *The Atoms Within Us*. New York: Columbia University Press.

Doolittle, R. F. 1985. Proteins. *Scientific American*, October.

Fruton, J. 1950. Proteins. *Scientific American* 182, 33–41.

Gold, M. 1986. *A Conspiracy of Cells*. Albany, NY: State University of New York Press.

Goldberg, J. 1988. *Anatomy of a Scientific Discovery*. New York: Bantam Books.

Gund, P., J. D. Andose, J. B. Rhodes, and G. M. Smith. 1980. Three-Dimensional Molecular Modeling and Drug Design. *Science* 208, 1425–31.

Hall, S. S. 1987. *Invisible Frontiers: The Race to Synthecize a Human Gene.* Boston: The Atlantic Monthly Press.

Hilts, P. J. 1982. *Scientific Temperaments: Three Lives in Contemporary Science*. New York: Simon and Schuster.

The Howard Hughes Medical Institute. 1990. *Finding the Critical Shapes*. Bethesda, MD.

Jevons, F. R. 1968. *The Biochemical Approach to Life*. New York: Basic Books.

Kendrew, J. C. 1961. The Three-dimensional Structure of a Protein Molecule. *Scientific American* 205, 96–110.

Lessing, L. 1969. The Life-Saving Promise of Enzymes. *Fortune*, March.

Marx, J. L. 1980. NMR Opens a New Window into the Body. *Science*, October.

Monod, J. 1971. *Change and Necessity: An Essay on the Natural Philosophy of Modern Biology*. New York: Alfred A. Knopf.

Perutz, M. F. 1964. The Hemoglobin Molecule. *Scientific American* 211, 64–76.

Salem, L. 1987. *Marvels of the Molecule*. New York: VCH Publishers.

Spilker, B., and P. Cuatrecasas. 1990. *Inside the Drug Industry*. Barcelona: Prous Science Publishers.

Stein, W. H., and S. Moore. 1961. The Chemical Structure of Proteins. *Scientific American* 205, February, 81–92.

Thomas, L. 1974. *The Lives of a Cell: Notes of a Biology Watcher*. New York: The Viking Press.

Watson, J. D. 1968. *The Double Helix*. New York: W. W. Norton and Co.

Weinberg, R. A. 1985. The Molecules of Life. *Scientific American*, October.

A great advantage in writing about science is that all work is painstakingly catalogued. When confusion reigned, I relied for clarification on the scientific literature. The following publications, either cited or referred to in the non-Starzl portions of the text, were of particular assistance:

Bierer, B. E., P. K. Somers, T. J. Wandless, S. J. Burakoff, and S. L. Schreiber. 1990. Probing Immunosuppressant Action with Nonnatural Immunophilin Ligand. *Science* 250, 556–59.

Boger, Joshua, et al. 1983. Novel renin inhibitors containing the amino acid statine. *Nature* 303, 81–84.

Handshumacher, R. E., M. W. Harding, J. Rice, and R. Drugge. 1984.

Cyclophilin: a specific cytosolic binding protein for Cyclosporine A. *Science* 226, 544–47.

Fischer, G., B. Wittmann-Liebold, K. Lang, T. Kiefhaber, and F. X. Schmid. 1989. Cyclophilin and peptidyl-prolyl cis-trans sis-trans isomerase are probably identical proteins. *Nature* 337, 476–78.

Friedman, J., and I. Weissman. 1991. Two Cytoplasmic Candidates for Immunophilin Action Are Revealed by Affinity for a New Cyclophilin: One in the Presence and One in the Absence of CsA. *Cell* 66, 799–806.

Fretz, H., M. W. Albers, A. Galat, R. F. Standaert, W. S. Lane, S. J. Burakoff, B. E. Bierer, and S. L. Schreiber. 1991. Rapamycin and FK506 Binding Proteins (Immunophilins). *J.Am.Chem.Soc.* 113, 1409–11.

Harding, M. W., A. Galat, D. E. Uehling, and S. L. Schreiber. 1989. A receptor for the immunosuppressant FK506 is a cis-trans peptidyl-prolyl isomerase. *Nature* 341, 758–60.

Lepre, C. A., J. A. Thomson, and J. M. Moore. 1992. Solution structure of FK506 bound to FKBP-12. *FEBS* Letters, May 4.

Liu, J., J. D. Farmer, Jr., W. S. Lane, J. Friedman, I. Weissman, and S. L. Schreiber. 1991. Calcineurin Is a Common Target of Cyclophilin-Cyclosporin A and FKBP-FK506 Complexes. *Cell* 66, 807–15.

Michnick, S. W., M. K. Rosen, T. J. Wandless, M. Karplus, and S. L. Schreiber. 1991. Solution Structure of FKBP, a Rotamase Enzyme and Receptor for FK506 and Rapamycin. *Science* 252, 836–39

Moore, J. M., D. A. Peattie, M. J. Fitzgibbon, J. A. Thomson, et al. 1991. Solution structure of the major binding protein for the immunosuppressant FK506. *Nature* 351, 248–50.

Navia, M. A., P. M. D. Fitzgerald, B. M. McKeever, C.-T. Leu, J. C. Heimbach, W. K. Herber, I. S. Sigal, P. L. Darke, and J. P. Springer. 1989. Three-dimensional structure of aspartyl protease from immunodeficiency virus HIV-1. *Nature* 337, 615–20.

Pauwels, R., K. Andries, J. Desmyter, D. Schols, M. J. Kukla, H. J. Breslin, A. Raeymaeckers, J. Van Gelder, R. Woestenborghs, J. Heykants, K. Schellekens, M. A. C. Janssen, E. De Clercq, and P. A. J. Janssen. 1990. Potent and selective inhibition of HIV-1 replication in vitro by a novel series of TIBO derivatives. *Nature* 343, 470–74.

Schreiber, S. L. 1991. Chemistry and Biology of the Immunophilins and Their Immunosuppressive Ligands. *Science* 251, 283–87.

———. 1992. Using the Principles of Organic Chemistry to Explore Cell Biology. *Chemical and Engineering News*, Oct. 26.

Siekierka, J. J., S. H. Y. Hung, M. Poe, C. S. Lin, and N. H. Sigal. 1989. A cytosolic binding protein for the immunosuppressant FK506 has peptidyl-prolyl isomerase activity but is distinct from cyclophilin. *Nature* 341, 755–57.

Takahashi, N., T. Hayano, and M. Suzuki. 1989. Peptidyl-prolyl cis-trans isomerase is the cyclosporin A-binding protein cyclophilin. *Nature* 337, 473–75.

Van Duyne, G. D., R. F. Standaert, P. A. Karplus, S. L. Schreiber, and J. Clardy. 1991. Atomic Structure of FKBP-FK506, an Immunophilin-Immunosuppressant Complex. *Science* 252, 839–42.

For the sections on Tom Starzl and transplantation immunology, I relied on many sources. Much of the work originally appeared as an article in *The New York Times Magazine* ("The Drug That Works in Pittsburgh," September 30, 1990), and though I was generally required by the University of Pittsburgh to be accompanied by a member of the staff when speaking with patients on my visits to the medical center, Starzl and all members of the staff were consistently open and receptive. I owe thanks to Cheryl Ackerman, Barbara Banner, Richard Cohen, Benjamin Eidelman, Anthony Demitris, Ashok Jain, Yukio Murase, Jerry McCauley, Mike Nalesnick, Camillo Ricordi, Raman Venkataramanan, and Vijay Warty. I also am indebted to Starzl's remarkable secretary, Terry Mangan.

Takenori Ochiai, the Japanese surgeon who pioneered the use of FK-506 in animals and whose role is often regrettably obscured by Starzl's, was extremely helpful in helping me to establish the early history of the drug. I owe him a special thanks.

Starzl's autobiography, *The Puzzle People* (1992, Pittsburgh, PA: University of Pittsburgh Press) was invaluable, as was *Many Sleepless Nights* by Lee Gutkind (1990, Pittsburgh, PA: University of Pittsburgh Press), especially for its descriptions of life on the transplant wards and of the Hardanger Vidda. The following books and articles were also of great use:

Billingham, R. E. 1966. Tissue Transplantation: Scope and Prospect. *Science* 153, 266–70.

Billingham, R. E., P. L. Krohn, and P. B. Medawar. 1951. Effect of Cortisone on Survival of Skin Homografts in Rabbits. *British Medical Journal*, May 26, 1158–63.

Foreman, J. 1987. Cracking the secrets of body's own "army." *Boston Globe*, 30–31, Oct. 18.

Medawar, P. 1957. *The Uniqueness of the Individual*. New York: Basic Books.

Silverstein, A. M. 1989. The History of Immunology. *Fundamental Immunology*, 21–37. New York: Raven Press.

Starzl, T. E. 1991. My Thirty-five Year View of Organ Transplantation. *History of Transplantation: 35 Recollections*, ed. P. I. Terasaki. Los Angeles: UCLA Tissue Typing Laboratory.

———. 1990. The Development of Clinical Renal Transplantation. *American Journal of Kidney Diseases* 16, 548–56.

Starzl, T. E., S. Todo, A. Tzakis, M. Alessiani, A. Casavilla, K. Abu-Elmagd, and J. J. Fung. 1991. The Many Faces of Multivisceral Transplantation. *Surgery, Gynecology and Obstetrics* 172, 335–44.

Thomas, E. D., 1987. Bone Marrow Transplantation in Hematologic Malignancies. *Hospital Practice*, 77–91, Feb. 15.

Thomas, E. D., H. L. Lochte, and J. Ferrebee. 1959. Irradiation of the Entire Body and Marrow Transplantation: Some Observations and Comments. *Blood* 14, 1–23, January.

Thompson, L. 1988. Jean-François Borel's Transplanted Dream. *Washington Post*, Nov. 15.

Four former senior research directors at Merck—Bob Denklewalter, Ralph Hirschmann, Eugene Cordes, and H. Boyd Woodruff—helped me immeasurably in characterizing the Tishler years. Woodruff, a protégé of Selman Waksman, tutored me in microbiology and soil screening. I also am grateful to Leon Gortler, who, along with John Heitmann, conducted a comprehensive interview of Tishler for the American Chemical Society's Eminent Chemists series, and to Peter Jacobi, chairman of the Chemistry Department at Wesleyan University, who directed me toward material about both Tishler and R. B. Woodward. Mary Feiser, a member of the chemistry faculty of Harvard for more than fifty years, also enriched my understanding, as did Don Ciappanelli, former director of Harvard's chemistry labs, who provided me with videotape of several of Woodward's lectures from the university archives.

I relied principally on two sets of archival material for information about medical research during World War II: the papers of A. N. Richards, which are at the University of Pennsylvania and which include the minutes of the Committee for Medical Research, and the personal papers of Vannevar Bush, which are at the Library of Congress.

The books and articles I found most useful for information

about the history of drug research, especially regarding Tishler, Merck, Woodward, and Harvard, are:

Barber, B. 1952. *Science and the Social Order*. Glencoe, IL: The Free Press.

Baxter, J. P., III. 1946. *Scientists Against Time*. Boston: Little, Brown.

Borkin, J. 1978. *The Crime and Punishment of I. G. Farben*. New York: The Free Press.

Braithwaite, J. 1984. *Corporate Crime in the Pharmaceutical Industry*. London: Routledge & Kegan Paul.

Browning, C. H. 1955. Emil Behring and Paul Ehrlich: Their Contributions to Science. *Nature* 175, 570–75.

Conant, J. B. 1970. *My Several Lives: Memoirs of a Social Inventor*. New York: Harper and Row.

Crosby, A. W., Jr. 1976. *Epidemic and Peace, 1918*. Westport, CT: Greenwood Press.

Dolphin, D. 1977. Robert Burns Woodward: Three Score Years and Then? *Aldrichimica Acta* 10, No. 1, 3–9.

DuBos, R. J. 1950. *Louis Pasteur: Free Lance of Science*. Boston: Little, Brown.

Engle, L. 1951. Cortisone and Plenty of It. *Harper's* 203, 56–62.

Epstein, S., and B. Williams. 1956. *Miracles from Microbes: The Road to Streptomycin*. New Brunswick, NJ: Rutgers University Press.

Galdston, I. 1943. *Behind the Sulfa Drugs: A Short History of Chemotherapy*. New York: D. Appleton Century.

Gallese, L. 1990. Venture Capital Strays Far from Its Roots. *New York Times Magazine* (The Business World), 24–39, April 1.

Garland, J. E. 1961. *Every Man Our Neighbor: A Brief History of the Massachusetts General Hospital*. Boston: Little, Brown.

Harris, R. 1964. *The Real Voice*. New York: Macmillan.

Hayes, P. 1987. *Industry and Ideology: IG Farben in the Nazi Era*. London: Cambridge University Press.

Hixson, J. 1976. *The Patchwork Mouse*. Garden City: NY: Anchor Press.

Hobby, G. L. 1985. *Penicillin: Meeting the Challenge*. New Haven: Yale University Press.

Kahn, E. J. 1981. *Jock: The Life and Times of John Hay Whitney*. Garden City: NY, Doubleday and Co.

Liebenau, J. 1987. *Medical Science and Medical Industry: The Formation of the American Pharmaceutical Industry*. Baltimore, MD: Johns Hopkins University Press.

Mahoney, T. 1959. *The Merchants of Life: An Account of the American Pharmaceutical Industry*. New York: Harper Brothers.

Merck and Co. 1992. *Values and Visions: A Merck Century*.

Merck, Sharp and Dohme. 1977. *Profiles in Discovery*.

Merck, Sharp and Dohme Research Laboratories. 1962. *By Their Fruits*.

Noble, D. F. 1977. *America by Design: Science, Technology and the Rise of Corporate Capitalism*. New York: Alfred A. Knopf.

Pearson, M. 1969. *The Million Dollar Bugs*. New York: G. P. Putnam's Sons.

Pfeiffer, J. 1939. Sulfanilamide: The Story of a Great Medical Discovery. *Harper's*, March.

Rettig, R. A. 1977. *Cancer Crusade: The Story of the National Cancer Act of 1971*. Princeton, NJ: Princeton University Press.

Richards, A. N. 1964. Production of Penicillin in the United States (1941–1946). *Nature* 201, 441–45.

Roberts, J. D. 1990. *The Right Place at the Right Time*. Washington, DC: American Chemical Society.

Roberts, R. 1989. *Serendipity: Accidental Discoveries in Science*. New York: John Wiley and Sons.

Roueché, B. 1955. Annals of Medicine: Ten Feet Tall. *The New Yorker*, Sept. 10.

Russell, F. 1957. A Journal of the Plague: The 1918 Influenza. *Yale Review*, December.

Sheehan, John C. 1982. *The Enchanted Ring: The Untold Story of Penicillin*. Cambridge, MA: The MIT Press.

Sheehan, J. C., and R. N. Ross. 1982. The Fire That Made Penicillin Famous. *Yankee*, 125–27, November.

Smith, F. R. 1947. Good Microbes Fight Bad Ones. *The New York Times Magazine*, 17–19, Aug. 10.

Soper, G. A. 1919. The Lessons of the Pandemic. *Science* 49, 501–5.

Sneader, W. 1985. *Drug Discovery: The Evolution of Modern Medicines*. Chichester: Wiley and Sons.

Sturchio, J. L. 1981. Chemists and Industry in Modern America: Studies in the Historical Application of Science Indicators. Graduate dissertation. University of Pennsylvania.

Swann, J. P. 1988. *Academic Scientists and the Pharmaceutical Industry: Cooperative Research in Twentieth Century America*. Baltimore, MD: Johns Hopkins University Press.

Talalay, P. (ed.) 1964. *Drugs in Our Society*. Baltimore, MD: Johns Hopkins University Press.

Temin, P. 1980. *Taking Your Medicine*. Cambridge, MA: Harvard University Press.

Tishler, M. 1974. Is Science Dead? New Brunswick Lecture, May 15.

———. 1969. The Siege of the House of Reason. *Science* 166, 192–95.

Todd, A. 1983. *A Time to Remember: The Autobiography of a Chemist*. London: Cambridge University Press.

Tuchman, B. W. 1978. *A Distant Mirror: The Calamitous 14th Century*. New York: Alfred A. Knopf.

Vogel, M. 1980. *The Invention of the Modern Hospital*. Chicago: University of Chicago Press.

Waksman, S. A. 1954. *My Life with the Microbes*. New York: Simon and Schuster.

———. 1949. *Streptomycin: Nature and Practical Applications*. Baltimore, MD: Williams and Wilkins.

Wilson, D. 1976. *In Search of Penicillin*. New York: Alfred A. Knopf.

Woodruff, H. B. 1981. A Soil Microbiologist's Odyssey. *Annual Review of Microbiology* 35, 1–28.

For historical information about Harvard, I relied primarily on Richard Norton Smith's *The Harvard Century: The Making of a University to a Nation* (1986, New York: Simon and Schuster) and Carl Vigeland's *Great Good Fortune: How Harvard Makes Its Money* (1986, Boston: Houghton Mifflin).

Management consultants Jim Feeney and Harutoshi Mayazumi taught me about the Japanese pharmaceutical industry and the ways of Japanese business. I found especially useful *The Samurai Factor: Japanese strategic thinking in industry*, a monograph coauthored by Mayazumi and Joseph Rudzinski. Other books and articles about Japan that I found useful were:

Christopher, R. C. 1983. *The Japanese Mind: The Goliath Explained*. New York: Linden Press.

Gibney, F. 1982. *Miracle by Design: The Real Reasons Behind Japan's Economic Success*. New York: Times Books.

Prestowitz, C. V., Jr. 1988. *Trading Places: How We Allowed Japan to Take the Lead*. New York: Basic Books.

Reich, R. 1983. *The Next American Frontier*. New York: Times Books.

Toland, J. 1970. *The Rising Sun*. New York: Random House.

The biotechnology industry is still too young to be the subject of many long-term histories. The best primer I found is Robert

Teitleman's *Gene Dreams* (1989, New York: Basic Books). To understand the early scientific, legal, and financial precepts behind the industry, I turned to the following articles:

Biddle, W. 1981. A Patent on Knowledge: Harvard goes public. *Harper's*, June.

Culliton, B. 1977. Harvard and Monsanto: The $23 Million Alliance. *Science* 195, 759–63.

Gurin, J., and N. Pfund. 1980. Bonanza in the Biolab. *The Nation*, Nov. 22.

Noble, D., and N. Pfund. 1980. Business Goes Back to College. *The Nation*, Sept. 20.

Wade, N. 1980. Gene Goldrush Splits Harvard, Worries Brokers. *Science* 210, 878–79.

INDEX